LESSONS

IN

ELEMENTARY CHEMISTRY:

INORGANIC AND ORGANIC.

LESSONS

IN

ELEMENTARY CHEMISTRY:

INORGANIC AND ORGANIC.

BY

HENRY E. ROSCOE, B.A. F.R.S.

PROFESSOR OF CHEMISTRY IN OWENS COLLEGE, MANCHESTER.

NEW EDITION.

NEW YORK:

WM. WOOD & CO., PUBLISHERS,

61 WALKER ST.

1870.

THE NEW YORK PRINTING COMPANY,
81, 83, *and* 85 *Centre Street*,
NEW YORK.

TABLE OF CONTENTS.

PREFACE.

—o—

In the following pages I have endeavored to arrange the most important facts and principles of Modern Chemistry in a plain but precise and scientific form, suited to the present requirements of elementary instruction.

For the purpose of facilitating the attainment of that exactitude in the knowledge of the subject, without which the introduction of physical science into the school system is worse than useless, I have added a series of Exercises and Questions upon the Lessons. The pupil must learn to work out accurately both the numerical and descriptive examples, and the teacher may find it advisable to add largely to their number. Particular attention should be given to the calculation of the relations between the weights of gases, and their volumes measured under varying circumstances of temperature and pressure.

The metric system of weights and measures, and the centigrade thermometric scale, are used throughout the work.

I have much pleasure in thanking my friend and assistant, Mr. Schorlemmer, for the aid which he has given me, especially in revising the proofs; and I have also to acknowledge the care and attention bestowed on the woodcuts by Mr. Dickes.

H. E. R.

MANCHESTER.

For further information concerning the absolute weights of the gases than is given on p. 26, a paragraph in the Appendix, page 373, may be consulted.

INTRODUCTION.

LESSON I.

BY *chemical action* we signify that which occurs when two or more substances so act upon another as to produce a third substance differing altogether from the original ones in properties; or when one substance is brought under such conditions that it forms two or more bodies differing from the original one in properties. Thus, if powdered sulphur and fine copper filings be well mixed together, the color of the sulphur as well as that of the copper will disappear, and to the unaided eye the mixture presents a uniform greenish tint; by the help of the microscope, however, the particles of copper may be seen lying by the side of the particles of sulphur; and we can wash away the lighter sulphur with water, leaving the heavier copper behind. Here no *chemical action* has occurred; the sulphur and copper were only *mechanically mixed.* If we next gently heat some of the mixture we see that it soon begins to glow, and on examining the mass we notice that both the copper and the sulphur have disappeared as such, that they cannot be distinguished even by the most powerful microscope, and that in their place we have formed a black substance possessing properties entirely different from those possessed either by the copper or by the sulphur. Here a *chemical change* has occurred; the copper and the sulphur are said to have *combined chemically* to form a *compound* out of which these two substances can be regained in exactly the quantities used. In like manner when a candle burns in the air a chemical change is going on; and although the candle gradually disappears, the materials of which it is made up are not destroyed or lost; they simply pass into a state in which they are invisible to our eyes, but their presence may be ascertained by other means. Thus, if we burn a candle for a few minutes in a clean bottle filled with air, and afterwards pour in some clear lime-water, we shall notice that the liquid, which remains clear in pure air, becomes at once milky, showing the presence of an invisible gaseous body produced by the burning of the candle, which possesses properties different from those of pure air. Although an apparent loss of matter occurs when a candle burns, it is easy to show by a simple experiment not only that this is not the case, but that on the contrary an increase of weight has occurred; this increase is occasioned by the constituent parts of the tallow or wax having united chemically with an invisible gas (called oxygen) present in the air. For this purpose a piece of glass tubing $\frac{3}{4}$ inch wide, and 10 inches long, is closed at each end with a cork; through the upper cork a bent glass tube passes, whilst through the lower one several holes are bored, and into one of these a small taper is fastened. The upper half of the tube is filled with pieces of caustic soda, a grid of perforated zinc being fixed inside the tube to keep them in their place. The tube thus arranged is hung at the end of one arm of a pair of scales, and exactly counterpoised by weights placed in the pan on the other arm. The top of the tube is now connected, by means of a piece of vulcanised caoutchouc tubing, with a closed vessel filled with water and furnished with a stop-cock through which the water can flow out; on opening this stop-cock, as the water flows out, air passes in to supply its place, through the holes in the perforated cork. This cork is then removed, the taper lighted, and the cork and taper quickly replaced; after the candle has burnt for three or four minutes, the vulcanised tubing is disconnected, and the glass tube allowed to hang freely. It is then seen that the weight of the tube is greater than it was before the candle was burnt, the pieces of caustic soda having absorbed the substances (carbonic acid and water) produced by the combination of the constituents of the taper (carbon and hydrogen) with the oxygen of the air.

By the careful examination of all the known cases of chemical action it has been satisfactorily proved that a *loss* of matter never takes place, that *matter is indestructible*, and that in chemical actions, such as that going on in the burning of the candle, a change of state and not an annihilation of matter occurs. The truth of this first great principle in chemical science has been gradually demonstrated by finding that the weights of the substances acting chemically upon one another, always remain the same after as before the chemical changes have occurred. For determining very accurately the weight

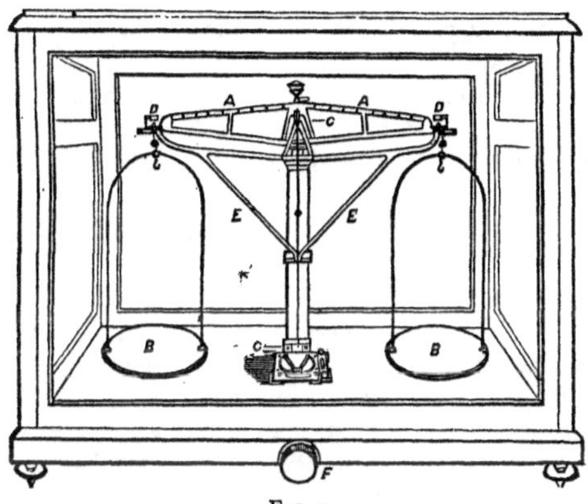

FIG. 1.

of substances, an instrument called the *chemical balance* is employed. Fig. 1 represents one form of chemical balance. It consists of a perforated brass beam (A A) vibrating about its centre, at which is fixed a triangular knife-edge of agate (C): this rests upon a horizontal agate plane attached to the upright brass pillar. To each end of the beam the light brass pans (B B) are attached, each pan hanging on by an agate plane upon an agate knife-edge fixed on the end of the beam at (D D). This mode of rest and support is to render the amount of friction as small as possible, and thus to insure delicacy in the instrument. In order to prevent the agate edges from being spoilt by constant wear on the agate planes, the beam and the ends (D D) are supported by the brass arm (E E) when the balance is not in use, so that the agate surfaces do not touch; the beam and pans are released when required by turning the handle (F). The substance to be weighed is placed in one pan, and weights added one by one to the other until the instrument is in equilibrium; this is ascertained by the long pointer (G) vibrating to an equal distance on each side of the central mark. A balance such as that represented in the figure will turn with $\frac{1}{10}$ of a milligramme when loaded with 100 grammes (*see* p. 21), or will indicate the one-millionth part of the substance weighed.

The aim of the chemist is to examine the properties of all substances with regard to their actions upon one another in producing bodies essentially differing from the originals. In order thoroughly to carry out his purpose he is obliged to resort to *experiment;* that is, he has to place the substances which he is examining under circumstances, perhaps not found in nature, which he can control and vary. Hence chemistry is called an *experimental science.* In thus investigating all the materials within his reach, whether solid, liquid, or gaseous, whether contained in the earth, sea, or air, whether belonging to the animal or to the vegetable creation, the chemist finds himself obliged to divide substances into two great classes: (1) COMPOUND SUBSTANCES—those which he is able to split up into two or more essentially different materials, and (2) ELEMENTS or SIMPLE SUBSTANCES—those which he is unable thus to split up, and out of which nothing essentially different from the original substances has been obtained.

Compound bodies are made up of two or more elementary substances chemically combined with each other; thus sulphur and copper are elementary bodies: out of each of these nothing different from sulphur or copper can be obtained; whereas, when the two bodies are heated together, a compound is formed from which both of the original elementary constituents can at any time be prepared. Water is a compound body,—it can be split up into two elementary gases, hydrogen and oxygen; common salt, again, is a compound of a gas (chlorine) with a metal (sodium); and limestone, clay, sugar, and wax may serve as examples of compound bodies: whilst phosphorus, charcoal, iron, mercury, and gold may be mentioned as belonging to the class of simple substances. The following experiment well illustrates the decomposition of a compound into two simple substances. A small quantity of the red powder called mercury oxide is introduced into a test tube, and heated in a gas flame; when hot, the oxide gradually decomposes, a grey deposit of metallic mercury in small globules collects upon the cooler parts of the glass, whilst the tube becomes filled with colorless oxygen gas, whose presence can be demonstrated by the rekindling of a glowing chip of wood plunged into the tube. On continuing the heat, the whole of the red powder is found to be split up into the two elements, mercury and oxygen, which together weigh exactly as much as the red oxide from which they were obtained.

The elementary bodies, for the sake of convenience, are arbitrarily divided into two classes, the *metals* and the *non-metals*. In the first are placed elements such as gold, iron, lead, mercury, tin; in the second, those elements which are gases at the ordinary temperature, such as oxygen, hydrogen, &c., together with some solid elements, as sulphur, charcoal, &c. The number of the metals is much larger than that of the non-metals; we are acquainted with forty-eight metals, and with only fifteen non-metals. These sixty-three elements constitute the material out of which the whole fabric of the science is built; every description of matter which has been examined is made up of these elements, either combined together to form compounds, or in the uncombined or free state. The science of chemistry has for its aim the experimental examination of the properties of the elements and their compounds, and the investigation of the laws which regulate their combination one with another. The applications of the prin ciples of chemical science to the arts and manufactures are of the highest importance and interest; they have exerted a most material influence upon the progress of civilization, and have greatly tended to the elevation and benefit of mankind; the instances are innumerable in which altogether new branches of industry have sprung up from the happy application of simple chemical principles, and there is scarcely an article in common use in the production of which some application of chemistry has not proved of essential value.

The following is a complete list of the elementary bodies known at present (1869). The names printed in large capitals, as BORON, are the non-metals; those in small capitals, as ALUMINIUM, are the more commonly occurring metals; those in small type, as Cadmium, are the rarer metals:

Names.	*Symbols.*	*Combining Weights.**
ALUMINIUM	Al	27·4
ANTIMONY	Sb	122
ARSENIC	As	75
BARIUM	Ba	137
BISMUTH	Bi	210
BORON	B	11
BROMINE	Br	80
Cadmium	Cd	112
Caesium	Cs	133
CALCIUM	Ca	40
CARBON	C	12
Cerium	Ce	92
CHLORINE	Cl	35·5
CHROMIUM	Cr	52·2
COBALT	Co	58·7
COPPER	Cu	63·5
Didymium	D	96
Erbium	E	112·6
FLUORINE	F	19
Glucinium	Gl	9·3
GOLD	Au	197

* For an explanation of these numbers see page 13.

Names.	*Symbols.*	*Combining Weights*
HYDROGEN	H	1
Indium	In	37·8
IODINE	I	127
Iridium	Ir	198
IRON	Fe	56
Lanthanum	La	92
LEAD	Pb	207
Lithium	Li	7
MAGNESIUM	Mg	24
MANGANESE	Mn	55
MERCURY	Hg	200
Molybdenum	Mo	96
NICKEL	Ni	58·7
Niobium	Nb	94·0
NITROGEN	N	14
Osmium	Os	199·2
OXYGEN	O	16
Palladium	Pd	106·6
PHOSPHORUS	P	31
PLATINUM	Pt	197·5
POTASSIUM	K	39·1
Rhodium	Rh	104·4
Rubidium	Rb	85·4
Ruthenium	Ru	104·4
SELENIUM	Se	79·5
SILVER	Ag	108
SILICON	Si	28
SODIUM	Na	23
STRONTIUM	Sr	87·5
SULPHUR	S	32
Tantalum	Ta	182
TELLURIUM	Te	128
Thallium	Tl	204
Thorium	Th	115·7
TIN	Sn	118
Titanium	Ti	50

Names.	*Symbols.*	*Combining Weights.*
Tungsten	W	184
Uranium	U	120
Vanadium	V	51·3
Yttrium	Y	61·6
ZINC	Zn	65·2
Zirconium	Zr	89·6

Some of these are very abundant and occur widely distributed, whilst others have only been found in such minute quantities, and in such rare fragments, that their properties have not yet been satisfactorily examined. Thus, for instance, oxygen occurs throughout the air, sea, and solid earth, in such quantities as to make up nearly half the weight of our planet. Whereas the compounds of yttrium, erbium, indium, &c., have only as yet been met with in most minute quantities.

The elements are distributed very irregularly throughout our planet: only four occur in the air, some thirty have been found in the sea; whilst all the known elements occur variously dispersed in the solid mass of the earth. The following table, giving the composition by weight of the primary rocks, shows that the bulk of the earth's solid body is made up of only eight elements, the remainder being found in much smaller quantities:

Composition of the Earth's Solid Crust in 100 *parts by weight.*

Oxygen	44·0 to 48·7	Calcium	6·6 to 0·9
Silicon	22·8 „ 36·2	Magnesium	2·7 „ 0·1
Aluminium	9·9 „ 6·1	Sodium	2·4 „ 2·5
Iron	9.9 „ 2·4	Potassium	1·7 „ 3·1

Doubtless other elements exist undiscovered in the earth in addition to the sixty-four now known, for we find that where, with the progress of science, new and more accurate methods of examining the composition of matter have been employed, the existence of new elements has frequently been brought to light; thus within the last four years, no less than four new elements have been discovered by the help of the new method of spectrum analysis (*see* p. 219). Whether any of the bodies now termed elementary may, by the application of more powerful means than we at present possess, at some future time be split up into simpler constituents, is a question which we cannot answer with certainty. Judging, however, from precedent, we may consider the occurrence of such a thing as possible, or even likely; for the alkalies potash and soda were believed to be elements until the year 1808, when Sir H. Davy proved that they were in reality compounds.

Our knowledge of the chemical composition of the heavenly bodies was restricted, until lately, to that gained from the examination of meteorites, in which no element has been found which is not known in the earth. Within the last few years the foundations of a solar and stellar chemistry have, however, been laid; and we are now able to ascertain the presence of many well-known chemical substances in the sun and far distant fixed stars with as great exactitude and certainty as we are able to prove their presence in terrestrial matter (*see* p. 223).

LESSON II.

NON-METALLIC ELEMENTS.

In the present work we shall consider the properties of the non-metals and their compounds in the following order :—

Oxygen,	Sulphur,
Hydrogen,	Selenium,
Nitrogen,	Tellurium,
Carbon,	Silicon,
Chlorine,	Boron,
Bromine,	Phosphorus,
Iodine,	Arsenic.
Fluorine,	

OXYGEN.

Symbol O. Combining Weight 16. *Density* 16.

Oxygen is a colorless invisible gas, possessing neither taste nor smell. It exists in the free state in the atmosphere, of which it constitutes about one-fifth by bulk, whilst, in combination with the other elements, it forms nearly half the weight of the solid earth, and eight-ninths by weight of water. Oxygen was discovered in the year 1774 by Priestley, and independently in 1775 by Scheele. Lavoisier first clearly pointed out in 1778 the part played by oxygen, and explained the chemical changes that go on when bodies burn in the air. The birth of the modern science of chemistry may be dated from the discovery of oxygen. Oxygen gas can be prepared from the air, but it is more easily obtained from many compounds which contain it in large quantities. Priestley prepared oxygen by heating red mercury oxide: this substance is made up of 200 parts by weight of mercury, and sixteen parts of oxygen; when strongly heated, it is decomposed, yielding metallic mercury and oxygen gas. Oxygen can be more cheaply obtained by heating potassium chlorate (commonly called chlorate of potash), a white salt which yields on heating 39·2 per cent. of its weight of this gas. In order to collect the oxygen thus given off, powdered potassium chlorate is placed in a small thin glass flask, furnished with a well-fitting cork, into which a bent tube is inserted. The lower end of the tube dips under the surface of water in a pneumatic trough, and the gas, on being evolved, bubbles out from the end of the tube, and is collected in jars or bottles filled with water, and placed with their mouths downwards in the trough. If a small quantity of manganese di-oxide (black oxide of manganese) be mixed with the potassium chlorate, the oxygen is given off from the chlorate at a much lower temperature, and thus the evolution of the gas is facilitated, but the manganese di-oxide undergoes no change whatever.

All the elements, with the single exception of fluorine, combine with oxygen to form *oxides.* In this act of combination, which is termed *oxidation*, heat is always, and light is frequently, given off. When bodies unite with oxygen, evolving light and heat, they are said to *burn*, or *undergo combustion.* All bodies which burn in the air, burn with increased brilliancy in oxygen gas; and many substances, such as iron, which do not readily burn in the air, may be made to do so in oxygen. A red-hot chip of wood, or a taper with glowing wick, is suddenly rekindled and bursts into flame when plunged into a jar of this gas. Sulphur, which in the air burns with a pale lambent flame, emits in oxygen a bright violet light; and a small piece of phosphorus, when inflamed and placed in oxygen, burns with a dazzling light. If the jars in which these experiments have been performed be afterwards examined, it is found that the substances produced by combustion in oxygen possess acid characters; they have the power of turning red certain vegetable blue coloring matters, such as litmus: owing to this fact, Lavoisier gave to oxygen the name it bears (from ὀξὺς, acid, γεννάω, I produce). Fine iron wire can be easily burnt in oxygen by tipping the end with burning sulphur, and then plunging the iron thus tipped into a jar of the gas; the oxide of iron, formed by the combustion, drops down in the molten state.

Many other substances may be employed for the preparation of oxygen; thus, if large quantities of the gas are needed, manganese di-oxide (a substance of frequent occurrence in nature) may be heated to redness in an iron bottle; 100 parts by weight of the oxide yield 12·3 by weight of oxygen. Another interesting decomposition by which oxygen is set free, is that effected by sunlight upon the carbonic acid gas contained in the air; this is accomplished by the green coloring matter of plants. Sunlight has the power, in presence of this green coloring matter, of decomposing carbonic acid; the carbon is taken up by the plant for its growth, whilst the oxygen is set free, and is afterwards used by animals for the support of the process of respiration. In the act of inspiration (filling the lungs) animals breathe in the oxygen of the air, whilst in that of expiration (emptying the lungs) they breathe out carbonic acid gas. Hence oxygen is necessary to animal life, wherefore this gas was formerly termed *vital air*. The chemical change which oxygen effects upon the body of the animal is in fact identical with that which goes on when a piece of charcoal burns in the air or oxygen; this may be rendered evident by a simple experiment. If some clear limewater be poured into a bottle of oxygen in which charcoal has been burnt, the limewater will become milky, owing to the formation of a compound of lime and carbonic acid (called chalk), this acid being produced by the combustion; if the air contained in the lungs be next blown through a piece of glass tubing into some more clear limewater, a turbidity (from the formation of chalk) will at once occur, proving that carbonic acid gas is given off from the lungs. This carbonic acid arises from the oxidation of the constituents of the body, and by this oxidation the heat of the body, which is greater than that of surrounding inanimate objects, is sustained. When this chemical process stops, the animal dies, and the temperature of the body sinks to that of the neighboring objects. Carbonic acid, nitrogen, and some other gases cause death when inhaled, because they do not contain free oxygen, and hence the process of oxidation in the body ceases. This cause of death is independent of any poisonous action of the gases. Other processes for preparing oxygen on a large scale will be mentioned in the lessons relating to bleaching powder, sulphuric acid, and barium di-oxide.

When the composition of a substance is determined by splitting the compound into its elementary constituents, a *chemical analysis* of that substance is said to have been made; and if the proportions by weight in which each of the constituents is present, be determined, a *quantitative* analysis of the substance has been made. When the composition is ascertained by bringing the constituent parts together, we are said to determine the composition by *synthesis.* If we analyze potassium chlorate we find that, from whatever source this salt may be derived, it always possesses the same unalterable composition. This is true of every definite chemical compound; indeed, were it not so, chemistry as a science could not exist. Potassium chlorate is made up of three elementary bodies, chlorine, potassium, and oxygen, combined together in the following proportions by weight:—

Chlorine	35·5	parts by weight.
Potassium . . .	39·1	"
Oxygen	48·0	"
Potassium Chlorate . . .	122·6	

When this salt is heated, the whole of the oxygen comes off as gas: 122·6 parts yield 48 parts of oxygen, while 74·6 parts of a white solid compound of chlorine and potassium, called potassium chloride, remain behind. Hence the weight of oxygen which can be obtained from any given weight of potassium chlorate, and *vice versâ*, can be calculated.

In order to express the composition of substances more conveniently than can be done by writing the names of the elementary constituents at full length, chemists use a kind of short-hand, or symbolic language, some of the principles of which must now be shortly explained. Instead of writing the whole name, the first letter or the first two letters of the name alone are employed to designate the element; sometimes using the English, sometimes the Latin or Greek name. Thus Cl stands for Chlorine, O for Oxygen, and K (from Kali, another name for Potash) for Potassium.

These letters, however, signify more than this: they stand not only for the elements in question, but they all have certain numbers belonging to them which indicate the proportions by weight in which the several elements are found to combine with each other. Thus Cl does not signify any weight of chlorine, but always exactly 35·5 parts by weight; K does not signify any weight of Potassium, but always 39·1 parts; while O signifies always 16 parts by weight of Oxygen. Hence it is evident that we may express by symbols not only the *qualitative* but also the *quantitative* composition of chemical substances. Thus, potassium chlorate consists of:—

Potassium . . .	39·1	or K.
Chlorine . . .	35·5	" Cl.
Oxygen	48·0=3×16	" O_3.

The symbol for potassium chlorate is, therefore, K Cl O_3.

In like manner each of the 63 elements has its particular symbol and number attached, signifying the proportion by weight in which it combines (*see* Table, page 6). The reasons which have led chemists to adopt these special numbers for the *combining weights* or *proportions* of the elements, and the laws which have been found to regulate their combination, will be explained as our stock of chemical facts gradually becomes larger.

The density or weight of a given volume of oxygen, compared with that of the same volume of hydrogen, is found to be sixteen, hydrogen, as the lightest body known, being taken as the standard. The specific gravity of oxygen, compared with the weight of the same volume of air taken as the unit, is found to be 1·1056. One litre of oxygen gas at 0° C., and under the pressure of 760 millimetres of mercury, weighs 1·4298 grammes.

OZONE.

Pure oxygen undergoes a remarkable modification when a series of electric discharges is passed through the gas: it thus attains more active properties; it is able to set free iodine from potassium iodide, and to effect oxidations which common oxygen is unable to bring about. This *allotropic* modification of oxygen has been termed *Ozone.* If a series of electric discharges be passed through pure oxygen, the gas becomes diminished in volume by about one-twelfth, and is partly transformed into ozone. If any substance be present, such as potassium iodide, capable of absorbing the ozone as it is formed, the whole of the oxygen can be transformed into this active modification. The peculiar smell which is observed when an electrical machine is worked, is caused by the presence of ozone; and if a paper, dipped in a solution of potassium iodide and starch paste, be held opposite a point on the conductor of the machine, the paper becomes blue, owing to the liberation of iodine and the formation of a blue compound of iodine and starch. Ozone can be obtained in several other ways; it is formed when a stick of phosphorus is allowed to hang in a bottle filled with moist air; it is produced in small quantities in the electrolytic decomposition of water (*see* p. 32); and it is formed by the action of strong sulphuric acid upon a salt called potassium permanganate.

HYDROGEN.

Symbol H. Combining Weight 1. *Density* 1.

Hydrogen is a colorless invisible gas, possessing neither

taste nor smell; it is the lightest body known, being 14·47 times lighter than air. It occurs free in small proportions in certain volcanic gases, and it has been lately shown to exist absorbed in certain specimens of meteoric iron; but it is found in much larger quantities, combined with oxygen to form water (ὕδωρ, water, and γεννάω, I produce), and it is by the decomposition of water, or of some other similar hydrogen compound, that the gas is always prepared. Hydrogen appears to have been first obtained by Paracelsus in the sixteenth century, but its properties were first exactly studied by Cavendish in 1781. One-ninth of the weight of water consists of hydrogen, and this gas can readily be obtained from it by the action of certain metals, which decompose the water, combining with the oxygen to form a metallic oxide, and liberating the hydrogen as a gas. The metals of the alkalies, potassium and sodium, decompose water at the ordinary temperature of the air; some other metals, as iron, are only able to do so at a red heat; whilst others, for instance silver and gold, are unable to decompose water at all. When a small piece of potassium is thrown into water, an instantaneous decomposition of the water ensues, potassium hydroxide (caustic potash) is formed, and the hydrogen of the water is liberated, so much heat being at the same time evolved that the hydrogen takes fire and burns. If the potassium, or, still better, sodium, be wrapped in a piece of wire gauze, and thus held in the water of the pneumatic trough, under the mouth of a cylinder, the hydrogen gas thus liberated may be collected, and its properties examined. Water consists of 2 parts by weight of hydrogen and 16 parts by weight of oxygen, and its chemical symbol is therefore H_2O. When potassium or sodium act upon water, half the hydrogen is liberated, the metal taking its place; this reaction can be represented by a *chemical equation*, as follows:—

$$\left.\begin{matrix}H\\H\end{matrix}\right\}O+K=\left.\begin{matrix}K\\H\end{matrix}\right\}O+H,^*$$

or water and potassium yield potassium hydroxide and oxy-

* The sign + used in chemical equations signifies "and" or "together with."

gen. This equation shows us that for every 1 part by weight of hydrogen which is liberated (H), 39·1 parts by weight of potassium (K) enter into combination. The hydroxide which is formed dissolves in the water, but its presence can easily be detected either by the peculiar caustic taste which the solution possesses (whence its name, caustic potash), or by its power of turning to a blue colour a solution of litmus which has been reddened by an acid.

To prepare hydrogen by the action of red-hot iron on water, a wrought-iron pipe, like a gun-barrel, filled with iron turnings, must be heated in a furnace, Fig. 3, and steam from a small flask or boiler passed over the red-hot metal through the tube ; hydrogen gas is given off, and oxide of iron left in the tube. The most convenient process of preparing pure hydrogen in quantity depends upon a property possessed by those metals, such as iron or zinc, which decompose water at a red heat ; namely, that these metals are able to evolve hydrogen from water at the ordinary temperature of the air if a dilute acid be present. For the purpose of thus obtaining hydrogen, a flask or bottle is provided with a cork and tube, some zinc clippings are introduced, and a mixture of one part of sulphuric acid and eight parts of water poured in through the tube funnel. After a few minutes a rapid effervescence commences, and the evolved gas is collected over water in bottles or cylinders as in the case of oxygen. Care must, however, be taken that all the air is expelled from the flask before the hydrogen is collected ; this is easily ascertained to be the case by filling a test tube with the gas, and trying whether it burns quietly when a lighted candle is brought to the mouth of the tube held downwards. If we concentrate by boiling the liquid remaining in the flask after the evolution of the hydrogen, we find that white crystals separate out when the liquid cools : these consist of zinc sulphate. A given weight of zinc (with sulphuric acid and water) can always be made to produce a certain weight of hydrogen, and a certain weight of zinc sulphate will always be formed. It is found by experiment that 2 parts by weight of hydrogen can be obtained by dissolving 65·2 parts of zinc with the formation of 161·2 parts of zinc sulphate. This can be represented by the equation—

$$H_2SO_4 + Zn = ZnSO_4 + H_2,$$

which not only indicates that sulphuric acid and zinc yield zinc sulphate and hydrogen, but also informs us as to the weights of the respective substances taking part in the reaction.

Hydrogen burns in the air when a light is brought to it with a very slightly luminous, although extremely hot flame ; and in the process the hydrogen combines with the oxygen of the air, forming water. The production of water by the combustion of hydrogen in the air may easily be shown by bringing a bright dry glass over the flame of hydrogen issuing from a fine jet, as in Fig. 5 ; the glass becomes at once dimmed, owing to

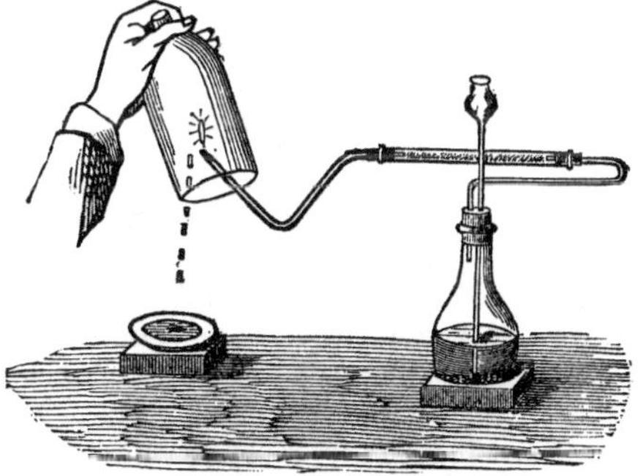

Fig. 5.

the condensation of water in small drops upon the cold dry surface. A number of these drops can be collected, and, upon examination, are found to consist of pure water. Hydrogen does not support the combustion of a candle, nor the life of an animal. If a burning taper is pushed up into a cylinder of this gas, held with its mouth downwards, the hydrogen burns at the mouth of the jar while the taper is extinguished; it can, however, be relit by the flame at the mouth. Hydrogen can be poured from one vessel to another in the air; but as it is lighter than air it must be poured upwards. The specific gravity of hydrogen, when air is taken as the unit, is found to be 0·0693; but for several reasons we shall find it more convenient to take hydrogen itself as our unit, and compare the weight of the same volumes of other gases with hydrogen instead of air. One litre of hydrogen gas at 0° C. and 760 mm. pressure weighs 0·08936 grams. Free hydrogen, like oxygen, has never been obtained in the liquid or solid state.

[The pupil must carefully work out the examples and exercises given for each Lesson at the end of the book, and thus test the accuracy of his knowledge.]

LESSON III.

PHYSICAL PROPERTIES OF GASES, &c.

It becomes now of importance to ascertain not merely the *weights* of oxygen and hydrogen capable of being evolved by using given weights of potassium chlorate or zinc, but likewise the *volume* of each gas thus obtained. Before we can enter into these calculations there are several important preliminary subjects, with the principles of which we must make ourselves acquainted.

The first of these is the *metric*, or French decimal system of weights and measures; the second is the mode of measuring temperature, and the construction and use of thermometers, together with the laws regulating the expansion of gases by heat; whilst the third relates to the measurement of atmospheric pressure by means of the barometer, and the laws regulating the changes which variations of pressure produce in the volumes of gases.

Metric System of Weights and Measures.

There are several distinct advantages to be gained by the adoption of this system, the chief of which is that the system is throughout a decimal one, and hence all calculations for reduction, such as occur in our old measures (from pennyweights to tons, or from inches to miles, for instance), are avoided. A second important consideration which renders our use of this system advisable, is that it is now generally adopted by men of science in all countries. The starting point of this system is the establishment of a unit of length called a *metre*, equal to rather more than our yard (more exactly, 39·37 English inches). This metre, like all other standards of length, is an arbitrary length : a standard metre was prepared, and of this copies are made for use.*

The metre is divided into tenths, hundredths, and thousandths ; these parts are termed respectively, *decimetres*, *centimetres*, and *millimetres*. The multiples of the metre, tens, hundreds, and thousands, are called *decametres*, *hectometres*, and *kilometres*. The measures of area, or square measure, and those of capacity, or cubic measure, are easily obtained ; we have square metres and square deci-, centi-, and milli-metres ; we have also cubic metres, and cubic deci-, centi-, and

* When the metre was first made, it was intended to give it a length which should have some reference to the earth's circumference, and a standard was made which had the length of the $\frac{1}{10,000,000}$ part of the distance from the equator to the pole, as measured by the French geometricians. Subsequent investigations have, however, proved that the measurement of the earth's circumference then made is not quite correct, and hence the metre turns out to be not quite (although very nearly) the $\frac{1}{10,000,000}$ part of the true distance of the pole from the equator. The value of the metric system does not at all depend upon this relation between the earth's circumference and the metre. *The metre* is the length of the bar of metal carefully preserved in Paris, from which copies have been taken for use.

milli-metres ; and we have the square and cubic measures derived from the multiples of the metre in the same way.

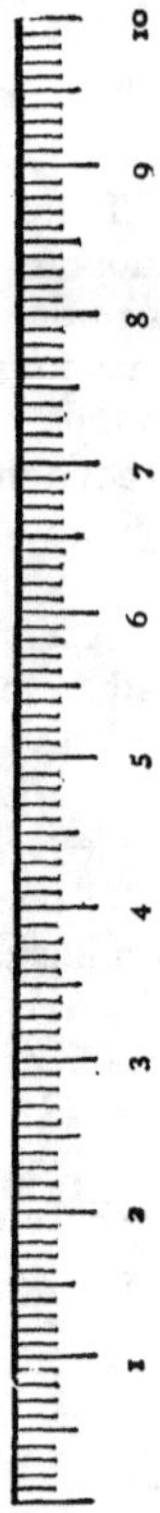

10 decimetres	1 metre.
100 centimetres	"
1,000 millimetres	"
100 square decimetres	1 square metre.
10,000 " centimetres	"
1,000,000 " millimetres	"
1,000 cubic decimetres	1 cubic metre.
1,000,000 " centimetres	"
1,000,000,000 " millimetres	"

The measure on the margin is one decimetre in length ; it contains 10 centimetres and 100 millimetres. For the sake of simplicity the word *litre* is used to signify 1 cubic decimetre (rather less than an English quart).

The French philosophers who arranged this metric system wished to have a simple relation between the measure of volume and that of weight, and they determined to take as their unit of weight the weight of 1 cubic centimetre of pure water of the temperature of 4° Centigrade, weighed at Paris. This weight is termed a *gramme*, or in English *gram.* It is divided like the metre into tenths, hundredths, and thousandths, called respectively deci-, centi-, and milli-gram ; whilst to the tens, hundreds, and thousands of grammes the names deca-, hecta-, and kilo-gramme are given. A table showing the relation between the weights and measures of the metric system and those commonly in use in this country is given in the Appendix.

Measurement of Temperature.—Thermometers.

Measurements of changes of temperature are always effected by ascertaining the expansion or contraction which bodies undergo by alteration of temperature. For this purpose liquids are generally used, as solids expand too little and gases too much to be convenient indicators. Mercury and alcohol are the liquids commonly employed, especially the former, because its rate of expansion is nearly uniform, and because the range of temperature which can be measured by a mercurial thermometer is large, this metal boiling at a very high temperature, and freezing at a comparatively low one. Alcohol is used when very low temperatures have to be measured, as this liquid has never yet been frozen. Air thermometers are only used in very delicate experiments in physics. In order to prepare a mercurial thermometer a straight piece of glass tubing, having a bore as uniform as possible throughout its whole length, is taken, and a bulb blown upon the end. This bulb, together with the whole of the tube, is then filled with mercury, and heated up to the highest temperature which the instrument is required to measure ; the open end of the tube is then completely closed, whilst full of mercury, by melting the glass before the blowpipe. The thermometer thus prepared requires graduating, in order that its indications may be compared with those of any other. This graduation is effected : 1. By plunging the bulb and stem in finely-powdered and melting ice, and marking on the stem the point where the mercury stands. 2. By immersing the bulb and stem in the steam given off from water boiling in a metallic vessel, and marking off the point where the mercury then stands. Care must be taken during this last experiment that the height of the barometer be observed ; the reasons for this precaution will be explained further on. Having obtained these two fixed points, it is easy to adapt a scale to the thermometer. Three scales, each of which is capable of being expressed in terms of the others, are at present in use : 1. The *Centigrade* scale. 2. *Fahrenheit's* scale. 3. *Réaumur's* scale. In the Centigrade scale (which we shall adopt, it being the one almost universally employed in scientific works, and in general use on the Continent) the space between these two points—called respectively the *freezing* and *boiling* points—is divided into 100 equal parts, each of which is called a *degree :* the Zero of the scale is placed at the freezing point, so that the boiling point is 100° C. These divisions are continued above the boiling and below the freezing points, and those below the freezing point are characterized by a minus sign, thus, —1°C., —2°C., etc. Fahrenheit divided the same space into 180 equal parts, each of which is called a degree Fahrenheit; he did not, however, commence his scale at the freezing point, as he erroneously thought that he had obtained the greatest possible degree of cold by making a mixture of snow and salt; the temperature of this mixture he found to be 32 of his degrees below freezing point; he therefore called the freezing point 32°. In Fahrenheit's scale, minus numbers are employed to denote degrees of temperature below the Zero of his scale; this scale is the one in common use in England, but is the most inconvenient one which we could adopt. Réaumur's scale (used in Russia and Sweden) resembles the Centigrade scale, except that the space between the freezing and boiling points is divided into 80 equal points; so that water boils at 80° Réaumur. The connection between these three scales is seen at a glance by reference to fig. 6. The relation between the degrees of Fahrenheit, Centigrade, and Réaumur, is expressed by the numbers 9, 5, 4. In converting from degrees Fahrenheit to Centigrade or Réaumur, we must remember first to subtract 32 and then reduce; whilst, when passing from degrees Centigrade and Réaumur to Fahrenheit, we must add 32 after the multiplication and division are completed.

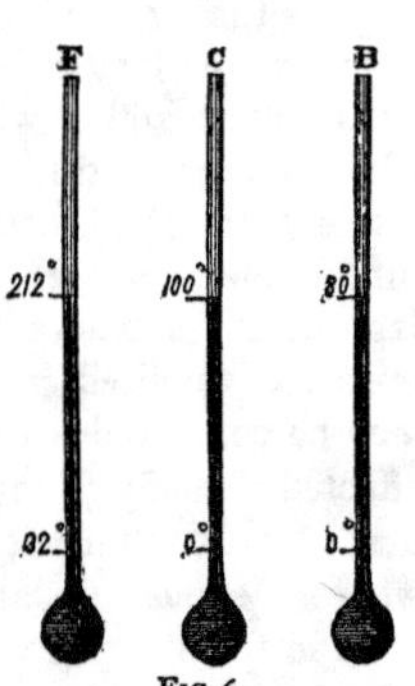

Fig. 6.

If very exact measurements are required, several precautions must be taken in the graduation and use of thermometers; thus, for instance, the tube must be calibrated—that is, the irregularities in the bore must be determined and allowed for, whilst any slight alteration in the position of the freezing point must from time to time be ascertained. Different mercurial thermometers often show slight differences in their indications, owing to the unequal expansion of different kinds of glass; hence it is necessary in exact experiments to have recourse to the air thermometer.

Expansion of Gases by Heat.

Solid and liquid bodies expand much less for equal increments of heat than gases; they also all expand differently, whilst all gases expand alike, or very nearly so. The expansion of solids and liquids is a subject with which, in elementary chemistry, we have little to do, whilst a knowledge of the laws regulating the expansion of gases is of more immediate importance. It has been found by exact and laborious experiment that all gases expand $\frac{1}{273}$ part of their volume at 0° C. for every increase in temperature of 1° Centigrade:

Thus	273	volumes of air	or hydrogen	at 0°C.
become	274	"	"	1° "
"	275	"	"	2° "
"	276	"	"	3° "
or	273+t	"	"	t° "

The decimal fraction corresponding to $\frac{1}{273}$ is 0·003665; 1 volume of air at 0° C. becomes 1·003665 volumes when heated to 1° C. This fraction is called the *co-efficient of the expansion of gases.** If we require to know the volume which 1,000 cubic centimetres of hydrogen measured at 0° C. will occupy when the temperature is raised to 20°, we must remember that the alteration in bulk takes place in the ratio of the numbers 273 to 273+20. Hence we multiply 1,000 by 293 and divide by 273. If we require to know what the volume 1,000 cbc.

* Regnault and Magnus have shown that hydrogen gas expands rather less than atmospheric air, whilst carbonic acid gas expands rather more than air. The coefficients of expansion from 0° to 100° obtained by these two renowned experimentalists are as follows:

	Regnault.	*Magnus.*
Hydrogen	0·36614	0·36556
Carbonic acid . . .	0·37099	0·36909

measured at 20° C. will occupy when the temperature sinks to 0°, we have to remember that the diminution in volume follows the same law, and that, therefore, 293 vols. at 20° will become 273 vols. at 0°. If we have 1,000 cbc. of gas at 20°, and desire to know the volume which it will occupy at 50°, we have in like manner to remember that 273+20, or 293 vols. at 20°, become 273+50, or 323 vols. at 50°; and then we can easily find the alteration in volume which the 1,000 cbc. of gas will undergo when heated from 20° to 50°.

Relation of Volume of Gases to Pressure.

When a gas is subjected to an increase of pressure, the volume of the gas becomes less; and when the pressure is withdrawn, the gas immediately expands again, and occupies exactly the same volume which it did before the pressure was increased. Solid and liquid bodies cannot be compressed in the same way. Gases are hence known as *compressible fluids*, and liquids as *incompressible fluids:* liquids, however, really are compressible, but only to a very slight extent: like gases, they recover their original volume on removal of the pressure. The law representing the relation between the volumes of a gas and the pressures to which the gas is subjected, is a very simple one: it is termed *Boyle's* or *Mariotte's Law*, from the names of the discoverers: it states that the *volume occupied by any gas is inversely proportional to the pressure to which it is subjected.* Thus, for instance, the volume 1 under pressure 1 becomes the volume 2 under the pressure $\frac{1}{2}$; the volume 3 under the pressure $\frac{1}{3}$, the volume $\frac{1}{2}$ under the pressure 2, and the volume $\frac{1}{3}$ under the pressure 3, and so on.* For a description of the experimental proof of this law, a work on Physics must be consulted.

The instrument which serves to measure the pressure

* This law, like many other physical laws, is only an approximation to the truth as ascertained by exact experiment. No gases obey the law exactly when high pressures are used, and many deviate perceptibly; still, as these deviations are but very slight, we may assume, for the purposes of our calculations, the absolute truth of the law of Boyle.

exerted by the air is termed a *barometer.* (Fig. 7.) This in its simplest form consists of a straight glass tube, about 800 mm. (33 inches) in length, closed at one end, and furnished with a millimetre scale. This tube is filled with dry mercury, and the open end placed downwards in a basin containing the same metal. It is then seen that the mercury sinks in the tube to a point about 760 mm. from the surface of the metal in the basin: it is sustained in this position by the pressure of the air. When this pressure increases, the height of the sustained column becomes greater; when it diminishes, the level of the mercury in the tube falls. All gases generated at the earth's surface are subject to this pressure, and their volumes increase or diminish according to the above law, when the superincumbent pressure becomes less or greater. In estimating the volume of hydrogen which can be collected from a given weight of zinc and sulphuric acid, it is clear that we require to know not only the temperature at which the gas is collected, but also the atmospheric pressure under which it is measured; and in order to be able to compare the bulks of two gases, we must always compare them under like conditions of temperature and pressure. For this purpose we agree to compare all the volumes of gases at the *standard temperature of* 0° C. and under the *standard pressure of* 760 *millimetres of mercury.* Suppose now that we desire to know what weight of potassium chlorate we need to take in order to fill with oxygen gas a gasholder having a capacity of 10 litres, the temperature of the room being 15° C. and the barometer standing at 752 mm. We know (1) that 122·6 parts by weight of potassium chlorate yield 48 of oxygen; (2) that a litre of oxygen at 0° C. and 760 mm. weighs 1·4298 grms. We must now ask, what will 10 litres of oxygen weigh if measured at 15° C. and under the pressure of 752 mm.? Now, 10 litres at 0° and 760 mm. will become

FIG. 7.

$\frac{10 \times 760 \times (273+15)}{752 \times 273}=10.661$ at 15° and 752 mm.; therefore, if 10 litres at 0° and 760 mm. weigh 14·298 grms. 10 litres at 15° and 752 mm. will weigh $\frac{14·298}{10·661}=13·411$ grms. Next we require to know how many grammes of chlorate will furnish this weight of oxygen; as every 122·6 parts of chlorate yield 48 parts of oxygen, we shall need $\frac{122·6 \times 13·411}{48}=34·254$ grms. of chlorate. In the same way we can calculate, for instance, the weight of zinc and sulphuric acid needed to inflate a balloon of the capacity of 150 cubic metres with hydrogen, when the thermometer stands at 11° C. and the barometer at 763 mm. [The student will do well to work out numerous examples of this kind, in order to familiarize himself with these methods of calculation (see Exercises at the end of the book).]

Diffusion of Gases.

Another physical property of gases is that of *diffusion.* Gases which, when mixed together, do not combine chemically, have the power of becoming intimately mixed together, even when differing in specific gravity, and when the heavier gas is placed at the bottom, and both remain at rest. This important property is called the *diffusive power of gases.* The rate at which gases diffuse varies greatly. Thus, a bottle filled with hydrogen lost 94.5 per cent. of this gas when left exposed to the air in the same time as that in which a bottle of carbonic acid lost only 47 per cent. of this gas in the same way. Gaseous diffusion goes on through the minute pores of certain solids, such as stucco, or thin plates of graphite; the different diffusive rates of air and hydrogen may be well seen by fixing a thin piece of stucco on to one end of a glass tube open at the other end, and filling this with hydrogen; on plunging the open end into water a steady rise of this liquid in the tube is noticed, and after some time the whole of the hydrogen is found to have disappeared, and the tube to contain only pure air. Experiments made upon this subject have shown that the velocity of diffusion of different gases is inversely proportional to the square roots of their densities; thus 4 volumes of hydrogen will pass through the diaphragm in the same time that 1 volume of oxygen is able to do so, oxygen being sixteen times as heavy as hydrogen. This property of gases has an important bearing upon the atmosphere of towns and dwelling-rooms, which is kept pure to a great extent by this diffusive power of gases.

The following table gives the rates of diffusion, as determined by Graham, of several gases, that of air taken to be equal to 1, compared with the inverse square roots of their densities, air also taken as the unit:—

	Density air=1.	$\frac{1}{\sqrt{\text{density}}}$.	Velocity of diffusion, air=1.
Hydrogen	0·06926	3·779	3·830
Nitrogen	0·9713	1·015	1·014
Oxygen	1·1056	0·9510	0·949
Carbon Dioxide . .	1·5290	0·8087	0·8120

LESSON IV.

OXIDES OF HYDROGEN.

WE are acquainted with two compounds of oxygen and hydrogen, namely:—

(1.) *Water* or *Hydrogen Mon-oxide.* Symbol H_2O. Combining Weight 18, Density 9.

(2.) *Hydrogen Di-oxide.* Symbol H_2O_2. Combining Weight 34.

Water, H_2O. When hydrogen burns in the air water is formed by the union of the former gas with oxygen. The discovery of the composition of water was made in 1781 by Mr. Cavendish, who showed that two volumes of hydrogen unite with one volume of oxygen to form water. In order to prove this, Cavendish made a mixture of these gases in this proportion by volume in a jar, and then allowed them to pass into a strong dry vessel (Fig. 8), from which the air had been pumped out. By means of two platinum wires melted through the glass (at B), an electric spark could be passed through the mixture of the two gases, causing their combination; dew was then seen to be deposited upon the sides of the vessel, and, when the stopcock was opened under water, this liquid rushed in, filling the whole space formerly occupied by the mixed gases. Cavendish weighed the glass before and after the ex-

FIG. 8.

plosion, and knowing the weight of gases taken, he found that the weight of the water produced was the same as that of the gases which combined. Since the above-mentioned year, the exact composition of water has been made the subject of careful synthetical experiment by many chemists, and the result has been to confirm by much more delicate methods this original conclusion. The most accurate of these methods of ascertaining the composition of water is a modification only of that originally used by Cavendish. We use for this purpose a long, accurately graduated, strong glass tube called a *Eudiometer* (*A*, Fig. 9), open at one end and closed at the other, whilst through the glass at the top are melted two platinum wires. This tube is first filled with mercury, and inverted mouth downwards over a trough filled with this metal (Fig. 9). Hydrogen gas is now allowed to enter the tube, and the volume admitted measured (suppose equal to 100 volumes); oxygen gas is next admitted, and the volume of the two mixed gases measured (suppose that 75 volumes of oxygen are added). In making this experiment, care must, however, be taken that the tube is not more than half full of the gaseous mixture, as great heat is evolved by the combustion, and hence a sudden expansion of volume occurs, for which reason it is necessary to press down the open end of the tube upon a plate of caoutchouc placed under the mercury. An electric spark is now passed through the gas along the platinum wires, when a flame is seen to pass down through the gas, showing that combination has occurred. As soon as the heat caused by the combustion has disappeared, the water produced will be deposited as dew upon the inside of the tube, and will then only take up about $\frac{1}{2000}$ part of the bulk which its constituent gases occupied, so that its volume may

FIG. 9.

be neglected; when the bottom of the eudiometer is opened the column of mercury in the tube rises, and we shall then find that only 25 volumes of gas remain, and this turns out to be pure oxygen. Thus we see that 100 volumes of hydrogen require exactly 50 volumes of oxygen for their complete combustion. By a modification of this experiment, it can be shown that the volume of the gaseous water formed, occupies exactly 100 volumes; or 2 volumes of hydrogen unite with 1 of oxygen to form 2 volumes of steam, hence the density of steam, or weight of 1 volume is $\frac{16+2}{2}=9$. The most striking method of demonstrating the composition of water analytically is by splitting it up into its constituent gases by means of a current of voltaic electricity. For this purpose we fill a glass vessel (Fig. 10) with water acidulated with sulphuric acid to enable it to conduct the electricity, and bring two test tubes filled with water and inverted into this vessel over two small platinum plates attached to wires of the same metal passing through the caoutchouc stopper at the bottom of the glass; on connecting these with the terminals of a battery of three

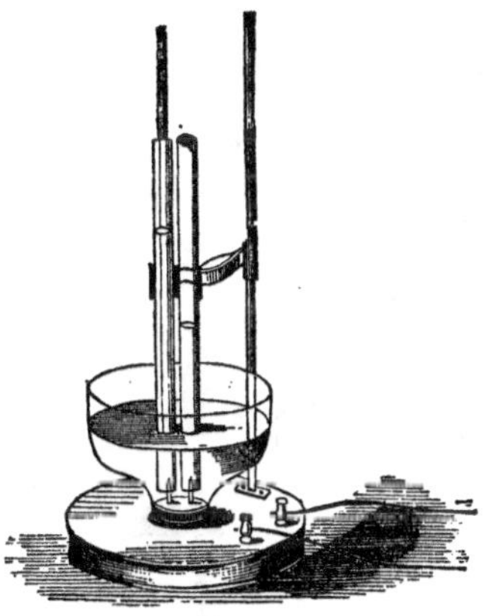

FIG. 10.

or four of Grove's elements, an evolution of gas from each plate is noticed, that disengaged from the plate in connection with the platinum end of the battery is found to be pure oxygen, whilst that coming off from the other plate, connected with the zinc end of the battery, is pure hydrogen gas. If the two tubes be graduated it will be seen that the volume of the hydrogen is a very little more than double that of the oxygen, for owing to the oxygen being rather more soluble in water than hydrogen, we do not thus get quite the exact proportions. In order to collect the detonating mixed gases evolved by this *electrolytic decomposition* of water, an apparatus represented in Fig. 11 may be employed.

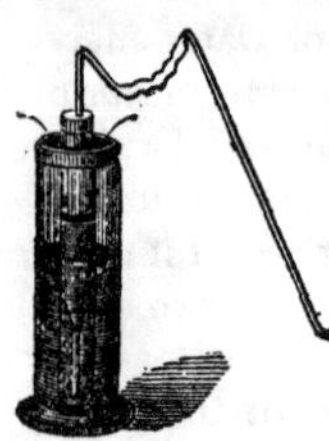
FIG. 11.

Oxygen being 16 times as heavy as hydrogen, and these gases combining to form water in the proportions by volume of one volume of the former to two of the latter, we now know that the proportions by weight in which these gases exist in water must be as 16 to 2. It is nevertheless most important that this calculation be verified by direct experiment. For this purpose use is made of the fact that copper oxide when heated alone, does not part with any of its oxygen, but when heated in the presence of hydrogen it parts with as much oxygen, as will, by combining with the hydrogen, form

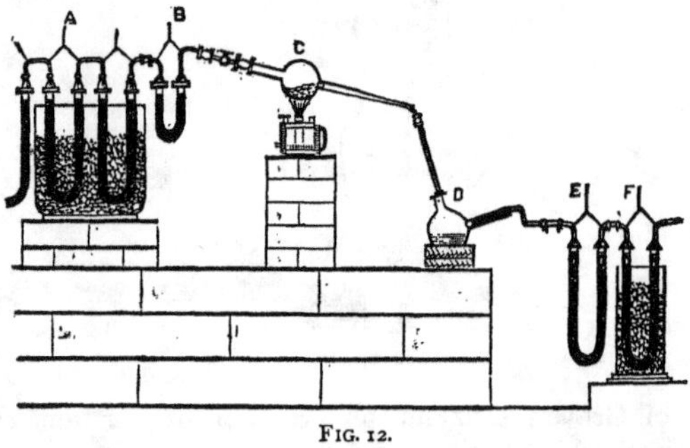

FIG. 12.

water, being itself wholly or partly reduced to metallic copper. If, therefore, we take a known weight of copper oxide, heat it, and pass pure hydrogen over it until it has parted with all its oxygen, and if we collect and weigh all the water thus formed, and likewise weigh the remaining metallic copper, we shall have made a *synthesis by weight of water*. For the loss in weight of the copper oxide is the weight of oxygen which has combined with hydrogen to form water ; and the difference between this weight and that of the water formed, is the weight of the hydrogen thus combined. The arrangement originally used for this determination is represented in Fig. 12. The hydrogen is purified from any trace of arsenic, sulphur, and moisture which it may contain by passing through the U tubes, containing absorbent substances, and placed in a freezing mixture. Some only of these (A) are represented in the figure. The tube B, containing a very hygroscopic substance, is weighed both before and after the experiment, and if no increase occurs, the dryness of the gas is insured. The gas then comes in a perfectly pure state into contact with the heated copper oxide contained in the bulb C. This first bulb, which is accurately weighed, is placed in connection with a second bulb D, in which the water formed by the reduction of the oxide collects ; any moisture which may escape condensation in this bulb is retained in weighed drying tubes, E and F, containing fragments of pumice moistened with sulphuric acid. Most careful experiments made according to this method, carried out with many precautions, which cannot here be detailed, have shown that 88·89 parts of oxygen by weight unite with 11·11 parts of hydrogen to form 100 parts of water.

Free oxygen and hydrogen combine together, when a light is brought in contact with them, with so much force that a violent and dangerous explosion occurs from the sudden expansion caused by the great heat evolved in combination. If we fill a strong soda-water bottle one-third full with oxygen and two-thirds with hydrogen, and then bring a flame to the mouth, the gases combine, producing a sudden detonation like the report of a pistol. Many fatal accidents have occurred to persons who have carelessly experimented with large volumes of this explosive mixture. In order to exhibit the great heat evolved by the combination of the two gases the oxyhydrogen blowpipe is employed; in this arrangement the gases are contained separately in two caoutchouc bags, being only brought together at the point at which the combination is desired, so that all danger of explosion is avoided. The flame thus produced is very slightly luminous, but its temperature is so high, that the most difficultly fusible metals, such as platinum, may be easily melted in it, whilst iron wire held in the flame burns with beautiful scintillations, forming an oxide of iron.

Water exists in nature in three forms: in the solid form as ice, in the liquid state as water, and in the gaseous form as steam. At all temperatures between 0° and 100° C. it takes the liquid form, and above 100° it entirely assumes the gaseous form (under the ordinary atmospheric pressure of 760 mm.). The melting point of ice is always found to be a constant temperature, and hence it is taken as the zero of the Centigrade scale; water may, however, under certain conditions, be cooled below 0° C. without becoming solid; still ice can never exist at a temperature above 0° C. In passing from the solid to the liquid state water becomes reduced in volume, and on freezing, a sudden expansion (from 1 volume to 1·099) takes place. That this expansion exerts an almost irresistible force is well illustrated by the splitting of rocks during the winter. Water penetrates into the cracks and crevices of the rocks, and on freezing widens these openings; this process being repeated over and over again, the rock is ultimately split into fragments. Hollow balls of thick cast-iron can thus easily be split in two by filling them with water and closing by a tightly fitting screw and then exposing them to a temperature below 0° C.

In the passage from solid ice to liquid water, we not only observe this alteration in bulk, but we notice that a very remarkable absorption, or disappearance of heat, occurs. This is rendered plain by the following simple experiment:—Let us take a kilogramme of water at the temperature 0°, and another kilogramme of water at 79° ; if we mix these, the temperature of the mixture will be the mean, or 39°·5 ; if, however, we take 1 kilogramme of ice at 0° and mix it with a kilogramme of water at 79°, we shall find that the whole of the ice is melted, but that the temperature of the resulting 2 kilogrammes of water is exactly 0° ; in other words, the whole of the heat contained in the hot water has just sufficed to melt the ice, but has not raised the temperature of the water thus produced. Hence we see that in passing from the solid to the liquid state a given weight of water takes up or *renders latent* just so much heat as would suffice to raise the temperature of the same weight of water through 79° C. ; the *latent heat of water* is therefore said to be 79 *thermal units*—a thermal unit meaning the *amount of heat required to raise a unit of water through* 1° C. When water freezes, or becomes solid, this amount of heat which is necessary to keep the water in the liquid form, and is therefore well termed the *heat of liquidity*, is evolved, or rendered sensible. A similar disappearance of heat on passing from the solid to the liquid state, and a similar evolution of heat on passing from the liquid to the solid form, occurs with all substances ; the amount of heat thus rendered latent or evolved varies, however, with the substance. A simple means of showing that heat is evolved on solidification consists in obtaining a saturated hot solution of Glauber's salt (sodium sulphate), and allowing it to cool. Whilst it remains undisturbed, it retains the liquid form, but if agitated, it at once begins to crystallize, and in a few moments becomes a solid mass. If now a delicate thermometer be plunged into the salt while solidifying, a sudden rise of temperature will be noticed. Similarly water at rest may be cooled down below 0° C. without solidifying, but if agitated, it at once solidifies, and the temperature of the whole mass instantly rises to 0° C.

When water is *heated* from 0° to 4°, it is found to *contract*, thus forming a striking exception to the general law, that bodies expand when heated and contract on cooling ; on *cooling* from 4° to 0° it *expands* again : above 4°, however, it follows this ordinary law, expanding when heated, and contracting when cooled. This peculiarity in the expansion and contraction of water may be expressed by saying that *the point of maximum density of water is* 4° C.; that is, a given bulk of water will at this temperature weigh more than at any other. Although the amount of contraction on heating from 0° to 4° is but small (1 volume of water at 4° becoming 1+0·00012 at 0°), it yet exerts a most important influence upon the economy of nature. If it were not for this apparently unimportant property, the climate of England would be perfectly arctic, and Europe would in all probability be uninhabitable. In order better to understand what the state of things would be if water obeyed the ordinary laws of expansion by heat, we may perform the following experiment. Take a jar containing water at a temperature above 4°, place a thermometer at the top and another at the bottom of the liquid. Now bring the jar into a place where the temperature is below the freezing point, and observe the temperature at the top and bottom of the liquid as it cools. It will be seen that at first the upper thermometer always indicates a higher temperature than the lower one; after a short time both thermometers mark 4°; and as the water cools still further, it will be seen that the thermometer at the top always indicates a lower temperature than that shown by the one at the bottom; hence we conclude that water above or below 4° is lighter than water at 4°. This cooling goes on till the temperature of the top layer of water sinks to 0°, after which a crust of ice is formed; but if the mass of the water be sufficiently large, the temperature of the water at the bottom is never reduced below 4°. In nature precisely the same phenomenon occurs in the freezing of lakes and rivers;* the surface-water is gradually cooled by cold winds, and thus becoming heavier, sinks, whilst lighter and warmer water rises to supply its place: this goes on till the temperature of the whole mass is reduced to 4°, after which the surface-water

* The point of maximum density of sea-water is considerably lower than that of fresh, and is in fact below 0° C.

never sinks, however much it be cooled, as it is always lighter than the deeper water at 4°. Hence ice is formed only at the top, the mass of water retaining the temperature of 4°. Had water become heavier as it is cooled down to the freezing point, a continual circulation would be kept up until the whole mass was cooled to 0°, when solidification of the whole would ensue. Thus our lakes and rivers would be converted into solid masses of ice, which the summer's warmth would be quite insufficient to melt thoroughly; hence the climate of our now temperate zone might approach in severity that of the arctic regions. Sea-water does not freeze *en masse*, owing to the great depth of the ocean, which prevents the whole from ever being cooled down to the freezing point; similarly, in England, very deep lakes never freeze, as the temperature of the whole mass never gets reduced to 4° C.

In passing from the liquid to the gaseous state, water exhibits several interesting and important phenomena. In the first place, when we heat water to 100° C. it begins to boil, or *enters into ebullition;* that is, a rapid disengagement of water-gas, or steam, from the lower or most heated surface takes place: this is well seen when water is heated in a glass globe over a gas flame. In this passage from the liquid to the gaseous state, a large quantity of heat becomes latent, the temperature of the steam given off being the same as that of the boiling water, as, like all other bodies, water requires more heat for its existence as a gas than as a liquid. The amount of heat latent in steam is roughly ascertained by the following experiment. Into 1 kilog. of water at 0°, steam from boiling water, having the temperature of 100°, is passed until the water boils: it is then found that the whole weighs 1·187 kilogs. or 0·187 kilogs. of water in the form of steam at 100°, have raised 1 kilog. of water from 0° to 100°; or 1 kilog. of steam at 100° would raise 5·36 kilogs. of ice-cold water through 100°, or 536 killogs. through 1°. Hence the *latent heat of steam is said to be* 536 *thermal units.*

Water, and even ice, constantly give off steam or aqueous vapor at all temperatures, when exposed to the air; thus we know that if a glass of water be left in a room for some days, the whole of the water will gradually evaporate. This power of water to rise in vapor at all temperatures is called the *elastic force*, or *tension*, of aqueous vapor; it may be measured, when a small quantity of water is placed above the mercury in a barometer, by the depression which the tension of the vapor thus given off is capable of exerting upon the mercurial column (as in Fig. 7). If we gradually heat the drops of water thus placed in the barometer, we shall notice that the column of mercury gradually sinks, and when the water is heated to the boiling point, the mercury in the barometer tube is found to stand at the same level as that in the trough, showing that the elastic force of the vapor at that temperature is equal to the atmospheric pressure. Hence *water boils when the tension of its vapor is equal to the superincumbent atmospheric pressure.* On the tops of mountains, where the atmospheric pressure is less than at the sea's level, water boils at a temperature below 100°; thus at Quito, where the mean height of the barometer is 527 mm., the boiling point of water is 99°·1; that is, the tension of aqueous vapor at 90°·1 is equal to the pressure exerted by a column of mercury 527 mm. high. Founded on this principle, an instrument has been constructed for determining heights by noticing the temperatures at which water boils. A simple experiment to illustrate this fact consists in boiling water in a globular flask, into the neck of which a stopcock is fitted: as soon as the air is expelled, the stopcock is closed, and the flask removed from the source of heat; the boiling then ceases; but on immersing the flask in cold water, the ebullition recommences briskly, owing to the reduction of the pressure consequent upon the condensation of the steam; the tension of the vapor at the temperature of the water in the flask being greater than the diminished pressure. All other liquids obey a similar law respecting ebullition; but as the tensions of their vapors are very different, their boiling points vary considerably.

When steam is heated alone, it expands according to the

law previously given for permanent gases, but when water is present, and the experiment is performed in a closed vessel, the elastic force of the steam increases in a far more rapid ratio than the increase of temperature. The following table gives the tension of aqueous vapor, as determined by experiment, at different temperatures measured on the air thermometer.

Tension of the Vapor of Water.

Temperature Centigrade.	Tension in millimetres of mercury.	Temperature Centigrade.	Tension in atmospheres, 1 atmosphere =760 mm. of mercury.
—20°	0·927	100°	1
—10	2·093	111·7	1·5
0	4·600	120·6	2
+5	6·534	127·8	2·5
10	9·165	133·9	3
15	12·699	144·0	4
20	17·391	159·2	6
30	31·548	170·8	8
40	54·906	180·3	10
50	91·982	188·4	12
60	148·791	195·5	14
70	233·093	201·9	16
80	354·280	207·7	18
90	525·450	213·0	20
100	760·000	224·7	25

We now see why the barometric height must be noticed in graduating a thermometer (Page 21); if the height differ from 760 mm. the temperature of the water boiling under that pressure will not be quite 100° C. A metal vessel is here employed, because it is found that water does not always boil at 100° in glass vessels, even though the atmospheric pressure be 760 mm., owing to some molecular action analogous to cohesion between the glass and water.

Water and ice, when seen in large masses, are found to possess a blue color; this is well seen in the glaciers and lakes of Switzerland. In order to obtain pure water the chemist is obliged to *distil* river or spring-water (that is, to boil the water and collect the water formed by the condensation of the steam thus produced), as all such water contains more or less *solid matter in solution* derived from the surface of the earth over which the water flows; this dissolved solid matter is left behind on boiling off the water. *Solid matter in suspension* can be got rid of by the simple process of *filtration* through paper, sand, &c. An arrangement for distillation used in laboratories is seen in Fig. 13. Rain-water is the

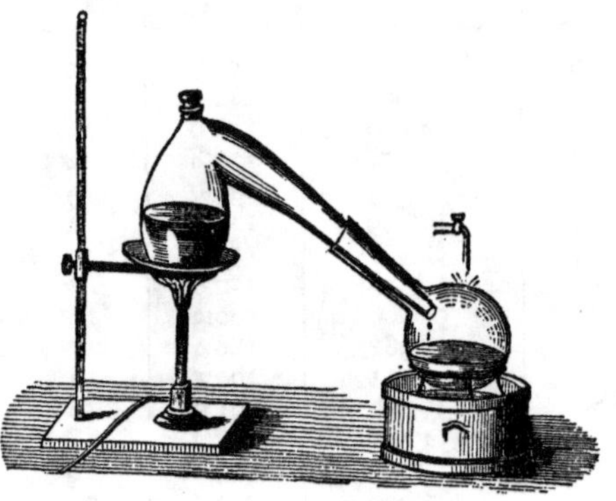

FIG. 13.

purest form of water occurring in nature, but even this contains impurities derived from the dust, &c. in the air, and no sooner does it touch the earth's surface than it dissolves some of the materials with which it comes in contact, and according to the nature of the ground over which it passes becomes more or less impure. All fresh water on the earth's surface has been derived from the ocean by a vast process of distillation, having been deposited in the form of rain or snow from the atmosphere.

Water is the most general solvent for chemical substances with which we are acquainted. Most salts are soluble to a greater or less extent in water, and are deposited again in crystals when the water is evaporated; we are unacquainted with any simple general law regulating the quantities of solids taken up by water; in most cases the solubility is greater in hot than in cold water. Water is also contained in the solid state in combination as *water of crystallization* in many salts; when this water is driven off by heat, the crystal falls to powder. Gases also dissolve in water in quantities varying with the nature of the gas, the temperature, and the pressure to which the gas is subjected. It is solely in consequence of the presence of oxygen derived from the air dissolved in the water of lakes, rivers, and seas, that fish are enabled to keep up their respiration; as the water passes through their gills the oxygen is taken up to purify their blood.

Hydrogen Di-oxide. Symbol $H_2 O_2$. This is a substance which does not occur in nature, but is artificially prepared by acting on barium di-oxide Ba O_2 with hydrochloric acid H_2 Cl_2; an exchange takes place between the barium and hydrogen, giving rise to hydrogen di-oxide and barium chloride. Thus

$$\begin{array}{c|c} Ba & O_2 \\ Cl_2 & H_2 \end{array}$$

It has received the name of oxygenated water, as it easily decomposes into oxygen and water: it is found to contain twice as much oxygen as water does, consisting of 2 parts by weight of hydrogen combined with 32 of oxygen; hence, if we represent water by the symbol H_2O, hydric peroxide will be written H_2O_2.

Hydrogen di-oxide acts as a powerful bleaching agent, owing to the ease with which it parts with half its oxygen, thus oxidizing the coloring matters; but owing to the difficulty attending its preparation, it has not been applied to this purpose on a large scale.

In addition to water and hydrogen di-oxide, we are acquainted with no compounds of oxygen and hydrogen.

LESSON V.

NITROGEN.

Symbol N. Combining Weight 14. *Density* 14.

NITROGEN exists in the free state in the air, of which it constitutes four-fifths by bulk; it occurs combined in the bodies of plants and animals in various chemical compounds, such as *nitre*, whence the gas derives its name (generator of nitre). It is best obtained from the air by taking away the oxygen with which it is mixed; for this purpose we may burn a piece of phosphorus in a bell-jar filled with air, the mouth of which is placed in a vessel full of water. White fumes of a compound of phosphorus and oxygen, phosphoric pentoxide, at first fill the jar, but these soon subside and dissolve in the water, leaving the nitrogen in a nearly pure state. One-fifth of the original volume of the air, consisting of oxygen, will have disappeared. Nitrogen may also be prepared by passing air over red-hot metallic copper, which combines with the oxygen, forming a solid compound, and leaving the gaseous nitrogen in a pure state. Nitrogen is also formed when a current of chlorine is passed through an excess of a solution of ammonia; nitrogen gas is evolved, and sal-ammoniac remains behind in solution. If the chlorine be present in excess, a most dangerous and explosive compound is formed. Nitrogen is a colorless and tasteless inodorous gas slightly lighter than air (specific gravity 0·972, air being 1.0). It does not combine readily with bodies, and is a very inert substance, neither supporting combustion nor animal life, nor burning itself: it has, however, no poisonous properties, and animals plunged into a jar of this gas die simply of suffocation from the want of oxygen. Nitrogen can be made to unite with both oxygen and hydrogen; when combined with the latter it forms a powerful alkaline base, *ammonia*, and united with both elements it forms a strong acid, *nitric acid.*

The Atmosphere.

The Atmosphere is the gaseous envelope encircling the earth ; and it constitutes the ocean of air at the bottom of which we live. We become aware of the existence of the air when we move rapidly, and experience the resistance offered to the passage of our bodies, and also when the air is in motion giving rise to a wind. We notice the pressure of the atmosphere if we withdraw the air from beneath the hand by a powerful air-pump, for we then find that the hand is pressed down with a force equal to 1·033 kilogs. on a square centimetre, or nearly 15 lbs. on every square inch. The total atmospheric pressure which the human body has to support hence amounts to several tons, but this pressure is not felt under ordinary circumstances, because the pressure is exerted equally in every direction. The instrument used for measuring the pressure of the air is termed a Barometer (*see* Fig. 7, p. 25), and the average pressure at the sea level is equal to that exerted by a column of mercury 760 mm. high. The air being elastic and having weight, it is clear that the lower layers of air must be more compressed than those above them, and hence the density of the air must vary at different heights above the sea level. The density of the air being thus dependent on the pressure to which it is subjected, the higher strata of air become extremely rarified, and it is hence difficult to say exactly whereabouts the air ceases, but it appears that the limit of the atmosphere is about 45 miles from the level of the sea. If the whole atmosphere were of the same density throughout as it is at the earth's surface, it would only reach to a height of a little less than 5 miles above the sea level. The weight of one litre of dry air at 0° and under 760 mm. of pressure is 1·2932 grammes.

Respecting the *chemical composition of the atmosphere*, we have to remark, in the first place, that the air is a mixture, and not a chemical compound of its constituent gases, although, as we shall see, these occur throughout the atmosphere in almost unvarying proportions. The grounds for coming to this conclusion are, first, if we bring oxygen and nitrogen together in the proportions in which they are found in air, no elevation of temperature or alteration in bulk occurs (as is invariably the case when gases combine), and yet the mixture acts in every way like air; secondly, that the relative quantities of the two gases present are, not those of their combining weights, nor of any simple multiples of these weights, and that although in general the proportions of the two gases are constant, yet instances not unfrequently occur in which this ratio is different from the ordinary one. The most convincing proof, however, that air is not a chemical compound, is derived from an experiment upon the solubility of air in water: when air is shaken up with a small quantity of water, some of the air is dissolved by the water; thus dissolved air is easily expelled again from the water by boiling, and on analysis this expelled air is found to consist of oxygen and nitrogen in the relative proportions of 1 and 1·87. Had the air been a chemical compound, it would be impossible to decompose it by simply shaking it up with water; the compound would then have dissolved as a whole, and on examination of the air expelled by boiling, it would have been found to consist of oxygen and nitrogen in the same proportions as in the original air, viz. as 1 to 4. This experiment shows, therefore, that the air is only a mixture, a larger proportion of oxygen being dissolved than corresponds to that contained in the atmosphere, owing to this gas being more soluble in water than nitrogen.

There are many ways of determining the amounts of oxygen and nitrogen contained in the air, the best of these being by the *eudiometer;** by means of which the composition by volume is ascertained. For this purpose the same arrangement is employed as that used in the eudiometric synthesis of water. (Fig. 14.) A quantity of air sufficient to fill the tube about one-sixth full is introduced into the eudiometer previously filled with mercury; the volume of this air is then accurately ascertained by reading off with a telescope the

* From εὔδιος, good, and μέτρον, a measure: a measure of the goodness or healthiness of the air, that is, of the quantity of oxygen which it contains.

number of the millimetre divisions on the tube to which the mercury reaches, whilst the height of the column of mercury in the tube above the trough, together with that of the barometer, and the temperature of the air, are also read off. Such a quantity of pure hydrogen gas is now added as is more than sufficient to combine with all the oxygen present; and the volume of this gas, and the pressure exerted upon it, are then determined as before. An electric spark is now passed through the mixture, care having been taken to prevent any escape of gas by pressing the open end of the eudiometer against a sheet of caoutchouc under the mercury in the

FIG. 14.

trough. After the explosion the volume is again determined as before, and is found to be less than that before the explosion, the whole of the oxygen and part of the hydrogen having combined to form water; the diminution, therefore, represents exactly the volumes of these gases which have united. We know, however, from our previous experiments upon the composition of water, that 2 vols. of hydrogen always unite with 1 vol. of oxygen to form water; hence one-third of the diminution in volume must represent the oxygen which has disappeared, and, therefore, the volume of oxygen contained in the air taken. An example may make this clearer. Suppose the volume of air taken amounted to 100 vols. and that after the addition of hydrogen the volume of the mixture was 150 vols.; after the explosion 87 vols. were found to remain, that is, 63 vols. had disappeared; then $\frac{63}{3}=21$ will be the volume of oxygen contained in 100 vols. of air.

Analyses of air collected in various parts of the globe thus made with the greatest care have shown that the relative quantities of oxygen and nitrogen remain the same, or very nearly the same, from whatever region the air may have been taken. So that whether the air be derived from the tropics, or the arctic seas, from the bottom of the deepest mine or from an elevation of 20,000 feet above the earth's surface, it contains from 20·9 to 21 vols. of oxygen per cent.

When we know the composition of air by volume and the relative densities of the two constituent gases (14 for nitrogen and 16 for oxygen), we can calculate its composition by weight; we thus find that in 100 grms. of air 23·16 grms. of oxygen are mixed with 76·84 grms. of nitrogen. It is important to control this calculation by experiment; for this purpose a large glass globe furnished with a stopcock is rendered vacuous by the airpump and then weighed; a tube of hard glass filled with copper turnings and also furnished with stopcocks is likewise weighed. This tube is then heated to redness in a long tube-furnace, and connected at one end with the empty flask, at the other with a series of tubes filled with caustic potash and sulphuric acid, for the purpose of completely freeing the air passing through them from carbonic acid and aqueous vapor; the cocks are then slightly opened and air allowed to pass slowly through the purifiers into the hot tube, where it is completely deprived of oxygen by the hot metallic copper which is thereby oxidized; the nitrogen passes on alone into the empty flask. After the experiment is concluded, the cooled tube is again weighed, and the increase over the former weighing gives the quantity of oxygen, whilst the increase in weight of the globe gives the nitrogen. The mean of a large number of experiments thus made shows that 100 parts by weight of air contained 23 parts by weight of oxygen and 77 of nitrogen.

In addition to the two above-mentioned gases, the air contains several other important constituents, especially carbonic acid gas, aqueous vapor, and ammonia gas. We have already noticed (page 11) the important part which the carbonic acid gas of the air plays in the phenomena of vegetation, this gas being the source from which plants obtain the carbon they need to form their tissues. The quantity of carbonic acid present in the air is very small compared with the quantities of oxygen and nitrogen, being only about 4 vols. to 10,000 of air ; nevertheless the absolute quantity of this gas contained in the whole atmosphere is enormously large (viz. about 3,000 billion kilogs.) The quantity of carbonic acid contained in the air can be found by drawing a known volume of perfectly dry air (not less than 20 litres) through weighed tubes containing caustic potash ; the increase in weight of the tubes gives the weight of carbonic acid contained in the air drawn through. Fig. 15 shows the arrangement of the

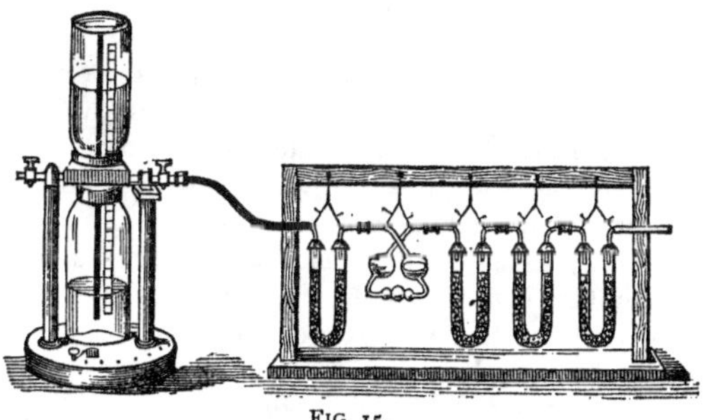

FIG. 15.

apparatus ; on the left is the aspirator, which by means of the flow of a known volume of water from the top to the bottom vessel causes the passage of an equal volume of air through the tubes ; the two tubes farthest from the aspirator contain pumice-stone steeped in sulphuric acid, and serve to dry the air completely before passing into the next tube and potash bulbs, in which the carbonic acid is absorbed by caustic potash : the tube nearest the aspirator also contains sulphuric acid and pumice to avoid a loss of moisture from the potash solution in the bulbs. The quantity of carbonic acid contained in the air in different localities and under different circumstances varies considerably—from 2 to more than 10 in 10,000 vols. of air. In houses and closed inhabited spaces, the quantity of carbonic acid present is often much larger, and the object of ventilation is to reduce the quantity of carbonic acid to as low a point as possible. Other methods for the estimation of carbonic acid are described in the larger manuals.

Aqueous vapor is contained in the air in quantities varying in different localities and at different times, and depending mainly upon the temperature of the air. Air at a given temperature cannot contain more than a certain quantity of moisture in solution, and when it has taken up this maximum quantity, it is said to be saturated with aqueous vapor. The higher the temperature of the air, the more water can it retain as vapor, and when air saturated with moisture is cooled, the water is deposited in the liquid form in very small globules, forming a mist, fog, or cloud. This is the cause of the fall of rain, snow, and hail ; when warm air heavily laden with moisture from the ocean passes into a higher and colder position, or meets with a current of air of lower temperature, it cannot any longer retain so much aqueous vapor, and a large quantity assumes the liquid form, falling as rain when the temperature is above the freezing point, or crystallizing as snow-flakes if the temperature be below that point. Hail is caused by the congelation of raindrops by passing through a stratum of air below the freezing point. The quantity of rain thus deposited is very large : 1 cubic metre of air saturated with moisture at 25° C. contains 22·5 grms. of water, and if the temperature of this air be reduced to 0° C., it will then be capable of retaining only 5·4 grms. of water vapor; hence 17·1 grms. of water will be deposited as rain. The air in England is often saturated with moisture, whilst the driest air observed on the coast of the Red Sea during a simoom contained only one-fifteenth of the saturating quantity. Instruments for ascertaining the degree of moisture or humidity of the air are termed *hygrometers.*

The deposition of dew is caused by the rapid cooling of the earth's surface by radiation after sunset, and by the consequent cooling of the air near the ground below the temperature at which it begins to deposit moisture.

The amount of aqueous vapor contained in the air at any time can be determined by the apparatus used for the estimation of the carbonic acid, for the moisture must be removed from the air before the carbonic acid can be absorbed, and the increase in weight of the tubes filled with pumice-stone moistened with sulphuric acid gives the weight of aqueous vapor. In general the air contains from 50 to 70 per cent. of the quantity necessary to saturate it. If the quantity be not within these limits the air is either unpleasantly dry or moist.

The next important constituent of the air is ammonia, which is a compound of nitrogen and hydrogen, and only exists in comparatively very minute quantities (about 1 part in 1,000,000 of air). Nevertheless it plays a very important part, as it is mainly from this ammonia that vegetables obtain the nitrogen which they need to form their seeds and fruit; for it appears that plants have not the power of assimilating the free nitrogen of the atmosphere. Other substances which occur in the atmosphere in very small quantities may be considered as accidental impurities. Amongst them, volatile organic matter is the most important, as probably influencing to a great extent the healthiness of the special situation. We become aware of the existence of such organic putrescent substances when entering a crowded room from the fresh air; and it is probable that the well-known unhealthiness of marshy and other districts is owing to the presence of some organic impurity. At present, however, we possess but little certain knowledge on this subject. Ozone is also present in fresh air, but generally absent in the close air of towns and dwelling-rooms, owing to its decomposition by the organic matter, &c., in such air; we do not know how it is formed in nature, unless it be by the discharge of atmospheric electricity.

LESSON VI.

COMPOUNDS OF NITROGEN WITH OXYGEN.

We are acquainted with five distinct chemical compounds of nitrogen with oxygen, viz.:

1	*Nitrogen Mon-oxide,*	containing	28	parts by weight	of N to	16 of O.
2	*Nitrogen Di-oxide,*	"	28	"	"	32 —
3	*Nitrogen Tri-oxide,*	"	28	"	"	48 —
4	*Nitrogen Tetr-oxide,*	"	28	"	"	64 —
5	*Nitrogen Pent-oxide,*	"	28	"	"	80 —

It will be seen that the oxygen contained in these compounds is in the proportion of the numbers 1, 2, 3, 4, 5, to one and the same quantity of nitrogen; and here, for the first time, we meet with a striking example of the *law of chemical combination in multiple proportion.* Thus, while 28 parts by weight of nitrogen combined with 16 parts of oxygen form 44 parts of nitrogen mon-oxide, we find that any other compounds of these two elements contain some simple multiple of 16 parts by weight of oxygen (thus, either 2×16, 3×16, 4×16, or 5×16), and that no compounds exist containing any intermediate quantity of oxygen.

This law of multiple proportions was first enunciated by Dalton, and is the expression of well-established experimental facts. Dalton endeavored to explain these facts by his celebrated *Atomic Theory.* He asked himself, Why do the elements combine only in multiples of their several combining proportions? and he answered the question by the following supposition.

Matter is made up of small indivisible portions, which he called *Atoms* (ἀ, privative, and τέμνω, I cut). These atoms do not all possess the same weights, but the relation between their weights is represented by that of the combining weights of the elements; thus he supposed the atom of oxygen to be 16 times as heavy as the atom of hydrogen, and the weights of the atoms of nitrogen and oxygen as 14 to 16. Dalton further assumed that chemical combination consists in the approximation of the individual atoms to one another; and, having made these assumptions, he was able to explain why compounds must contain their constituents in the combining proportions, or in multiples of them, and in no intermediate proportion. Let us take, for example, the compounds of nitrogen and oxygen: the lowest of these consists of one single atom of oxygen combined with 2 atoms of nitrogen, or with one double atom of nitrogen, as it contains 16 parts of oxygen to 28 of nitrogen; thus, (N)(N)(O); and we therefore write its formula, N_2O, and call it nitrous oxide. The next compound that can be formed must be produced by the addition of another atom of oxygen to this; thus we get (N)(N)(O)(O) $=N_2O_2$, or nitrogen di-oxide. The next must be formed by the attachment of another atom of oxygen, and thus we get (N)(N)(O)(O)(O) $=N_2O_3$, or nitrogen tri-oxide. The next possible compound is (N)(N)(O)(O)(O)(O) $=N_2O_4$, or nitrogen tetr-oxide;* and the next (N)(N)(O)(O)(O)(O)(O)

* If nitric oxide and nitric tetr-oxide be considered to be represented respectively by the formula N_2O_2 and N_2O_4, they will be exceptions to the law mentioned on page 27, respecting the density of compound gases or vapors, as instead of having their densities represented by the halves of their combining proportions, they will have them represented by the quarters of these numbers.

$=N_2O_5$, or nitrogen pent-oxide. We thus see that an atom being indivisible, no intermediate compounds can be formed. In considering this subject, we shall do well to remember that the law of multiple proportions being founded on experimental facts, stands as a fixed bulwark of the science, which must ever remain true; whereas the Atomic Theory, by which we now explain this great law, may possibly in time give place to one more perfectly suited to the explanation of new facts.

Combining Volumes of Gases.

The relation existing between the *volumes of gases* when they combine together has been found to be a very simple one, inasmuch as *the densities of all the elements known in the gaseous state are identical with their atomic weights;* or, what is the same thing, the atoms in the gaseous state all occupy the same space.*

Thus the density and combining weight of oxygen are alike 16; or, oxygen is 16 times heavier than hydrogen: the density and combining weight of nitrogen are alike 14; or, nitrogen is 14 times heavier than hydrogen; the density of chlorine is 35·5, that of sulphur vapor 32, and so on. Remembering this fact, it is easy to calculate the absolute weight of a given volume—say one litre of these different gases—when we know that one litre of hydrogen at the standard pressure and temperature weighs 0·08936 grams. Thus 1 litre of oxygen, under the same circumstances, weighs

		$16 \times 0{\cdot}08936 = 1{\cdot}430$	grams
1 litre of nitrogen	weighs	$14 \times 0{\cdot}08936 = 1{\cdot}251$	"
" chlorine	"	$35{\cdot}5 \times 0{\cdot}08936 = 3{\cdot}172$	"
" sulphur vapor	"	$32 \times 0{\cdot}08936 = 2{\cdot}860$	"

* Certain notable exceptions to this law occur in the case of phosphorus and arsenic, whose vapors possess a density twice as great as that required to be in accordance with the above law, and also of a few volatile metals, such as zinc and mercury, whose density is only half their atomic weight.

With respect to compounds, we find that *the density of a compound gas is one-half its molecular weight;* or the molecule of a compound gas occupies the space of 2 atoms of hydrogen.*

Thus the density of water-gas, or steam, H_2O, is $\frac{18}{2}$ or 9; that is, it is nine times heavier than hydrogen; the density of hydrochloric acid, H Cl, is $\frac{36 \cdot 5}{2}$, or 18·25; that of ammonia, N $\frac{17}{2}$ or 8·5; that of carbonic acid, CO_2, $\frac{44}{2}$ or 22.

Hence the weights of 1 litre of these compounds (estimated at 0° C and 760 mm.) are as follow:—

1 litre of steam	weighs	9 × 0·08936	gram.
" ammonia	"	8·5 × 0·08936	"
" hydrochloric acid	"	18·25 × 0·08936	"
" carbonic acid	"	22 × 0·08936	"

The symbol for water, H_2O, therefore, not only indicates that it is composed of 2 parts by weight of hydrogen and 16 of oxygen, but also that 2 volumes of hydrogen have united with 1 volume of oxygen to form 2 volumes or one molecule of water gas. The symbol NH_3 denotes that 3 volumes of hydrogen and 1 volume of nitrogen have united to form 2 volumes, one molecule, of ammonia, whilst the symbol HCl shows that 2 volumes of hydrochloric acid gas contain 1 volume of chlorine and 1 of hydrogen.

We have seen that 28 parts by weight of nitrogen unite with 32 parts of oxygen to form nitrogen di-oxide; the density of this compound is, however, found by experiment to be 15: hence its molecular weight is 30, consisting of 14 parts by

* The exceptions to this law are mentioned under the several compounds.

weight of nitrogen to 16 of oxygen, or 1 volume of each constituent, and its formula must, therefore, be N O.

Nitrogen and oxygen do not readily combine together, but under certain circumstances they are found to do so; thus, if a series of electric sparks are passed through a glass vessel filled with dry air, the presence of red colored vapors, possessing a peculiar acrid smell, is soon noticed. These consist of nitrogen tri-, and tetr-oxides, formed by the union of the nitrogen and oxygen of the air. If an alkali, such as potash, be present in the air through which the sparks are passed, a new substance, called nitre, or potassium nitrate, is formed; and from this an important compound, called nitric acid, can be prepared. This substance is formed when flashes of lightning pass through the air, being carried down to the earth's surface in the rain. Nitric acid may be considered as a compound of nitrogen pentoxide with water; and as all the other nitrogen oxides can be prepared from it, we shall first consider its properties and mode of preparation.

Nitric Acid, or *Hydrogen Nitrate.*

Symbol HNO_3, Molecular Weight 63. Nitre, or potassium nitrate, is generally formed by the gradual oxidation of nitrogenous animal matter in presence of the alkali potash. Spring water, especially the surface well-water of towns, frequently contains nitrates in solution, owing to water passing through soil containing decomposing animal matters, which by oxidation yield nitrates. For this reason, water containing nitrates is unfit for drinking purposes. Potassium Nitrate, $K\,NO_3$ (commonly called saltpetre) occurs as an incrustation on the soil in various localities, especially in India; and Sodium Nitrate, $NaNO_3$, or Chili saltpetre, is found in large beds on the coast of Chili and Peru. Nitric acid, $H\ NO_3$, is the hydrogen salt, and hence is termed Hydrogen nitrate: it may be regarded as water, $\left.\begin{matrix}H\\H\end{matrix}\right\}O$, in which one atom of hydrogen is replaced by the group NO_2, thus, $\left.\begin{matrix}H\\NO_2\end{matrix}\right\}O$; it is obtained by heating nitre, $K\ NO_3$, with sulphuric acid, or hydrogen sulphate, $H_2\ SO_4$, when nitric acid, $H\ NO_3$, and hydrogen potassium sulphate, $HK\ SO_4$, are formed. The decompositions here effected may serve as the type of a very large number of chemical changes classed as *double decompositions.* These may all be represented as consisting in an exchange between two elements, or groups of elements; thus in the case in question, one atom of the hydrogen in the sulphuric acid changes place with one atom or its *equivalent* of potassium in the nitre. These double decompositions may be represented in the form of an equation, in which one side signifies the arrangement and relative weights of the elements before combination, theother the arrangement and relative weights of the same elements after the chemical change has taken place, thus—

$$KNO_3 + H_2SO_4 = H\ NO_3 + HK\ SO_4{}^*$$

or, Nitre and Sulphuric Acid give Nitric Acid and Hydric Potassium Sulphate.

The relative weights of the elements and compounds entering into the decomposition is easily ascertained when we remember that each symbol expresses not merely the nature of the element, but also the relative weight with which it combines, and that the combining weight of a compound is the sum of the combining weights of its constituents. The numbers expressed by the above equation are

$$\begin{matrix} K & & N & & O_3 & + & H_2 & & S & & O_4 & = & H & & N & & O_3 & + & H & & K & & SO_4 \\ 39{\cdot}1 & + & 14 & + & 48 & + & 2 & + & 32 & + & 64 & = & 1 & + & 14 & + & 48 & + & 1 & + & 39{\cdot}1 & + & 32 + 64 \\ & & 101{\cdot}1 & & & + & & & 98 & & & = & & & 63 & & & + & & & 136.1 & & \end{matrix}$$

* The sign + placed between two symbols signifies "and" or "together with."

We may express these double decompositions perhaps more clearly by representing by a curved line the actual exchange of hydrogen for potassium, thus—

$$\begin{array}{ll} \mathrm{H} & \mathrm{H\,SO_4} \\ \mathrm{K} & \mathrm{NO_3} \end{array}$$

or by a straight line, thus—

$$\begin{array}{r|l} \mathrm{H} & \mathrm{H\,SO_4} \\ \mathrm{NO_3} & \mathrm{K} \end{array}$$

This signifies that if we require 63 parts by weight of nitric acid we shall require to take exactly 101·1 parts of nitre and 98 parts of sulphuric acid, and that we shall have 136·1 parts of hydrogen potassium sulphate formed. Knowing these numbers,

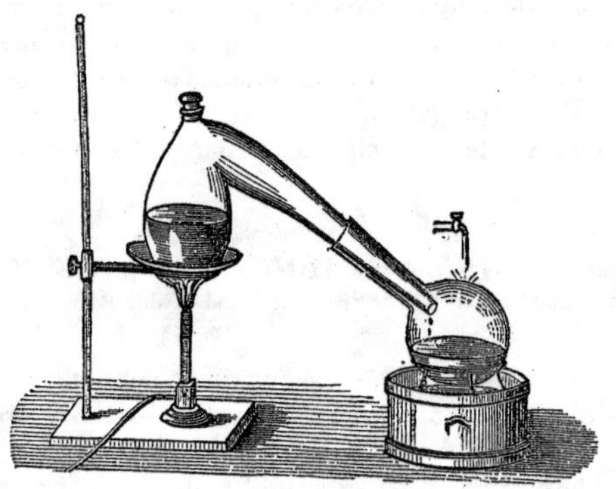

FIG. 16.

it is easy to calculate the proportions of ingredients needed to produce any given quantity of nitric acid.

Nitric acid is prepared on a small scale by placing equal weights of nitre and sulphuric acid in a stoppered retort, which is gradually heated by a Bunsen's burner, as in Fig. 16:

the nitric acid formed distils over, and may be collected in a flask cooled with water. On a large scale this substance is prepared in iron cylinders, into which the charges of nitre and acid are brought, the nitric acid being collected in large stoneware bottles.

Nitric acid thus obtained is represented by the formula $H\ NO_3$; it is a strongly fuming liquid, colorless when pure, but usually slightly yellow from the presence of lower oxides of nitrogen. Its specific gravity is 1·51 at 18°; it does not possess a constant boiling point, as it gradually undergoes decomposition by boiling and becomes weaker: if mixed with water, and distilled under the ordinary atmospheric pressure, the residual acid is found at last to attain a fixed composition, boiling constantly at 120°5, containing 68 per cent. of $H\ NO_3$, and possessing a specific gravity of 1·414. When mixed with less water, a stronger acid than this comes over; when mixed with more water, a weaker one distils over till this constant composition is attained. Nitric acid contains 76 per cent. of oxygen, with some of which it easily parts; hence it acts as a strong oxidizing agent; this is seen when we bring a small quantity of copper or tin into this liquid diluted with a little water; red fumes are immediately given off, and the metals are oxidized; for the same reason nitric acid bleaches indigo solution, oxidizing, and therefore destroying, the coloring matter: this reaction, and the formation of red fumes in presence of metallic copper, &c., serve as modes of detecting the presence of nitric acid. One of the most delicate tests for this acid consists in adding to the liquid to be tested an equal volume of strong sulphuric acid, well cooling the mixture, and then carefully pouring on to its surface a solution of ferrous sulphate, $Fe\ SO_4$: a black ring is produced where the two layers of liquid meet if any nitric acid be present. Nitric acid forms, with metallic oxides, by the process of double decomposition, a numerous family of salts called nitrates: these are nearly all soluble in water, and many of them are largely used in the arts for various purposes. They will be mentioned under the several metals.

In nitric acid we have the first example of a series of important compounds known as *acids.* Most of the acids are soluble in water; they possess an acid taste, and have the property of turning blue litmus-solution red. All acids contain hydrogen, combined either with an element, or with a group of elements, which almost always contains oxygen, and in this case the substances are termed *oxi-acids.* These acids may be regarded as water $\left.\begin{matrix}H\\H\end{matrix}\right\}O$, in which part of the hydrogen is replaced by the oxygenated group of atoms; thus nitric acid may be represented as $\left.\begin{matrix}NO_2\\H\end{matrix}\right\}O$. When the rest of the hydrogen of an acid is replaced by a metal, as for instance when sulphuric acid acts upon zinc, the acid character of the substance disappears, and a *salt,* called zinc sulphate, is formed, thus:—

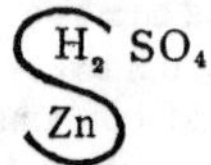

Salts are likewise produced when certain hydroxides and oxides are brought into contact with acids; thus if the solution of potassium hydroxide (caustic potash), obtained by the action of the metal potassium on water, is added to nitric acid, the alkaline or caustic properties of the hydroxide as well as the sour taste of the nitric acid disappear at a certain point; the solution becomes *neutral,* that is, it does not change the color of either blue or red litmus, and the salt potassium nitrate is contained in the liquid: thus—

$$\left.\begin{matrix}H\\K\end{matrix}\right\}O+\left.\begin{matrix}NO_2\\H\end{matrix}\right\}O=\left.\begin{matrix}H\\H\end{matrix}\right\}O+\left.\begin{matrix}NO_2\\K\end{matrix}\right\}O.$$

The soluble hydroxides which thus act upon acids are termed *alkalies,* and have the power of turning red litmus solution blue. In the same way many metallic oxides, called *basic oxides* or *bases,* act upon acids to form salts; thus silver oxide dissolves in nitric acid, and neutralizes its acid character, forming soluble silver nitrate, thus—

$$\left.\begin{matrix}Ag\\Ag\end{matrix}\right\}O+2\left.\begin{matrix}NO_2\\H\end{matrix}\right\}O=\left.\begin{matrix}H\\H\end{matrix}\right\}O+2\left.\begin{matrix}NO_2\\Ag\end{matrix}\right\}O.$$

Nitrogen Pentoxide, or *Nitric Anhydride.*—Symbol $N_2\ O_5$, or $\left.\begin{matrix}NO_2\\NO_2\end{matrix}\right\}O$. This oxide of nitrogen cannot be prepared directly from liquid nitric acid; but if dry chlorine gas be passed over silver nitrate, silver chloride is formed, oxygen is given off, and a white crystalline substance produced, which on analysis is found to be nitrogen pentoxide. The decomposition is thus represented—

$$2\ Ag\ NO_3 + 2\ Cl. = N_2O_5 + O + 2\ Ag\ Cl.$$

Nitrogen Pentoxide melts at + 30° and boils at + 45°; it very easily undergoes decomposition, and unites with great energy with water, forming nitric acid $N_2O_5 + H_2O = 2\ NO_3H$: this may be represented as a double decomposition, in which one atom of hydrogen changes place with NO_2, thus :—

$$\left.\begin{matrix}NO_2\\NO_2\end{matrix}\right\}O + \left.\begin{matrix}H\\H\end{matrix}\right\}O = \left.\begin{matrix}NO_2\\H\end{matrix}\right\}O + \left.\begin{matrix}NO_2\\H\end{matrix}\right\}O.$$

The fact that the composition of nitric pentoxide is represented by the formula N_2O_5, may be ascertained experimentally by determining the quantity of nitrogen contained in 100 parts of nitrogen pentoxide, which is first converted into nitric acid by the aid of water as above, and then into lead nitrate by treatment with lead oxide, thus :—

$$Pb, O + 2\ NO_3H = Pb, 2\ NO_3 + H_2O.$$

We thus find the nitrogen to weigh 25·93 parts, and hence the oxygen 100–25·93, or 74·07 parts. We then wish to know what is the simplest relation in which the combining weights of nitrogen and oxygen are contained in this compound; in other words, what is the ratio of the number of atoms of nitrogen present to the number of those of oxygen. This is ascertained by dividing the above numbers by the respective combining weights of these two elements ; thus—

$$\frac{25·93}{14} = 1·852 \text{ and } \frac{74·07}{16} = 4·63:$$

hence the ratio of the number of atoms of nitrogen present to the number of atoms of oxygen, is that of the numbers 1·852 to 4·6294, or that of 2 to 4·999. Hence we conclude that the exact relation between the number of atoms of nitrogen and oxygen respectively is that of 2 to 5, the slight difference which is noticed being due to the unavoidable errors which accompany every experimental inquiry, and are, therefore, termed *errors of experiment.* All the other oxides of nitrogen may be obtained from nitric acid by depriving it of its hydrogen, and more or less of its oxygen.

LESSON VII.

Nitrogen Mon-oxide, or *Nitrous Oxide.*

Symbol N_2O, Molecular Weight 44, Density 22, is obtained by heating ammonium nitrate, NH_4NO_3 or $\left.\begin{matrix}NH_4\\NO_2\end{matrix}\right\}O$, in a flask such as that used for the production of oxygen, and is best collected over warm water (*see* Fig. 17). The salt de-

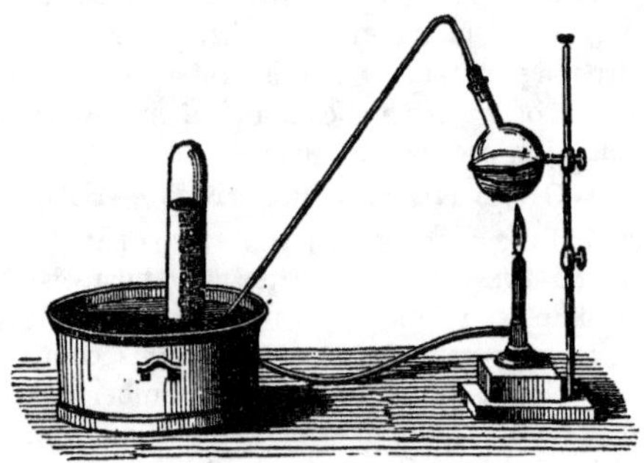

FIG. 17.

composes on heating, into nitrogen mon oxide and water: $NH_4NO_3 = N_2O + 2H_2O$; or ammonium nitrate yields nitrous oxide and water. Nitrous oxide is a colorless inodorous gas possessing a slightly sweet taste; it is somewhat soluble in cold water, one volume of water at 0° dissolving 1·305 volumes of the gas, whilst one volume of water at 24°

dissolves only 0·608 volume. Nitrogen mon-oxide differs from all the gases which we have previously considered, inasmuch as it liquefies when exposed either to great pressure or to an intense degree of cold. Thus, if it be brought under a pressure of about 30 atmospheres at 0°, or if it be cooled down to −88° under the ordinary pressure, it forms a colorless liquid (in other words the tension of nitrous oxide vapor or gas is 1 atmosphere at −88°, and 30 atmospheres at 0° C.). If this liquid be cooled below −115° it solidifies to a transparent mass. By the rapid evaporation of this liquid in vacuô, the lowest artificial temperature hitherto known has been attained, viz., about −140° C.

A glowing chip of wood when plunged into nitrous oxide rekindles, and the wood continues to burn with a brighter flame than in the air, whilst phosphorus on burning in this gas evolves nearly as much light as in pure oxygen ; a feeble

Fig. 18.

flame of sulphur is, however, extinguished on bringing it into this gas, but if burning strongly it also continues to burn brightly. This is owing to the fact that the gas has to be decomposed into nitrogen (1 volume) and oxygen (½ a volume) before bodies can burn in it ; and to effect this decomposition a tolerably high temperature is necessary,—the same products of combustion are produced as if the combustion went on in the air. When inhaled, nitrous oxide produces a peculiar intoxicating effect on the human frame ; hence the name Laughing gas. The composition of nitrous oxide may be determined as follows : a bent tube (Fig. 18) is filled with the dry gas over mercury up to a certain mark on the tube, a small pellet of metallic potassium having been previously introduced into the bent part of the tube ; this is then heated by a spirit lamp, or Bunsen's burner, while the open end of the tube is closed with the thumb under the mercury, to prevent a loss of gas by sudden expansion caused by the combustion. The potassium burns in the gas, uniting with the oxygen to form solid potassium hydro-oxide, or potash, whilst the nitrogen remains in the tube. On removing the thumb and allowing the tube to cool, it will be seen that the volume of nitrogen is exactly the same as the volume of nitrous oxide taken ; hence this gas contains its own volume of nitrogen. But we know by experiment that the weight of one volume of the gas is 22, so that if we subtract from this the weight of one volume of nitrogen (viz. 14) we shall obtain the weight of oxygen (8) contained in one volume of nitrogen mon-oxide. Hence we see that two volumes of nitrous oxide are composed of two volumes of nitrogen and one volume of oxygen, or 44 parts by weight contain 28 of nitrogen and 16 of oxygen. The specific gravity of nitrous oxide (air= 1) is 1·527 : 1000 cbc. at 0° and 760 mm. weigh 1·972 grm.

Nitrogen Di-oxide or *Nitric Oxide.*—Symbol NO ; Molecular Weight 30 ; Density 15 ; a colorless gas obtained by acting upon copper turnings with nitric acid thus :—$3\ Cu. + 8\ HNO_3 = 3\ (Cu.\ 2\ NO_3) + 2\ NO + 4\ H_2O$. Copper and nitric acid give copper nitrate, nitrogen di-oxide, and water.

This substance has not been condensed to a liquid ; in contact with oxygen it combines directly with this latter gas, forming red fumes which are readily soluble in water, and by this property it may be distinguished from all other gases. Although nitric oxide contains half its volume of oxygen, and more oxygen in proportion by weight than nitrous oxide, it does not easily support combustion, as it requires a high temperature for its decomposition.

The composition of this gas may be determined according

to the method described under nitrogen mon-oxide; one volume of nitric oxide yields ½ a volume of nitrogen; as the weight of one volume of nitric oxide is 15, the weight of oxygen contained in one volume of this gas is 15—7=8: or two volumes of nitric oxide weigh 30, and are composed of one volume of nitrogen weighing 14, and one of oxygen weighing 16. Hence, in accordance with the law mentioned on p. 27 respecting the densities of compound gases, the formula of nitric oxide should be NO and not N_2O_2: the physical properties of the gas likewise, compared with those of nitrous oxide, seem to indicate that this latter has a more complicated constitution; thus nitric oxide has not as yet been seen in the liquid form, does not condense to a liquid at temperatures and pressures at which nitrous oxide readily liquefies; nitric oxide is more difficultly decomposed by heat than nitrous oxide, and therefore supports combustion less easily; and it is a general law that in a series of similar bodies the more complicated be the constitution of one member, the more readily does it condense to the liquid form, and the more easily does it decompose.

The specific gravity of nitric oxide (air = 1) is 1·038, and 1,000 cbc. of this gas at 0° and 760 mm. weigh 1·343 grm.

Nitrogen Trioxide.—Symbol N_2O_3; Molecular Weight 76; Density 38. This substance is prepared by mixing four volumes of dry nitrogen di-oxide with one volume of oxygen, and cooling the mixture to −18°; the two gases combine to form red fumes, which condense to a volatile indigo-blue colored liquid; the same blue body is obtained by adding water to nitrogen tetroxide and drying the distillate over calcium chloride. When mixed with water nitrogen trioxide decomposes, nitric oxide and nitric acid being formed, thus:

$$3\,N_2O_3 + H_2O = 2\,HNO_3 + 4\,NO.$$

As Nitric Pentoxide is connected with a series of salts called *Nitrates*, by the fact that all these salts may be represented as nitric pentoxide, in which one of the groups NO_2 has been replaced by a metal, thus :

$$\text{Nitric pentoxide } \left.\begin{matrix} NO_2 \\ NO_2 \end{matrix}\right\} O \text{ ; potassium nitrate } \left.\begin{matrix} K \\ NO_2 \end{matrix}\right\} O \text{ ;}$$

$$\text{hydric nitrate } \left.\begin{matrix} H \\ NO_2 \end{matrix}\right\} O \text{ ;}$$

so also is nitric trioxide connected with a parallel series of salts called *Nitrites ;* thus :

$$\text{Nitric trioxide } \left.\begin{matrix} NO \\ NO \end{matrix}\right\} O \text{ ; potassium nitrite } \left.\begin{matrix} K \\ NO \end{matrix}\right\} O \text{ ;}$$

$$\text{hydric nitrite (unknown) } \left.\begin{matrix} H \\ NO \end{matrix}\right\} O.$$

As Hydric Nitrate is known by the name nitr*ic* acid, so Hydric Nitrite is referred to as nitr*ous* acid ; it being a general rule in chemical nomenclature that if the specific name of an acid or hydrogen salt end in "ous," the names of the corresponding metallic salts end in "ite ;" while if the name of the acid end in "ic," the names of the corresponding metallic salts end in "ate ;" thus nitric acid and potassium nitrate, nitrous acid and potassium nitrite, are respectively analogous.

Potassium nitrite $K\,NO_2$ is obtained by heating nitre or potassium nitrate KNO_3, when the third atom of oxygen is evolved ; and also by passing N_2O_3 into solutions of KHO.

Nitrogen Tetroxide.—Symbol NO_2, Molecular Weight 46, Density 23. This substance forms the greater part of the reddish-brown fumes evolved when nitric oxide gas escapes into the air ; it is however best prepared by heating lead nitrate in a hard glass retort ; lead oxide, oxygen, and nitrogen tetroxide are produced by the decomposition of the nitrate, thus :

$$2\ (Pb\ 2\ NO_3) = 2\ PbO + 4\ NO_2 + O2.$$

Nitrogen tetroxide NO_2 solidifies at — 9° to long prisms ; these on fusing yield a yellow liquid, boiling at 22°. Owing to the fact that the density of nitrogen tetroxide is 23, its formula is considered to be NO_2 and not N_2O_4.

COMPOUNDS OF NITROGEN WITH HYDROGEN.

But one is as yet known in the free state, viz.:

Ammonia.—Symbol NH_3, Molecular Weight 17, Density 8·5. Nitrogen and hydrogen do not readily unite when brought together alone, but do so under certain circumstances, especially when water is evaporated; the nitrogen of the air then combines with the elements of the water, forming small quantities of the compound ammonium nitrite

$$\text{Thus } N_2 + 2\,H_2\,O = N_2\,H_4\,O_2, \text{ or } NH_4, NO_2.$$

Ammonia is chiefly obtained from the decomposition of animal or vegetable matter containing nitrogen and hydrogen, being formed either gradually at the ordinary temperature, or quickly under the influence of heat: thus when horns, or clippings of hides, or coal, is heated, ammonia is given off; hence ammonia was known as *spirits of hartshorn.* The name ammonia is derived from the fact that a compound containing ammonia, called salammoniac, was first prepared by the Arabs in the deserts of Libya, near the temple of Jupiter Ammon, by heating camels' dung. Guano, the dried excrement of sea birds, and the urine of animals, likewise contain large quantities of ammonia. Ammonia and its compounds are now, however, mainly obtained from the ammoniacal liquors of the gasworks: coal contains about 2 per cent. of nitrogen, which, when the coal is heated in close vessels, mostly comes off in combination with the hydrogen of the coal as ammonia. Ammonia may also be formed by the action of nascent hydrogen on dilute nitric acid; thus, when this acid is placed in contact with metallic zinc or iron, ammonia is formed, thus:

$$9\,HNO_3 + 4\,Zn = 4\,(Zn\,2NO_3) + 3H_2O + H_3N.$$

Ammonia gas is best prepared by heating in a glass flask one part of salammoniac, or ammonia hydrochlorate, $NH_3\,H\,Cl$ or $N\,H_4Cl$, and two parts of powdered quicklime, thus

$$CaO + 2\,NH_3\,HCl = CaCl_2 + 2NH_3 + H_2O.$$

Quicklime and salammoniac give calcium chloride, and ammonia and water. Ammoniacal gas is colorless, and possesses a most pungent and peculiar smell, by means of which it can be readily recognized; it is lighter than air, its specific gravity (air=1) being 0·59, and it may be collected by displacement, the neck of the bottle intended to receive the gas being turned downwards, as in Fig. 19. It may also be collected over

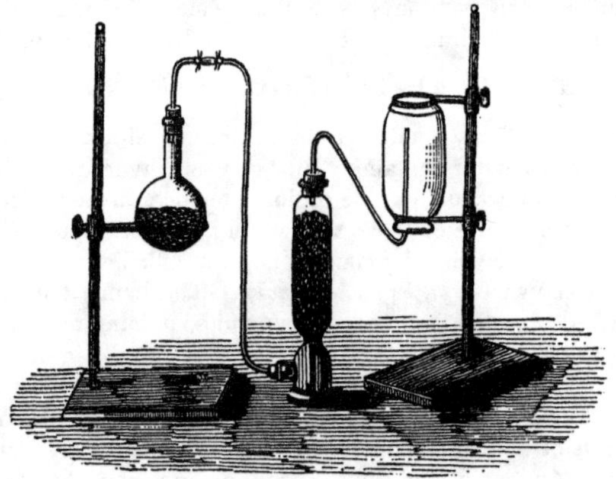

FIG. 19.

mercury, but not over water, as it is extremely soluble in this liquid, one gramme of water at 0° absorbing 0·877 grms., or 1149 times its volume of ammonia, under a pressure of 760 mm.; whilst at 20° the same weight of water absorbs 0·520 grm., or 681·1 times its volume, under the same pressure. The solution of ammonia gas in water is the common Liquor Ammoniæ of the shops, which has a specific gravity of about 0·880. Ammonia gas, as well as the aqueous solution, possesses a strong alkaline reaction, turning red vegetable colors blue: it unites with the most powerful acids, forming compounds called the Salts of Ammonia (*see* p. 167), which closely resemble the salts of the alkaline metals; hence the name of the Volatile Alkali has been given to Ammonia. The action of ammonia gas on nitric acid may be thus represented :

$$NH_3 + NO_3H = NH_4NO_3 \text{ ; or } \left.\begin{matrix} NH_4 \\ NO_2 \end{matrix}\right\} O.$$

On exposure to a pressure of seven atmospheres at the ordinary temperature of the air (about 15° C.), ammonia condenses to a colorless liquid, boiling at −38·5° ; and this liquid, if cooled below −75°, freezes to a transparent solid ; it has at 0° the specific gravity 0·62. An elegant application of the principle of the latent heat of vapors has recently been made in the case of ammonia in M. Carré's freezing machine,

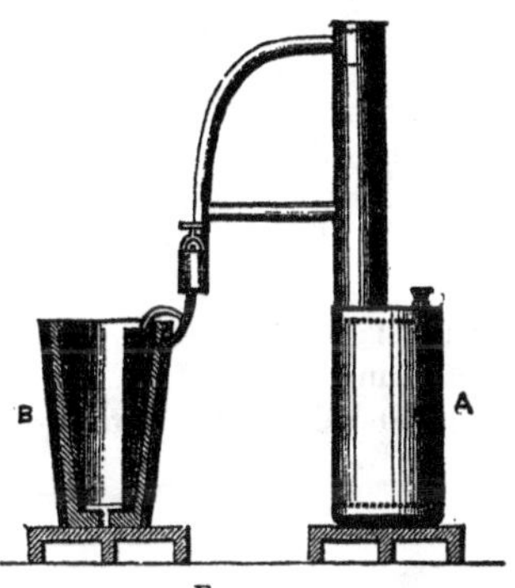

FIG. 20.

Fig. 20. This consists essentially of two strong iron vessels connected in a perfectly air-tight manner by a bent pipe ; one of these vessels contains an aqueous solution of ammonia saturated with the gas at 0°. When it is desired to procure ice, the vessel A containing the ammonia solution (which we will term the retort) is gradually heated over a large gas burner, the other vessel B (the receiver) being placed in a bucket of cold water : in consequence of the increase of temperature, the gas cannot remain dissolved in the water, and passes into the receiver, where, as soon as the pressure amounts to about 10 atmospheres, it condenses in the liquid form. When the greater part of the gas has thus been driven out of the water, the apparatus is reversed, the retort (A) being cooled in a current of cold water, whilst the liquid it is desired to freeze is placed in the interior of the receiver (B). A re-absorption of the ammonia by the water now takes place, and a consequent evaporation of the liquefied ammonia in the receiver; this evaporation is accompanied by an absorption of heat, which becomes latent in the gas; hence the receiver is soon cooled far below the freezing point, and ice is produced around it.

The composition of ammonia may be ascertained by leading the gas through a red-hot tube, or passing a series of electric sparks through the gas, when it will be decomposed into nitrogen and hydrogen, which will be found to occupy together a volume twice as large as the ammonia taken, and to be mixed together in the proportions of three volumes of hydrogen to one volume of nitrogen. Hence the formula $N H_3$ is given to the gas.

The Salts of Ammonia will be described together with those of Potassium and Sodium (page 167). The compound ammonias will be noticed under Organic Chemistry.

LESSON VIII.

Carbon. *Symbol C. Combining Weight* 12.

Carbon is the first solid element which we have to notice; it is not known in the free state, either as a liquid or as a gas. Carbon is remarkable as existing in three distinct forms, which, in outward appearance or physical properties, have nothing in common, whilst their chemical relations are identical. These three *allotropic* forms of carbon are (1) Diamond,

(2) Graphite or Plumbago, (3) Charcoal: these substances differ in hardness, color, specific gravity, &c., but they each yield, on combustion in the air or oxygen, the same weight of the same substance, Carbonic Acid, or carbon dioxide;* 12 parts by weight of each of these forms of carbon yielding 44 parts by weight of carbon dioxide. Carbon is the element which is specially characteristic of animal and vegetable life, as every organized structure, from the simplest to the most complicated, contains carbon: if carbon were not present on the earth, no single vegetable or animal body such as we now know could exist. In addition to the carbon which is found free in these three forms, and that contained combined with hydrogen and oxygen in the bodies of plants and animals, it exists combined with oxygen as free carbonic di-oxide in the air, and with calcium and oxygen as calcium carbonate in limestone, chalk, marble, corals, shells, &c. The fact has already been noticed that plants are able, when exposed to sunlight, to decompose the carbonic di-oxide of the air, liberating the oxygen, and taking the carbon for the formation of their vegetable structure, whilst all animals living directly or indirectly upon vegetables absorb oxygen, and evolved carbonic di-oxide. Thus the sun's rays, through the medium of plants, effect deoxidation or reduction, while animals act as oxidizing agents with respect to carbon.

The element carbon not only combines directly with oxygen, but also with hydrogen, forming a compound called Acetylene, C_2H_2. Carbon forms with oxygen, hydrogen, and nitrogen a series of more or less complicated compounds very much more extended than the series formed with these bodies by any other element; so that these compounds are considered as a separate branch of the science under the name of *Organic Chemistry*, or the *Chemistry of the Carbon Compounds.* The properties of the majority of these compounds will be examined in a subsequent chapter, owing

* Although the term Acid strictly denotes a hydrogen salt, yet the word has been applied so long to a few other compounds containing no hydrogen, such as carbon di-oxide, &c., that these bodies are universally known by the names carbonic acid, &c

to their complexity; hence till then it will be better to postpone the consideration of several of the properties of carbon.

The Diamond was first found to consist of pure carbon by Lavoisier, in 1775–6, by burning it in oxygen, and collecting the carbon dioxide formed; it occurs crystallized in certain sedimentary rocks and gravel in India (Golconda), Borneo, and the Brazils. Diamond occurs crystallized in forms (Fig. 21), derived by a symmetrical geometric operation from a regular Octahedron, known as belonging to the regular system

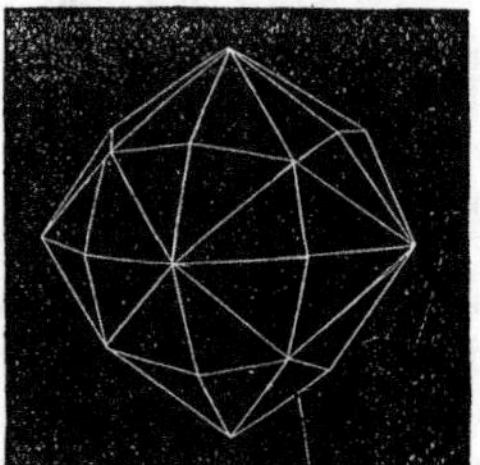

FIG. 21.

of Crystallography (*see* p. 150). The specific gravity of Diamond varies from 3·3 to 3·5; it is the hardest of all known bodies, and when cut possesses a brilliant lustre, and a high refractive power. In addition to its employment as a gem, the diamond is used for cutting and writing upon glass. We are altogether unacquainted with the mode in which the diamond has been formed; it cannot, however, have been produced at a high temperature, because when heated strongly in a medium incapable of acting chemically upon it, the diamond swells up, and is converted into a black mass resembling coke.

Graphite, or *Plumbago*, crystallizes in six-sided plates which have no relation to the form in which the diamond crystallizes. Graphite occurs in the oldest sedimentary formations, and in granitic or primitive rocks: it is found in Bor·

rowdale in Cumberland, and in large quantities in Siberia and Ceylon. It has a black metallic appearance (whence the familiar name black lead), and leaves a mark when drawn upon paper. The specific gravity of graphite is 2·15 to 2·35. Coarse impure graphite may be purified by heating the powder with sulphuric acid and potassium chlorate; a compound is thus obtained which, on being heated strongly, decomposes, leaving pure graphite in a bulky and finely-divided powder: this powder when strongly compressed forms a coherent mass, from which pencils and other articles can be made. Graphite is used for polishing surfaces of iron-work, and also for giving a protecting varnish to grains of gunpowder. Graphite is produced in the manufacture of iron; it occasionally separates from the molten pig-iron in the form of scales.

Charcoal is the third allotropic modification of carbon. It is obtained in a more or less pure state whenever animal or vegetable matter is heated to redness in a vessel nearly closed; the volatile matters (compounds of carbon, hydrogen, and oxygen) are thus driven off, and the residue of the carbon, together with the ash or mineral portion of the organism, remains behind.

The purest form of charcoal carbon is found in lampblack; it also occurs as wood charcoal, coal, coke, and animal charcoal. This form of carbon does not crystallize, and is hence termed *amorphous*; it is much lighter than either of the other two forms, the specific gravity of powdered coke varying from 1·6 to 2·0. Charcoal appears at first sight to be lighter than water, as a piece of it floats on the surface of this liquid; this is, however, due to the porous nature of the charcoal, for if it be finely powdered it sinks to the bottom of the water. This porous nature of charcoal enables it to exert a remarkable absorptive power, of which much use is made in the arts. Charcoal is thus able to absorb about ninety times its own volume of ammonia gas, and about nine volumes of oxygen. In the process of sugar-refining, use is made of the property of charcoal to absorb the coloring matters present in the raw sugar: the kind of charcoal best suited to this purpose is that obtained by heating bones in a closed vessel. Charcoal is also used as a disinfectant in hospitals and dissecting rooms, &c. It appears that the putrefactive gases when absorbed by the charcoal undergo a gradual oxidation from contact with the oxygen of the air taken up by the charcoal, and are thus rendered harmless.

Coal is a form of carbon less pure than wood charcoal. It consists of the remains of a vegetable world which once flourished on the earth's surface: the original woody fibre has undergone a remarkable transformation in passing into coal, having been subjected to a process similar, in a chemical point of view, to that by which wood is transformed intc

Composition of Fuels (ash being deducted).

Description of Fuel.	Percentage Composition.		
	Carbon.	Hydrogen.	Nitrogen and Oxygen.
1 Woody Fibre	52·65	5·25	42·10
2 Peat from the Shannon .	60·02	5·88	34·10
3 Lignite from Cologne . .	66·96	5·25	27·76
4 Earthy Coal from Dax .	74·20	5·89	19·90
5 Wigan Cannel	85·81	5·85	8·34
6 Newcastle Hartley . . .	88·42	5·61	5·97
7 Welsh Anthracite . . .	94·05	3·38	2·57

charcoal. It has not, however, lost the whole of its hydrogen and oxygen, and it has at the same time become bitumenized; so that for the most part all the vegetable structure has disappeared. There are many different kinds of coal, containing more or less of the oxygen and hydrogen of the original wood: cannel coal and boghead coal contain the most hydrogen, and anthracite coal the least. The alteration in composition which wood has undergone in passing into the various forms of coal is seen from the above table.

COMPOUNDS OF CARBON WITH OXYGEN.

Carbon forms two compounds with oxygen, viz.:

Carbon Monoxide, or C O.

Carbon Dioxide, or CO_2.

Carbon Dioxide, or *Carbonic Acid.* Symbol CO_2, Molecular Weight 44, Density 22.

Carbon dioxide is always formed when carbon is burnt in excess of air or oxygen. It is best prepared by acting upon marble, chalk, or other form of calcium carbonate, with hydrochloric acid. On pouring some of this acid upon pieces of marble contained together with some water in a flask, a rapid effervescence from the disengagement of carbon dioxide gas at once occurs, calcium chloride being left behind in solution in the flask: the decomposition is thus represented:

$$Ca\,CO_3 + 2HCl = CO_2 + H_2O + Ca\,Cl_2.$$

Calcium carbonate and hydrochloric acid give carbon dioxide, water, and calcium chloride.

Carbon dioxide occurs free in the air, and in the water of many mineral springs. The quantity of this gas present in the air is nearly constant, and amounts to about 4 volumes per 10,000 of air; this quantity, though relatively small, is, taken altogether, very large, being about 3 billions of tons in weight, as can be easily calculated if we know the weight of the atmosphere and the density of carbonic acid.

Owing to the evolution of carbon dioxide in respiration and in the burning of coal-gas, &c., this gas is always found in larger quantities in dwelling-rooms than in the open air. When the air of a room contains 0·10 per cent. of this gas it is certainly unfit for continued respiration, not only on account of the deleterious effects produced by carbon dioxide, but also because, together with this gas, volatile putrescible matters are given off from the skin and lungs of animals, and these matters act in a prejudicial manner upon the health; hence the necessity for attention to the ventilation of dwelling-rooms and public buildings. Carbon dioxide gas is also given off in the process of fermentation; it occurs frequently at the bottom of old wells, and forms the *chokedamp* of the coal mines. Compounds of carbon dioxide with lime or magnesia, such as limestone or calcium carbonate, $\left.\begin{matrix}Ca\\CO\end{matrix}\right\}O_2$, and magnesian limestone, &c., occur plentifully in nature, sometimes forming whole mountain chains. Calcium carbonate also constitutes the main portion of coral, a substance of which whole continents are being built up in the Pacific Ocean.

Carbon dioxide gas is colorless and inodorous, but possesses a slightly acid taste; it is 1·529 times heavier than air, and is tolerably soluble in water, one volume of water at 0° dissolving 1·797 volumes of this gas, whilst at 20° only 0·901 volumes is absorbed. The volume of this gas absorbed by water at the same temperature is found to remain the same, under whatever pressure the gas may be measured. As the volumes occupied by any given quantity of gas measured under different pressures vary inversely as these pressures, it is clear that the weights of carbon dioxide thus absorbed must be proportional to the pressures. Thus, for instance, under the pressure of 1 atmosphere and at the ordinary temperature of the air 1 cbc. of water dissolves 1 cbc. or 1·529 milligrams of carbon dioxide. So under a pressure of 2 atmospheres 1 cbc. of water will at the same temperature dissolve 1 cbc. (measured under the pressure of 2 atmospheres) or $2 \times 1{\cdot}529 = 3{\cdot}058$ milligrams of carbon dioxide. The increased quantity of absorbed carbonic acid under increased pressure is seen when a bottle of soda-water or champagne is opened; the pressure being diminished by removal of the cork, a brisk effervescence and escape of the dissolved gas occurs. The same relation is found to hold good when many other gases are dissolved in water under varying pressures.

The aqueous solution of carbon dioxide reddens blue litmus paper, and when placed in contact with a metallic oxide, such as calcium oxide or lime, CaO, gives rise to the formation of salts such as calcium carbonate: this aqueous solution may be considered to contain a true acid, the real carbonic acid, $\left.\begin{matrix}H_2\\CO\end{matrix}\right\} O_2$ (which, however, has never yet been isolated), and the reaction which then takes place may be thus represented:

$$\left.\begin{matrix}H_2\\CO\end{matrix}\right\} O_2 + Ca\,O = \left.\begin{matrix}Ca\\CO\end{matrix}\right\} O_2 + H_2O.$$

Carbonic acid and calcium oxide give calcium carbonate and water. The red color produced by the acid on litmus paper disappears on drying, owing to the decomposition of this true carbonic acid into carbon dioxide and water, thus:

$$\left.\begin{matrix}H_2\\CO\end{matrix}\right\} O_2 = CO_2 + H_2O.$$

Carbon dioxide can be condensed to a liquid by the application of great pressure, or by cooling the gas to a very low temperature: liquid carbon dioxide is a colorless and very mobile liquid, which is remarkable as being found to expand by heat more than the gaseous form of the same substance, 100 volumes of this liquid at 0° becoming 106 volumes at 10°, while 100 volumes of the gas at 0° must be heated to 16·4° before they expand to 106 volumes; hence this body is an exception to the rule that liquids expand by heat less than gases, and at the same time forming an excellent illustration of the fact, that liquids expand proportionally much more when submitted to a high pressure than when under a low one; thus, the expansion of water above 100° is much greater than that below 100°. The boiling point of carbon dioxide is -78°. At a still lower temperature it freezes to a colorless, ice-like solid. At 0° the tension of its vapor is 35·5 atmospheres; and at 30°, 73·5 atmospheres. The liquefaction of carbon dioxide gas can be effected either by evolving the gas in a strong closed vessel, so that it is condensed by its own pressure, or by pumping the gas by means of an ordinary forcing syringe into a strong wrought-iron receiver, kept during the process at a temperature of 0°. As soon as the volume of gas pumped in amounts to about 36 times the volume of the receiver, each stroke of the syringe produces a condensation of the gas which is pumped in, and thus the receiver can easily be filled with liquid. If the stopcock be then opened so that the liquid is forced out, a portion at once assumes the gaseous state; and so much heat is absorbed by this sudden transition from the liquid to the gaseous form, that a portion of the liquid is solidified and deposited in the form of white snow-like flakes, which can be collected by allowing the stream of liquid to flow into a thin brass box, with perforated sides. Solid carbon dioxide thus obtained, is a light, snow-like substance, which, owing to the bad conducting power for heat of the gas which the solid substance is constantly giving off, may be handled without damage, although its temperature is below -78° C. If, however, the solid be forcibly pressed between the fingers, so that the substance really comes in contact with the skin, a sharp pain will be felt, and a blister like one produced by touching a hot iron will be produced. This solid carbon dioxide is much used for the production of very low temperatures; for this purpose it is mixed with ether, and the mixture brought into the vacuum of the air-pump, whereby a temperature as low as -100° C. can be obtained, and large quantities of mercury may easily be frozen. Carbon dioxide gas does not support the combustion of bodies in general, such as wood, sulphur, or phosphorus; but certain metals—for instance, potassium and magnesium—heated in the gas, are able to decompose it, burning in it, and

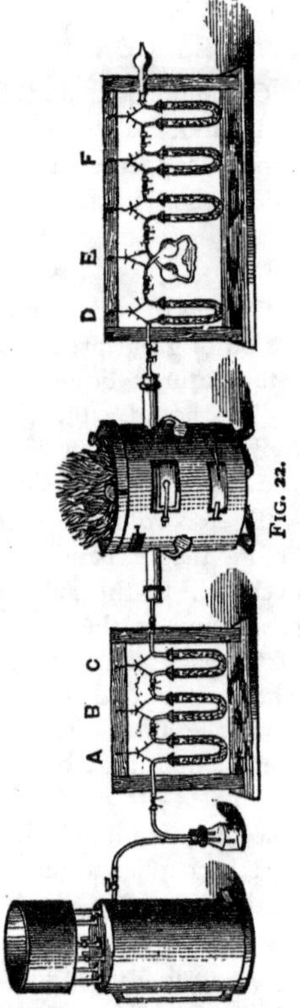

FIG. 22.

uniting with the oxygen to form oxides, while the carbon is liberated.

The composition of carbon dioxide may be ascertained with great exactness by burning a known weight of pure carbon, such as the diamond, in a current of pure oxygen gas, and weighing the carbon dioxide produced. The apparatus for this synthesis of this gas is represented in Fig. 22. The weighed quantity of diamond placed in a small platinum boat, is pushed into the porcelain tube, which can be strongly heated in the furnace. One end of this tube is connected with a gasholder and drying tubes, A, B, C, by means of which pure and dry oxygen gas is supplied. The other end is connected, as is seen, with a number of tubes and bulbs destined to absorb the carbon dioxide formed by the combustion; the tube D and the bulbs E contain solution of caustic potash, and the other tubes F are filled with pumice-stone and sulphuric acid. The bulbs and tubes are carefully weighed, and then the apparatus is filled with pure oxygen, and the tube slowly brought to a red heat. The gas passes gradually through the system of tubes, and carries along with it the carbon dioxide formed by the combustion of the diamond: this gas is wholly absorbed by the potash in the tube and bulbs, whilst any moisture which might be given off from the bulbs is taken up by the tubes F. The oxygen gas is dried as it enters and also as it leaves the apparatus; so that the gain in weight which the tubes have experienced gives exactly the weight of carbon dioxide formed by the combustion of the carbon of the diamond. Usually the diamond contains a small quantity of ash, or inorganic matter, and this weight must be subtracted from the original weight of the diamond, in order that we may know the exact weight of pure carbon burnt; for this reason the diamond is placed in a platinum boat, which can be withdrawn and weighed after the experiment, and thus the amount of ash determined. Another precaution that must be taken is, to fill the greater part of the red-hot tube with porous copper oxide, in case any trace of carbon mon-oxide gas (C O) should be formed by the incomplete combustion of the carbon; this gas would pass unabsorbed through the potash if not oxidized to carbon dioxide by the copper oxide. In this way it has been shown that 100 parts of carbon dioxide consist of

Carbon	27·27
Oxygen	72·73
Carbon dioxide . .	100·000

If we divide 27·27 by the combining weight of carbon, and 72·73 by that of oxygen, we have $\frac{27 \cdot 27}{12} = 2 \cdot 273$ and $\frac{72 \cdot 73}{16} = 4 \cdot 545$; or, the relation between the number of atoms of carbon and that of those of oxygen is that of 1 to 2; hence the formula of carbon dioxide is CO_2. Hence the gas should contain its own volume of oxygen; for 44 parts by weight of carbon dioxide occupying a volume equal to that occupied by 2 parts by weight of hydrogen, contain 32 parts by weight of oxygen, which likewise occupy a volume equal to that of 2 parts of hydrogen. That this is the case can be experimentally proved by burning charcoal in a known volume of oxygen in excess, when it is observed that when the gas has cooled after the combustion, no alteration in its volume has occurred; hence the volume of carbon dioxide formed must be precisely equal to that of the oxygen used in its formation.

LESSON IX.

Carbon Mon-oxide, or *Carbonic Oxide Gas.*

Symbol CO. Molecular Weight 28. *Density* 14.

When carbon burns with a limited supply of oxygen, carbonic oxide is formed. The production of this gas in an ordinary red-hot coal fire is often observed; oxygen of the air, which enters at the bottom of the grate, combines with the carbon of the coal, forming carbon dioxide; this substance then passing upwards over the red-hot coals, parts with half its oxygen to the red-hot carbon; thus:—

$$CO_2 + C = 2\,CO.$$

This carbon mon-oxide on coming out at the top of the fire meets with atmospheric oxygen, with which it at once combines, burning with a lambent blue flame, and reforming carbon dioxide. Carbon mon-oxide gas in the pure state can be prepared by passing a slow current of carbon dioxide over pieces of charcoal heated to redness in a tube by means of a furnace, as represented in Fig. 23: it may likewise be obtained in the pure state from several compounds of carbon. Thus, if crystallized oxalic acid be heated with strong sulphuric acid, a mixture of equal volumes of carbon mon-oxide and carbon dioxide gases is evolved; this latter can be easily separated from the former by shaking the mixed gas up with caustic soda solution, when sodium carbonate will be formed,

FIG. 23.

half the volume of the gas will disappear, and the remainder will be found to be pure carbon mon-oxide. This decomposition of oxalic acid results from the fact that sulphuric acid has a strong tendency to abstract water, or the elements of water, from the bodies with which it comes into contact; thus the oxalic acid, which may be represented as $C_2\,H_2\,O_4$ (*see* p. 289), being deprived of the elements of one molecule of water, which are taken up by the sulphuric acid, yields a compound,

C_2O_3, which cannot exist alone, and immediately splits up into CO_2 and CO.

Carbon mon-oxide can also be prepared by heating Formic acid CH_2O_2 (*see* p. 271), with sulphuric acid: here, as with oxalic acid, the elements of water are removed, and pure CO is thus evolved. Carbon mon-oxide is a colorless, tasteless gas, which has not been condensed to a liquid; it is a little lighter than air, its specific gravity being 0·969 (air = 1): it is but very slightly soluble in water. It acts as a strong poison, producing death when inhaled even in very small quantities, the fatal effects often observed of the fumes from burning charcoal, or from limekilns, being due to the presence of carbonic oxide. When heated in contact with oxygen, carbon mon-oxide takes fire, burning with a characteristic lambent blue flame and forming carbon dioxide. In contact with caustic potash at a high temperature, carbon mon-oxide produces potassium formate thus:

$$\left.\begin{matrix}H\\K\end{matrix}\right\}O + CO = CHKO_2.$$

Caustic potash and carbon mon-oxide give potassium formate.

The composition of this gas can be ascertained by combustion in the Eudiometer with oxygen. 100 volumes of carbon mon-oxide and 75 volumes of oxygen yield on passing the electric spark 125 volumes, of which 100 are found to be absorbed by caustic potash, and hence are carbon dioxide, the remaining 25 volumes being unaltered oxygen. Hence the volume of carbon dioxide produced is equal to that of the carbon mon-oxide taken, whilst the volume of oxygen needed is half as large. But as carbon dioxide contains its own volume of oxygen, carbon mon-oxide must contain half its volume of oxygen; or two volumes of this gas weighing 28 contain one volume of oxygen weighing 16, and hence 12 parts of carbon by weight: therefore its formula is CO.

COMPOUNDS OF CARBON WITH HYDROGEN.

These compounds are very numerous; they are known in the gaseous, liquid, and solid forms. The most important and numerous parts of these will be considered under Organic

Chemistry; we now have only to describe some of the simplest of these compounds.

Methyl Hydride, Light Carburetted Hydrogen, or *Marsh gas.* Symbol CH_4; Molecular Weight 16, Density 8.

This is a colorless, tasteless, inodorous gas, which has not been condensed to a liquid. It is found in coal mines, and known under the name of *firedamp;* it also occurs in stagnant pools, being produced by the decomposition of dead leaves, whence the name marsh gas; it is one of the constituents of coal gas, &c., and is evolved in many volcanic districts. Marsh gas may also be artificially prepared by heating sodium acetate (*see* p. 275) with caustic soda, thus:

$$\left.\begin{matrix} Na \\ C_2H_3O \end{matrix}\right\} O + \left.\begin{matrix} H \\ Na \end{matrix}\right\} O = \left.\begin{matrix} Na_2 \\ CO \end{matrix}\right\} O_2 + CH_4.$$

Sodium acetate and caustic soda give sodium carbonate and marsh gas.

Marsh gas burns with a bluish-yellow non-luminous flame, forming carbon dioxide and water; with a limited supply of air it yields several products, amongst which is *acetylene,* C_2H_2. If mixed with ten times its volume of air, and twice its volume of oxygen, it ignites with a sudden and violent explosion on the application of a light, and hence the great damage produced by the escape of this gas in coal mines. The composition of marsh gas is ascertained by exploding it with oxygen in the eudiometer; 1 volume of this gas and 3 volumes of oxygen yield 2 volumes after passage of the spark. On absorbing by potash the carbon dioxide produced, 1 volume of oxygen is found to remain. Hence of the 2 volumes of oxygen needed to burn the 1 volume of marsh gas, 1 has gone to unite with the carbon, and 1 to form water with the hydrogen. It is thus seen that 2 volumes of marsh gas contain 4 volumes of hydrogen weighing 4 (as water contains 2 volumes of hydrogen and 1 of oxygen), and as much carbon as is contained in 2 volumes of carbon dioxide, viz., 12 parts by weight: and hence the formula CH_4 is given to this gas.

Acetylene.

Symbol C_2H_2.—This gas is formed by the direct union of carbon and hydrogen at a very high temperature. For this purpose the carbon terminals of a powerful galvanic battery are brought together in an atmosphere of hydrogen. At the high temperature thus evolved, a direct union of carbon and hydrogen takes place, and acetylene is formed. Acetylene is a colorless gas, which burns with a bright luminous flame, and possesses a disagreeable and very peculiar odor; it is produced in all cases of incomplete combustion, and its smell may be noticed when a candle burns with a smoky flame. Acetylene combines with certain metals, such as copper and silver; and the compounds thus formed are distinguished by the ease with which they undergo explosive decomposition. This gas likewise unites directly with hydrogen, forming the next substance, ethylene, $C_2H_2 + H_2 = C_2H_4$.

Ethylene, Heavy Carburetted Hydrogen or *Olefiant Gas.* Symbol C_2H_4, Molecular Weight 28, Density 14.

This gas is obtained on the destructive distillation of coal, and is an important constituent of coal gas. It is obtained in the pure state by heating 1 part of alcohol (spirits of wine) C_2H_6O, with 5 or 6 parts by weight of strong sulphuric acid; as in formation of carbon mon-oxide from formic acid, the elements of water are separated by the sulphuric acid, and C_2H_4 is evolved as a gas. This gas is colorless, but possesses a sweetish taste; by exposure to a high pressure at a temperature of—110° it has been condensed to a colorless liquid. On bringing it in contact with a light, in the air, it burns with a luminous smoky flame, forming carbon dioxide and water. When mixed with three times its bulk of oxgyen and fired, it detonates very powerfully; 1 volume of olefiant gas requires 3 volumes of oxygen to burn it completely, and yields 2 volumes of carbon dioxide; so that 1 volume of oxygen is needed to combine with the hydrogen. Hence this gas contains twice as much carbon as marsh gas, with the same quantity of hydrogen; we must therefore write its formula C_2H_4.

Olefiant gas combines directly with its own volume of chlorine gas, forming an oily liquid, and owing to this property it has received the above name.

Coal Gas.

The gas so largely used for illuminating purposes, and obtained by the destructive distillation of coal (*i. e.* by heating the coal in large closed retorts so as to decompose or destroy the coal, the volatile products of this decomposition being condensed and collected), is not a simple chemical compound, but a mixture of a large number of distinct substances. In order to prepare coal gas of good quality, cannel, or some highly bitumenized coal is heated in a closed retort ; volatile bodies are thus formed and expelled, while a residue of (impure) carbon is left behind as coke. The volatile products of this decomposition may be distinguished as tar, ammonia, water, and gas. The tar contains a great variety of substances, from some of which the well-known Aniline colors are produced (*see* p. 322) ; and the ammonia derived from the nitrogen in the coal is our chief source of ammoniacal salts (*see* p. 174). The gas which comes off consists of a mixture of various substances, some of which are useful for illuminating or heating purposes, whilst some are hurtful and must be removed. Amongst those which burn with a luminous flame, are olefiant gas, and other hydrocarbons having an analogous composition, as $C_3 H_6$ and $C_4 H_8$ (where the number of atoms of hydrogen is double that of those of carbon). The gases which serve to dilute these luminous hydrocarbons, and burn themselves with non-luminous flames, are hydrogen, carbonic oxide, and marsh gas. The impurities consist of carbon dioxide, hydrogen sulphide (sulphuretted hydrogen), and the vapor of carbon di-sulphide ; and these substances are almost always withdrawn from the gas by a system of purification before it is sent out from the gasworks. The relative proportion of these three ingredients present in coal gas varies greatly according to the kind of coal employed, and according to the heat to which the coal is subjected.

The value of coal gas, as regards its illuminating power, is ascertained by comparing the light given off by the gas burning at a certain rate, usually 5 cubic feet per hour, with that of a sperm candle burning 120 grains per hour. Thus the cannel gas is said to be equal to 34·4 candles, and the coal gas to be equal to 13 candles.

It will be convenient here to mention the nature and structure of flame, and the principle of the Davy lamp. Flame consists of gas in a high state of ignition. When a jet of burning hydrogen is plunged into oxygen, the flame of *hydrogen in oxygen* is seen. This is caused by the ignition of the particles of hydrogen and oxygen, owing to the heat evolved in their combination. A similar flame of *oxygen in hydrogen* is seen when a jet of the former gas is lit in an atmosphere of hydrogen. The temperatures of flames differ as much as their illuminating powers, and the hottest flames do not necessarily give off much light; thus the oxyhydrogen flame, which is so hot as to burn iron or steel wire like tinder, can scarcely be seen in bright daylight. In order that a flame shall give off much light, it must contain solid matter, which becomes heated up to whiteness. If a piece of lime be held in the oxyhydrogen flame, it becomes strongly heated, and gives off an intense light; so also if we bring solid matter, such as powdered charcoal, into the colorless flame of hydrogen, it becomes luminous. The difference between the non-luminous flame of marsh gas and the luminous flame of olefiant gas is due to the fact that in the latter, carbon is separated out in the solid form, whereas in the former all the carbon is at once burnt to a carbonic acid. The flame of a candle consists of three distinct parts—(1), the dark central zone or supply of unburnt gas surrounding the wick; (2), the luminous zone or area of incomplete combustion; and (3), the non-luminous zone or area of complete combustion. If we bring one end of a small piece of glass tubing (Fig. 24) into the dark central zone (1), the unburnt gases will

FIG. 24.

pass up the tube, and may be ignited at the other end, where they escape into the air. In the luminous part of the flame the gases are not completely burnt, and carbon is separated out in the solid state; and it is to the presence of this carbon that the flame owes its luminous power. In the outer zone the supply of oxygen is greater, all the carbon is at once burnt to carbon dioxide, and the flame here becomes non-luminous.*

The effect of allowing a complete combustion to proceed at once throughout the flame is well seen in the small Bunsen gas-lamp, now universally employed in laboratories. In this lamp (Fig. 25) the coal gas issues from a small central burner (*a*), and passing unburnt up the tube (*e*) draws air up with it through the holes (*d*); the mixture of air and gas thus made can be lighted at the top of the tube, where it burns with a

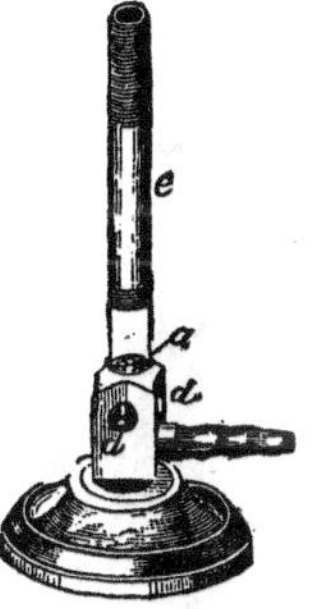

FIG. 25.

FIG. 26.

non-luminous, perfectly smokeless flame; if the holes (*d*) be closed, the gas alone burns with the ordinary bright smoky flame. The blow-pipe flame (Fig. 26) may also be divided into two distinct parts—the oxidizing flame (*a*), where there is excess of oxygen, and the reducing flame, (*b*) where there is excess of carbon; and these are distinguished by the same

* The optical difference between these two classes of flame is pointed out in the paragraph on Spectrum Analysis, (*see* p. 222.)

properties as the outer and inner mantle of the candle flame Every mixture of gases requires a certain temperature to inflame it; and if the temperature be not reached, the mixture does not take fire; we may thus cool down a flame so much that it goes out, by placing over it a small coil of cold copper wire, whereas, if the coil be previously heated, the flame will continue to burn. The same fact is well shown with a piece of wire gauze containing about 700 meshes to the square inch: if this be held close over a jet of gas, and the gas lit, it is possible to remove the gauze several inches above the jet, and yet the inflammable gas below does not take fire, the flame burning only above the gauze (Fig. 27). The metallic wires in this case so quickly conduct away the heat that the temperature of the gas at the lower side of the gauze cannot rise to the point of ignition. This simple principle was made use of by Sir Humphry Davy in his safety-lamp for coal mines. It consists of an oil lamp (Fig. 28), the top of which is inclosed in a

FIG. 27.

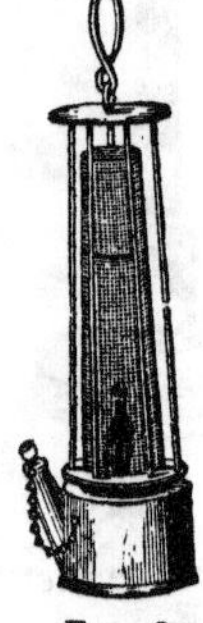

FIG. 28.

covering of wire gauze; the air can enter through the meshes of the gauze, and the products of combustion of the oil can escape, but no flame can pass from the inside to the outside of the gauze; and hence, even if the lamp be placed in a most inflammable mixture of fire-damp and air, no ignition is possible, although the combustible gas may take fire and burn inside the gauze. It is, however, then advisable that the miner should withdraw, to avoid risk of explosion of the gas from the gauze thus becoming overheated and inflaming the fire-damp which surrounds it. The compounds of carbon being generally of a more complicated nature than the preceding ones, they will be more completely considered under the head of Organic Chemistry.

CARBON AND NITROGEN.

Cyanogen Compounds.—Carbon and nitrogen do not unite together; but if nitrogen gas be passed over a white-hot mixture of charcoal and potassium carbonate, a remarkable compound, termed Potassium Cyanide, KCN, is formed, thus:

$$K_2CO_3 + N_2 + 4C = KCN + 3\ CO.$$

From this substance a large number of bodies can be prepared, all of which contain the group of atoms CN, and possess characteristic and peculiar properties; to this group the name Cyanogen is given, from its forming a number of blue compounds (κύανος, blue, and γεννάω, I produce). Cyanogen combines with metals to form *Cyanides*, and in this respect resembles chlorine.

Gaseous Cyanogen $\left.\begin{matrix}CN\\CN\end{matrix}\right\}$* can be easily obtained as a colorless gas by heating mercury cyanide. It is best collected over mercury, as it is soluble in water. It condenses to a colorless liquid when exposed to a pressure of about four atmospheres; it is inflammable, and burns with a beautiful purple flame, forming carbonic acid (dioxide CO_2) and free nitrogen. Cyanogen compounds are prepared on a large scale for various purposes by heating nitrogenous organic matter, such as clippings of hides, hoofs, &c.; with iron and potashes, a double cyanide containing iron and potassium,

* The reason for giving the double formula to gaseous cyanogen will be explained further on.

called Potassium Ferrocyanide, or yellow Prussiate of Potash (*see* Organic Chemistry, p. 298), is formed.

Cyanogen forms with hydrogen a compound analogous in composition to hydrochloric acid HCl, and called *Hydrocyanic Acid*, or commonly, *Prussic Acid*, HCN. This important substance is prepared by acting on potassium cyanide with dilute sulphuric acid in a retort. Hydrocyanic acid mixed with water distils over, leaving potassium sulphate in the retort, thus:

$$2\,KCN + H_2SO_4 = K_2SO_4 + 2\,HCN.$$

It is prepared pure and free from water by passing sulphuretted hydrogen gas, H_2S, over dry mercury cyanide, hydrocyanic acid, and mercury sulphide being formed, thus:

$$Hg\left\{\begin{matrix}CN\\CN\end{matrix}\right. + H_2S = 2\,(HCN) + HgS.$$

Mercuric cyanide and sulphuretted hydrogen yield hydrocyanic acid and mercury sulphide. Hydrocyanic acid thus prepared is a volatile liquid, boiling at 26·5°, and solidifying at — 15°; it is the most poisonous substance known, one drop of the pure substance being sufficient to produce fatal results; much care must therefore be taken in its preparation not to inhale the vapor which, even in small quantity, may produce death. It possesses a peculiar and characteristic smell of bitter almonds, and occurs in small quantities in the kernels and leaves of many plants.

Cyanogen forms a large number of compounds, some of them of a very complicated constitution, and connected with other carbon compounds, under which they will be considered.

LESSON X.

We now pass to the consideration of a group of elements which resemble each other closely, and possess strongly-marked and active properties: viz. *Chlorine, Bromine, Iodine, and Fluorine.*

CHLORINE. *Symbol* Cl. *Atomic Weight* 35·5.
Density 35·5.

Chlorine was discovered in the year 1774 by Scheele: it does not occur free in nature, but can easily be prepared from its compounds. It is found combined with metals forming chlorides; of these sodium chloride, sea or rock-salt, is the most common: to obtain chlorine from this, it must be heated with sulphuric acid and manganese dioxide, thus:

$$2\,NaCl + 2\,SO_4\,H_2 + MnO_2 = 2\,Cl + Na_2\,SO_4 + MnSO_4 + 2H_2O.$$

Sodium chloride, sulphuric acid, and manganese dioxide, give chlorine, sodium sulphate, manganese sulphate, and water.

If one part of salt to one part of manganese dioxide be mixed with two parts of sulphuric acid and two of water, and the mixture brought into a large flask, the chlorine gas is given off regularly upon the application of a very slight heat; in order to obtain the gas pure, it may be passed through the water contained in a wash-bottle before it is collected for use. Chlorine is a green-yellow gas, possessing a most disagreeable and peculiar smell, which, when the gas is present in small traces only, resembles that of seaweed, but when present in large quantities acts as a violent irritant, producing inflammation of the mucous membrane, and even causing death when inhaled. Chlorine gas when submitted to a pressure of five atmospheres at the ordinary temperature, is condensed to a heavy yellow liquid, but it has not yet been solidified. This gas cannot be collected over water or mercury, as it is soluble in the former (1 volume of water dissolving 2·37 volumes of chlorine at 15°), and combines directly with the latter, forming mercuric chloride. It can, however, be easily collected by displacement, as it is nearly 2·5 times as heavy as air. If metals in a freely divided state are brought into chlorine gas, they take fire spontaneously, forming metallic chlorides: thus powdered arsenic, antimony, or thin copper leaf, burn when thrown into the gas.

The most remarkable property of chlorine is its power of combining with hydrogen to form hydrochloric acid: when these two gases are brought together in equal volumes, they combine with explosion on bringing a flame into contact with them, or on exposing the mixture to sunlight. Chlorine is even able to decompose water in the sunlight, combining with the hydrogen and liberating the oxygen. Several experiments illustrative of this property of chlorine may be mentioned: if a burning candle be plunged into this gas, the taper continues to burn, but with a very smoky flame, the hydrogen alone of the wax entering into combination with the chlorine, whilst the carbon is given off as smoke or soot: the same effect is produced when a paper moistened with turpentine (a compound of carbon and hydrogen) is held in a jar of chlorine gas; the hydrogen of the turpentine at once combines with the chlorine, forming hydrochloric acid, and the carbon is liberated; so much heat is given off by this action that the paper frequently takes fire.

The well-known bleaching action of chlorine also depends upon its power of combining with the hydrogen of water and liberating the oxygen. Dry chlorine gas does not bleach; we may inclose a piece of cotton cloth or paper colored by a vegetable substance, as madder or litmus, in a bottle of dry chlorine, and no change of color takes place, even after the lapse of many weeks: if, however, a few drops of water are added, the coloring matter is immediately destroyed, and the cotton or paper is bleached. Here the chlorine combines with the hydrogen of the water, and the oxygen at the moment of its liberation (when it is said to be *nascent*) combines with the vegetable coloring matters, forming compounds destitute of color. Ordinary free oxygen has not this power—not at least to any great extent; it is a frequent observation that bodies in this *nascent state* have more active properties than the same bodies when free and alone. Chlorine is unable to bleach mineral colors; the difference between printer's ink, colored by lampblack or carbon, and writing ink, a vegetable black, is well illustrated by placing a sheet of paper having characters written and printed upon it in a solution of chlorine in water. Chlorine gas is largely used for bleaching purposes in the cotton, linen, and paper manufactures. It is sometimes used in the form of a gas, but more usually in combination with calcium and oxygen, forming the article called chloride of lime (mixture of calcium chloride, $CaCl_2$, and calcium hypochlorite, $CaCl_2O_2$) or Bleaching Powder. Chlorine is also largely employed as a disinfectant and deodorant, its action on organic putrefactive substances being similar to that upon organic coloring matters.

CHLORINE WITH HYDROGEN.

Hydrochloric Acid, or *Hydrogen Chloride.* Symbol HCl. Molecular Weight 36·5. Density 18·25.

This substance, the only known compound of chlorine and hydrogen, is obtained when equal volumes of chlorine and hydrogen are mixed and exposed to the diffused light of day; the gases then combine, and form an unaltered volume of hydrochloric acid gas. If the light be strong, this combination takes place so rapidly that a violent explosion occurs. Hydrochloric Acid may, however, be more easily prepared by heating common salt (sodium chloride) and sulphuric acid in a flask.

$$NaCl + So_4 H_2 = HCl + Na\ H\ So_4.$$

Sodium chloride and sulphuric acid give hydrochloric acid and hydrogen sodium sulphate.

Hydrochloric Acid is a colorless gas 1·269 times heavier than air; it fumes strongly in damp air, combining with the moisture, and has a strongly acid reaction. It is very soluble in water, one volume of this liquid at 15° dissolving 454 volumes of the gas; this solution is the ordinary Hydrochloric or Muriatic Acid of the shops. Under a pressure of 40 ats. the gas forms a limpid liquid. The gas can be collected over mercury, and its solubility in water strikingly shown by allowing a few drops of water to ascend to the surface of the mercury in contact with the gas: a rapid rise of the mercury in the jar immediately occurs. A saturated solution of hydrochloric acid in water has the specific gravity of 1·21; it fumes strongly in the air, and when heated in a retort, loses at first hydrochloric acid gas, but after a time an aqueous acid distils over at the ordinary atmospheric pressure, containing 20·22 per cent. of HCl, and boiling constantly at 110°. If the distillation be conducted under a diminished pressure, the acid boils constantly at a lower temperature, and attains a composition which is different for each boiling point; hence the constant acids thus obtained by boiling the solution of hydrochloric acid gas in water cannot be considered as definite compounds of HCl and water. This fact holds good for many other aqueous solutions of acids, &c., viz.: that residues constantly boiling at the same temperature, and having constant compositions, are obtained on distillation, the composition and boiling point varying, however, with the pressure under which the distillation has been conducted.

Enormous quantities of Hydrogen Chloride (commonly called Muriatic Acid, from *muria*, sea-salt) are obtained as a bye-product in the manufacture of sodium carbonate (*see* p. 162). More than 1,000 tons of this acid are made every week in the South Lancashire district alone. The acid thus produced is, however, very impure, having a yellow color, and containing iron, arsenic, organic matter, and sulphuric acid in solution.

The arrangement represented in Fig. 29 is adapted to show, that when gaseous hydric chloride is passed over heated manganese dioxide, water and chlorine gas are formed, thus:

$$4\ HCl + MnO_2 = Cl_2 + 2(H_2O) + MnCl_2.$$

If the gas is allowed to pass over the oxide contained in the first bulb before it is heated, no formation of water is noticed, and the red litmus paper in the bottle remains colored; as soon as the oxide is heated, moisture is at once seen to be deposited in the second bulb, and the paper becomes bleached, showing the presence of chlorine. The exact composition of hydrogen chloride is best determined by decomposing the aqueous acid by means of a current of voltaic electricity, by an arrangement similar to that shown

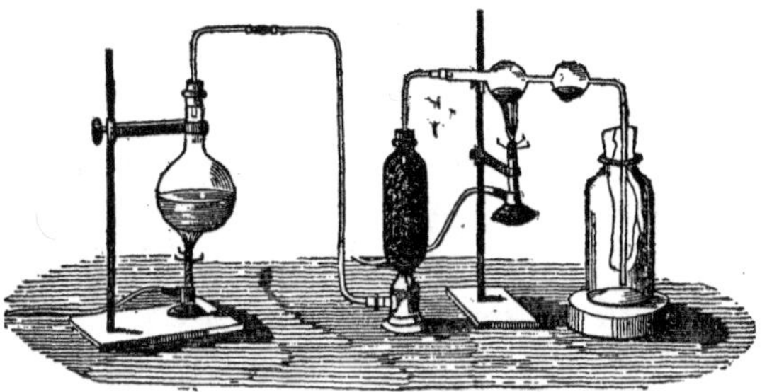

FIG. 29.

in Fig. 11, and collecting the gases (hydrogen and chlorine) evolved in a long tube after allowing the decomposition to go on for some time. If the tube thus filled be opened in the dark under a solution of potassium iodide, the solution will rise in the tube, the iodine being liberated, the chlorine combining with the potassium, until exactly half the tube is filled with liquid; the remaining gas is found to consist of hydrogen. If the mixture of electrolytic gases, which can with care be sealed up in a strong tube, be exposed to the action of daylight, or of a bright artificial light, immediate combination of the two gases will ensue, and on opening the tube under water, this liquid will completely fill the whole of the tube, showing that the component gases were present in exactly the proportion needed to form hydrochloric acid gas, which dissolves in water.

Nitro-hydrochloric Acid, or Aqua Regia.

Certain metals, such as gold and platinum, and many metallic compounds, such as certain sulphides, which do not dissolve in either nitric or hydrochloric acid separately, are readily soluble in a mixture of both of these acids, especially upon warming. This mixture has been termed Aqua Regia (because it dissolves the noble metals), and its solvent action depends upon the fact that it contains free chlorine, liberated by the oxidizing action of nitric acid on the hydrogen of the hydrochloric acid. The metals combine directly with this free chlorine to form soluble chlorides, and the sulphides are decomposed by it. The nitric acid is reduced to nitrogen dioxide, and this combines with a portion of the chlorine to form the compounds N O Cl and N O Cl_2, which are liberated as yellowish gases, condensing to a dark yellow, very volatile liquid when they are led into a freezing mixture. The same compounds are formed by direct combination when two gases, nitrogen dioxide and chlorine, are mixed together. If chlorine is present in excess, N O Cl_2 is formed; whilst N O Cl is produced when the oxide of nitrogen is present in largest quantity.

COMPOUNDS OF CHLORINE WITH OXYGEN.

Chlorine and oxygen do not unite directly, but they may indirectly be made to form the following compounds, viz.:—

Compounds of Chlorine and Oxygen.	Corresponding Acids.	
(1) *Chlorine Mon-oxide,* Cl_2 O, yielding	*Hypochlorous Acid* or *Hydrogen Hypochlorite*	HClO
(2) *Chlorine Tri-oxide,* Cl_2O_3, yielding	*Chlorous Acid* or *Hydrogen Chlorite*	$HClO_2$
(3) *Chlorine Tetroxide,* Cl_2 O_4 . . .	No corresponding Acid is known.	
(4) No corresponding Oxide is known	*Chloric Acid* or *Hydrogen Chlorate*	$HClO^3$
(5) No corresponding Oxide is known .	*Perchloric Acid* or *Hydrogen Perchlorate*	$HClO_4$

The oxides Cl_2O_5 and Cl_2O_7, corresponding respectively to the two last acids, have not yet been prepared in the free state.

Chlorine Mon-oxide.

Symbol Cl_2O. *Molecular Weight* 87. *Density* 43·5.

This body is obtained by the action of chlorine upon mercury oxide—the chlorine combining not only with the metal, but also with the oxygen, thus:

$$HgO + 2\,Cl_2 = Cl_2O + HgCl_2.$$

Mercury oxide, and chlorine, give chlorine mon-oxide, and mercuric chloride.

It is a colorless gas, which may be condensed by means of a freezing mixture to a red liquid, which is very explosive.

It is very soluble in water, yielding a yellow solution, which bleaches powerfully, and destroys vegetable coloring matter more rapidly than chlorine, inasmuch as twice as much oxygen is liberated from one molecule of chlorine mon-oxide as from one molecule of chlorine, thus:

$$\left.\begin{matrix}Cl\\Cl\end{matrix}\right\} + \left.\begin{matrix}H\\H\end{matrix}\right\}O = 2\left.\begin{matrix}H\\Cl\end{matrix}\right\} + O.$$

$$\left.\begin{matrix}Cl\\Cl\end{matrix}\right\}O + \left.\begin{matrix}H\\H\end{matrix}\right\}O = 2\left.\begin{matrix}H\\Cl\end{matrix}\right\} + O_2.$$

If a current of chlorine gas be led into a dilute and cold solution of caustic soda, a mixture is formed of sodium chloride and a compound having the formula $NaClO$, known as sodium hypochlorite, thus:

$$2\,NaHO + Cl_2 = NaClO + NaCl + H_2O.$$

If slaked lime is employed instead of caustic soda the chlorine is rapidly absorbed, and a substance well known as bleaching powder (or chloride of lime) is formed.

Bleaching powder consists of a mixture of calcium hypochlorite and calcium chloride: it is used in enormous quantities for bleaching purposes, and is prepared on the large scale by passing chlorine gas, generated from manganese di-oxide and hydrochloric acid in large stone tanks, into spacious chambers, on the floors of which a layer of slaked lime two inches thick is laid; the gas is all absorbed, and bleaching powder formed. The reaction may be thus explained:

$$2\ Ca\ H_2O_2 + 2\ Cl_2 = 2\ H_2O + Ca\ Cl_2 + Ca\ Cl_2\ O_2.$$

Slaked lime and chlorine give water, calcium chloride, and calcium hypochlorite.

Hypochlorous Acid, or *Hydrogen Hypochlorite.* Symbol HClO.

This body may be obtained in solution in water by distilling dilute solutions of sodium hypochlorite or calcium hypochlorite with dilute nitric acid. Thus:

$$Na\ OCl + HNO^3 = Na\ NO_3 + H\ OCl.$$

Sodium hypochlorite and nitric acid give sodium nitrate and hypochlorous acid.

It may be regarded as hypochlorous oxide in which one atom of chlorine has been replaced by one of hydrogen. When this atom of chlorine is replaced by metals, a series of salts called Hypochlorites is obtained. These are formed (together with chlorides) whenever chlorine gas acts upon the solution of certain metallic oxides in water, or on the damp solid oxides.

Hypochlorous acid stands therefore in the same relation to chlorine monoxide as nitric acid to nitrogen pentoxide, or as the carbonates to carbon dioxide. Hydrochloric acid decomposes hypochlorous acid with evolution of chlorine thus:

$$\left.\begin{matrix}Cl\\H\end{matrix}\right\}O + \left.\begin{matrix}H\\Cl\end{matrix}\right\} = \left.\begin{matrix}H\\H\end{matrix}\right\}O + \left.\begin{matrix}Cl\\Cl\end{matrix}\right\}.$$

Hence neither this acid nor sulphuric acid, which liberates hydrochloric acid from calcium chloride, can be used for the preparation of hypochlorous acid from the hypochlorites; but they are employed in the process of bleaching for the decomposition of the bleaching powder, to liberate the chlorine in the fibre of the cloth. This is accomplished by first dipping the goods to be bleached in a solution of bleaching powder, and then passing them through dilute hydrochloric or sulphuric acid, whereby the chlorine is liberated in the fibre of the cloth: hence the bleaching effect is only visible after the goods have been "soured," or dipped in the acid.

Chlorine Tri-oxide, Symbol $Cl_2 O_3$, is produced by the deoxidation of chloric acid, $HClO_3$; it is connected with a series of salts called chlorites, just as hypochlorous oxide is with the hypochlorites; thus, Sodium chlorite is $NaClO_2$.

Chlorine Tetroxide, Symbol Cl_2O_4, is an explosive gas obtained by the action of sulphuric acid on potassium chlorate.

Chloric Pentoxide, Symbol Cl_2O_5, has not been as yet isolated; but the acid derived from it is known, viz.:

Chloric Acid. Symbol $H\ ClO_3$. Certain metallic salts corresponding to this body (and termed chlorates) are obtained by the action of chlorine in excess upon the warm solution of the metallic oxide, thus:

$$3\ Cl_2 + 6\ KHO = KClO_3 + 5KCl + 3H_2O.$$

Chlorine, and potash, give potassium chlorate, potassium chloride, and water.

The potassium chlorate can be easily separated from the more soluble chloride by crystallization. Chloric acid itself can be prepared by decomposing potassium chlorate by hydrofluosilicic acid, whereby an insoluble potassium compound is precipitated, and chloric acid remains in solution; or by adding sulphuric acid to barium chlorate, insoluble, barium sulphate being precipitated, thus:

$$Ba\ 2\ ClO_3 + H_2\ So_4 = Ba\ So_4 + 2\ (H\ ClO_3).$$

Chloric acid solution may be concentrated to a syrup, but it decomposes on further evaporation; it acts as a powerful oxidizing agent, and when dropped upon paper it produces ignition, parting with its oxygen. The chlorates also yield up all their oxygen on heating, and the potassium salt is used as a most convenient source of this gas. The composition of chloric acid is ascertained by determining the weight of oxygen, which potassium chlorate yields on heating (*see* ante, p. 12), together with the amount of chlorine contained in the residual potassium chloride.

Chlorine Heptoxide, Symbol Cl_2O_7, is unknown as yet in the free state. Its corresponding acid is known, viz.:

Perchloric Acid. Symbol $HClO_4$. Molecular Weight 100·5 When potassium chlorate is heated, it first fuses and begins to give off oxygen: at a certain point, however, the whole mass solidifies; if the decomposition be stopped at this stage a new salt will be found to be contained in the residue, together with chloride and unaltered chlorate, thus: $2KClO_3 = KClO_4 + KCl + O_2$. This new salt is termed Potassium perchlorate, and its composition is $KClO_4$. It may be easily separated from the chlorate by the action of hydrochloric acid, which decomposes this latter, but has no action on the perchlorate. Perchloric acid, $HClO_4$, can easily be prepared from the potassium salt by the action of strong sulphuric acid. If a mixture of one part of the dry perchlorate and four of sulphuric acid be distilled in a retort, a colorless fuming liquid condenses in the receiver; this is perchloric acid, $HClO_4$. It has a specific gravity of 1·78 at 15°·5, and does not solidify at −35°. Perchloric acid is one of the most powerful oxidizing agents known; when thrown upon wood or paper it produces instant ignition, and when dropped on to charcoal it decomposes with a loud explosion. It combines with water to form a crystalline hydrate, $HClO_4 + H_2O$, which when further diluted with water forms a thick oily liquid, boiling constantly at 203°, and containing 72·3 per cent. of $HClO_4$, and thus not corresponding to any definite hydrate. Perchloric acid is by far the most stable of the acids derived from chlorine.

COMPOUNDS OF CHLORINE AND NITROGEN.

Chlorine combines with nitrogen, though only indirectly, to form a very remarkable compound, the composition of which has not been as yet determined. If chlorine gas is passed into a solution of ammonia, nitrogen, as we have seen, is liberated; if an excess of chlorine be employed, drops of an oily liquid are seen to form, which, on being touched, explode with fearful violence, so that the greatest caution must be used in manipulating even traces of this body. The explosive nature of this compound arises from the fact that its constituent elements are very loosely combined, and separate with sudden violence.

COMPOUNDS OF CHLORINE AND CARBON.

Chlorine and carbon unite to form four different compounds, which will be discussed under Organic Chemistry.

LESSON XI.

BROMINE. *Symbol* Br. *Combining Weight* 80. *Density* 80.

This element, which closely resembles chlorine in its properties and compounds, was discovered by Balard, in 1826, in the salts obtained by the evaporation of sea-water. It does not occur free in nature, and is, like chlorine, found combined with sodium and magnesium in the waters of certain mineral springs. In order to obtain pure bromine, use is made of the fact that free chlorine liberates bromine from its combinations with metals, forming a metallic chloride. The bromine thus set free may be separated by shaking the liquid up with ether, which dissolves the bromine, forming a bright red solution. On adding caustic potash to this ethereal solution, the color at once disappears, the bromine becomes combined, forming potassium bromide, and bromate: on evaporation of the ether these salts remain, and after ignition (to decompose the bromate), the bromine can again be liberated by the action of sulphuric acid and manganese dioxide, exactly as in the case of chlorine.

$$2\,BrK + 2H_2SO_4 + MnO_2 = 2Br + K_2SO_4 + MnSO_4 + 2H_2O.$$

Potassium bromide, and sulphuric acid, and manganese

dioxide, give bromine, potassium sulphate, manganese sulphate, and water.

Bromine is a dark, reddish-black, heavy liquid (the only element liquid at ordinary temperatures besides mercury): its specific gravity at 4° is 2·966; it freezes at — 22° to a black solid, and boils at 63°. It possesses a very strong, irritating smell, resembling that of chlorine (βρῶμος a stink), and, when inhaled, acts as a strong poison: one part of bromine dissolves in about 30 parts of water at 15°; and this solution possesses bleaching powers, feebler, however, in action than those of chlorine. This bleaching action is caused by the oxidation of the coloring matter, bromine combining with the hydrogen of the water to form an acid called hydrobromic acid, corresponding in mode of formation and properties to hydrochloric acid.

Hydrobromic Acid, or *Hydrogen Bromide*. Symbol HBr. Molecular Weight 81. Density 40·5.

Hydrogen and bromine do not unite together, even when placed in the sunlight, but they combine to form hydrobromic acid when passed through a red-hot porcelain tube. Hydrobromic acid is prepared by the action of acids (phosphoric acid) on the bromides; it is a colorless gas, having a strong acid reaction, and fumes strongly in moist air: it is very soluble in water. When concentrated, the aqueous acid boils (under 760 mm. pressure) at 126°, and contains 47·8 per cent. of HBr. Two volumes of this gas contain one of bromine united with one of hydrogen. The aqueous acid neutralizes bases, forming the bromides and water. The gas liquefies at - 73°.

Oxides and Oxi-acids of Bromine.

These are analogous to those of Chlorine, although not so numerous.

Bromine Mon-oxide, Br_2O, is not known, but the corresponding *Hypobromous Acid*, HBrO, is known, though only in aqueous solution. It is obtained by the action of bromine water upon mercuric oxide.

Thus, $HgO + 2\ Br_2 + H_2O = 2\ H\ BrO + Hg\ Br_2$.

Like hypochlorous acid, it bleaches vegetable coloring matters by oxidation, hydrobromic acid being formed. *Bromic Pentoxide*, Br_2O_5, has not yet been isolated, but the corresponding *Bromic Acid*, $HBrO_3$, can be obtained by the action of chlorine upon bromine water, thus:

$$Br + 3H_2O + 5Cl = 5HCl + HBrO_3.$$

In both its properties and composition it corresponds to chloric acid. Certain metallic bromates can be obtained, like the corresponding chlorates, by the action of bromine on the metallic oxides in aqueous solution. The best method of obtaining the bromates of the alkaline metals (potassium and sodium) consists in saturating a concentrated solution of the metallic carbonate with chlorine until carbonic acid begins to escape, and then adding bromine; all the chlorine escapes, and solution of pure bromate remains. Hence it appears that bromine can displace chlorine from its compounds with oxygen, whilst chlorine can liberate bromine from its compound with hydrogen. The bromates are decomposed by heat in the same way as the chlorates.

Perbromic Acid, or *Hydrogen Perbromate*, $HBrO_4$, has been obtained by the action of bromine upon perchloric acid, $HClO_4$.

IODINE. *Symbol I. Combining Weight* 127. *Density* 127.

Iodine likewise occurs in sea-water, and is obtained from kelp, the ash of certain seaweeds, in which it is found combined with sodium and magnesium, forming compounds called Iodides. It was discovered in 1812 by Courtois. Iodine is obtained from kelp by exactly the same process as that by which chlorine and bromine are obtained from chlorides and bromides, viz. by heating with sulphuric acid and manganese dioxide. Iodine is thus liberated in the form of a deep violet-colored vapor, which condenses to a dark gray solid, with bright metallic lustre. Iodine melts at 115°, and boils above 200°, and has a specific gravity of 4·95; it gives off a perceptible amount of vapor at the ordinary temperature, and possesses a faint, chlorine-like smell. Water dissolves a very small quantity of iodine; but, in presence of a soluble iodide, it is freely dissolved, forming a deep red or brown solution: it is easily soluble in alcohol and other organic liquids. Iodine does not possess such active properties as either of the preceding elements, its solution does not bleach organic coloring matters, and it is liberated from its compounds by both bromine and chlorine. Free iodine forms a remarkable compound with starch, of a splendid blue color, and by this means the minutest trace of this substance can be detected. To apply this test, one drop of potassium iodide solution is added to starch paste largely diluted with water; no blue color is observed until the iodine is set free by the addition of a drop or two of chlorine-water, when a deep blue coloration is instantly perceived. Iodine acts as a powerful poison, but, given in small quantities, it is much used as medicine.

Hydriodic Acid, or *Hydrogen Iodide*. Symbol HI. Molecular Weight 128. Density 64. Iodine and Hydrogen may be made to unite with each other by heating them together: hydriodic acid is liberated when dilute sulphuric acid acts on an iodide. This substance is, however, best prepared by acting upon the phosphorus iodide with water, thus:

$$PI_3 + 3H_2O = 3HI + H_3PO_3.$$

Phosphorus tri-iodide and water produce hydriodic acid and phosphorous acid.

Hydriodic acid is a colorless gas, possessing a strong acid reaction, and fuming strongly in the air: it is very soluble in water, yielding a solution which boils at 127°, and contains 57 per cent. of HI. The gas liquefies under pressure, and solidifies at - 55°. An analysis of this gas shows that hydriodic acid (like hydrochloric) is composed of one volume of hydrogen and one of iodine, forming two volumes of hydriodic acid.

Oxides and Oxi-acids of Iodine.

Iodine, when treated with caustic alkaline solutions, does not yield bleaching liquors ; nor is there any compound corresponding to hypochlorous acid known in the iodine series. It forms, however, two important acids, iodic and per-iodic acids, corresponding respectively with chloric and perchloric acids ; the oxide, I_2O_7, corresponding with the latter, is also known. *Iodine pentoxide*, I_2O_5, is obtained as a white crystalline solid by heating iodic acid, HIO_3, to 170°.

Iodic Acid, or *Hydrogen Iodate.* Symbol HIO_3. Molecular Weight 176. This acid, which corresponds closely with chloric acid, may be obtained by the direct oxidation of iodine by nitric acid, and also by acting upon iodine water with chlorine, thus :

$$I + 3H_2O + 5Cl = HIO_3 + 5HCl.$$

Iodine, water, and chlorine, yield iodic acid and hydrochloric acid.

The alkaline iodates are formed (together with the iodides of the metals employed) like the chlorates and bromates, by dissolving iodine in the caustic alkalies.

$$6I + 6HKO = KIO_3 + 5KI + 3H_2O.$$

Iodine and caustic potash give potassium iodate, potassium iodide, and water.

The whole of the iodine is converted into iodate if chlorine gas be passed into the solution, thus :

$$I + 6KHO + 5Cl = KIO_3 + 5KCl + 3H_2O.$$

Iodine, caustic potash, and chlorine, yield potassium iodate, potassium chloride, and water.

Hence we see that oxygen combines with iodine to form an iodate in preference to forming a chlorate with chlorine. The iodates of the alkaline metals decompose on heating like the corresponding chlorates, yielding oxygen and an iodide, whereas the iodates of the heavy metals yield the metallic oxides, iodine, and oxygen.

Iodine Heptoxide, I_2O_7, is prepared by heating periodic acid

Periodic Acid, HIO_4, or *Hydric Per-iodate*, can be obtained from the corresponding perchloric acid by the addition of iodine.

Iodine and Nitrogen.

The three atoms of hydrogen in ammonia can be wholly or partly replaced by iodine; the resulting compounds are black powders, which, when touched in the dry state, suddenly decompose with a loud report, and sometimes even explode spontaneously. The pure iodide of nitrogen is prepared by the action of a strong alcoholic solution of iodine upon aqueous ammonia, thus:

$$6I + 4NH_3 = NI_3 + 3NH_4I.$$

FLUORINE. *Symbol F. Combining Weight* 19.

This element occurs combined with the metal calcium, forming calcium fluoride CaF_2, or fluorspar, a mineral crystallizing in cubes and found in Derbyshire; it also exists in large quantities in cryolite ($3\ NaF + AlF_3$), a mineral found in Greenland, whilst it has been detected in minute quantities in the teeth, and even in the blood of animals. Fluorine is remarkable as forming no compounds with oxygen, and as being extremely difficult to prepare in a pure state. Many unsuccessful attempts have been made to obtain fluorine, but none of the methods which yield chlorine, bromine, or iodine give any result. By the action, however, of dry iodine upon dry silver fluoride, it appears that fluorine has been isolated, and it is found to be a colorless gas which does not act upon glass, and is absorbed by caustic potash with the formation of potassium fluoride and hydric di-oxide.

$$2\ KHO + F_2 = 2\ KF + H_2O_2.$$

Hydrofluoric Acid, or *Hydrogen Fluoride.* Symbol HF Molecular Weight 20. Density 10.

This gas corresponds in composition to the hydrogen compounds of the three preceding elements, and may be obtained in an exactly similar manner by the action of sulphuric acid upon calcium fluoride, thus:

$$H_2\,SO_4 + CaF_2 = 2\,HF + Ca\,SO_4.$$

Sulphuric acid, and calcium fluoride, give hydrofluoric acid and calcium sulphate.

Hydrofluoric acid gas must be prepared in a leaden or platinum vessel, as glass is rapidly attacked by the vapor. The colorless gas thus obtained fumes strongly in the air; if it be passed into a metallic tube (Fig. 30) placed in a freezing mixture of the temperature of $-20°$, a liquid is formed; this liquid is strong aqueous hydrofluoric acid: it appears doubtful whether the dry acid HF has been obtained in the liquid state. The strong acid acts very violently upon the skin, producing painful wounds; and the fumes of the gas are likewise dangerous from their corrosive power. When brought into contact with water the strong acid dissolves with a hissing noise; this aqueous acid attains a constant boiling point under the ordinary atmospheric pressure, when the liquid contains 37 per cent. of HF.

The most remarkable property of hydrofluoric acid is its power of etching upon glass; this arises from the fact that fluorine forms, with the silicon contained in the glass, a volatile compound called Silicic Fluoride (*see* p. 123). This etching serves as a very delicate test of the presence of fluorine, and is effected in a very simple manner by covering a watch-glass with a thin coating of wax, removing a portion by means of a sharp point, and then exposing the glass for a short time to the vapor of hydrofluoric acid given off by heating the materials in a small leaden saucer; on removing the wax with a little turpentine, the marks on the glass will be distinctly visible. The solution of hydrofluoric acid in water is also used for the purpose of etching on glass. Fluorspar is used in metallurgic operations as a flux, whence its name (*fluo*, to flow).

The members of the foregoing group of elements exhibit certain remarkable relations among themselves, especially a gradation in properties. Thus chlorine is a gas, bromine a liquid, and iodine a solid, at ordinary temperatures; the specific gravity of liquid chlorine is 1·33, of bromine 2·97, and of iodine 4·95; liquid chlorine is transparent, bromine but slightly so, and iodine is opaque. The combining weight, and, therefore, the density of bromine, is nearly the mean of those of chlorine and iodine, $\frac{35 \cdot 5 + 127}{2} = 81 \cdot 25$; and in its general chemical deportment bromine stands half-way between the other two elements The property which distinguishes these substances from the rest of the elements is the power of forming, with hydrogen, compounds containing equal volumes of the constituent gases united without condensation.

LESSON XII.

SULPHUR. *Symbol S. Combining Weight* 32. *Density* 32.

Sulphur occurs both free and combined in nature; it is

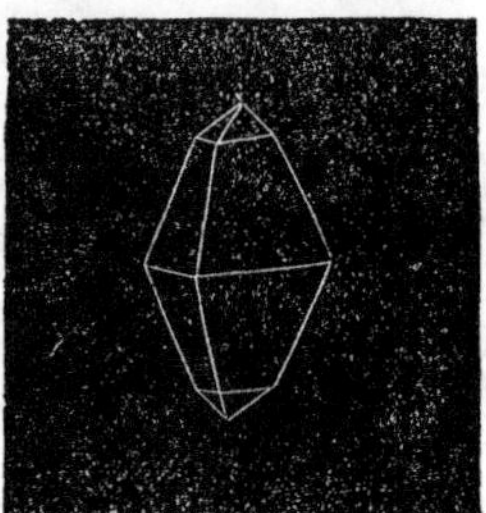

FIG. 31.

found free in certain volcanic countries, especially in Sicily and Iceland, and occurs crystallized in yellow transparent crystals in the form of rhombic octahedra (Fig. 31); it exists in combination with many metals, forming compounds termed Sulphides, which constitute the common ores from which the metals are usually obtained. Thus PbS, lead sulphide, or *Galena;* Zn S, zinc sulphide, or *Blende;* and Cu S, copper sulphide, are the substances from which those metals are generally procured. Sulphur also is found in nature, combined with metals and oxygen, to form a class of salts called Sulphates; of these, calcium sulphate or gypsum, barium sulphate or heavy spar, sodium sulphate or Glauber's salt, occur in the largest quantity. Sulphur likewise occurs combined with hydrogen as a gas called Sulphuretted Hydrogen, H_2S, in the waters of certain springs, as at Harrogate. In order to obtain pure sulphur, the mineral containing the crude substance mixed with earthy impurities is heated in earthenware pots (Fig. 31); the sulphur distils over

FIG. 31.

in the form of vapor, which is condensed in similar pots placed outside the furnace. When brought to this country, the sulphur thus obtained is refined or purified by subjecting it to a second distillation. If the vapor of sulphur is quickly cooled below its melting-point, it solidifies in the form of a fine crystalline powder called Flowers of Sulphur, exactly as aqueous vapor, when cooled down below the freezing point of water, deposits as snow. When sulphur is gently heated it melts, and may be cast into sticks, and is then known as brimstone or roll sulphur.

Sulphur exists in three modifications: the first is that in which sulphur crystallizes in nature, and the other two are obtained by melting sulphur. If melted sulphur be allowed to cool slowly, it crystallizes in long, transparent, needle-shaped, prismatic crystals, which are quite different in form from the natural crystals of sulphur, and have the specific gravity of 1·98; whereas the specific gravity of the crystals of native sulphur is 2·07. These transparent crystals become opaque after exposure to the air for a few days, owing to each crystal splitting up into several crystals of the natural, octahedral, or permanent form. The third allotropic modification of sulphur is obtained by pouring melted sulphur heated to 230° into cold water; the sulphur thus forms a soft tenacious mass resembling caoutchouc, and has a specific gravity of 1·96. This form of sulphur is, however, not permanent; in a few hours, at the temperature of the air, the mass assumes the ordinary brittle form of the element, while, if heated to 100°, it instantly changes to the brittle form, and thereby evolves so much heat as to raise its temperature up to 111°. These peculiar modifications become apparent when sulphur is heated; thus, sulphur melts, to begin with, at 115°, and forms an amber-colored mobile liquid; as the temperature rises, the liquid becomes dark-colored, and attains the consistency of thick treacle, so that at about 230° it can scarcely be poured out of the vessel: heated above 250° it again becomes fluid, and remains as a dark reddish-black-colored thin liquid, until the temperature rises to 490°, when it begins to boil and gives off a red-colored vapor.

Sulphur is an inflammable substance, and when heated in the air or in oxygen, burns with a bluish flame, combining with the oxygen to form sulphur dioxide (often called sulphurous acid), SO_2, which is given off as a gas possessing the peculiar and well-known suffocating smell which is evolved when a common lucifer match is burnt. Sulphur is insoluble in water and most organic liquids, but both the natural octahedral variety, and the other crystalline (or prismatic) variety, dissolve freely in carbon di-sulphide, CS_2, whilst the tenacious form of sulphur is insoluble in this liquid. When deposited from solution in carbon di-sulphide, sulphur crystallizes in the ordinary natural or octahedral form.

COMPOUNDS OF SULPHUR AND OXYGEN.

Two compounds of sulphur and oxygen are known in the free state, viz., the *di-oxide*, SO_2, and the *tri-oxide*, SO_3. These each unite with a molecule of water to form important acids :

H_2SO_3, Hydrogen Sulphite, or Sulphurous Acid, and

H_2SO_4, Hydrogen Sulphate, or Sulphuric Acid.

Besides these, five other oxi-acids of sulphur are known, with whose corresponding oxides we are unacquainted. The following table shows the composition of seven sulphur oxacids; the first three on the list are important compounds, the remaining bodies are but little known, and do not as yet serve any purpose in the arts or manufactures. These compounds exhibit in a striking manner the law of multiple combining proportions enumerated by Dalton. (*See ante*, p. 50.)

Sulphurous Acid	H_2SO_3
Sulphuric Acid	H_2SO_4
Hyposulphurous Acid	HS_2O_3
Dithionic Acid	$H_2S_2O_6$
Trithionic Acid	$H_2S_3O_6$
Tetrathionic Acid	$H_2S_4O_6$
Pentathionic Acid	$H_2S_5O_6$

Sulphur Di-oxide or *Sulphurous Acid.* Symbol SO_2. Molecular Weight 64. Density 32.

This gas is obtained when sulphur is burnt, and it is given off in large quantities from volcanic craters ; it may be prepared more conveniently on the small scale by removing the elements of water, and one additional atom of oxygen from sulphuric acid by heating along with it the metals copper or mercury, thus:

$$Cu + 2H_2SO_4 = SO_2 + CuSO_4 + 2H_2O.$$

Copper, and sulphuric acid, yield sulphur dioxide, copper sulphate, and water.

The gas thus given off may be washed to purify it and then collected over mercury or by displacement. It is a colorless gas, possessing a suffocating smell of burning sulphur; it is 2·247 times heavier than air, and may be condensed to a colorless liquid by cooling down to -10° under the ordinary atmospheric pressure; when cooled below -76° the liquid freezes to a transparent solid. The arrangement for liquefying the gas is seen in Fig. 32: it consists of the usual evolution flask and washing bottle, which is connected with a spiral glass tube surrounded by a freezing mixture of salt and pounded ice. The gas condenses in this tube and falls down into the small flask placed below, which is also placed in a freezing mixture. When a sufficient quantity of the liquid has been collected, the neck of the flask can be sealed up by the

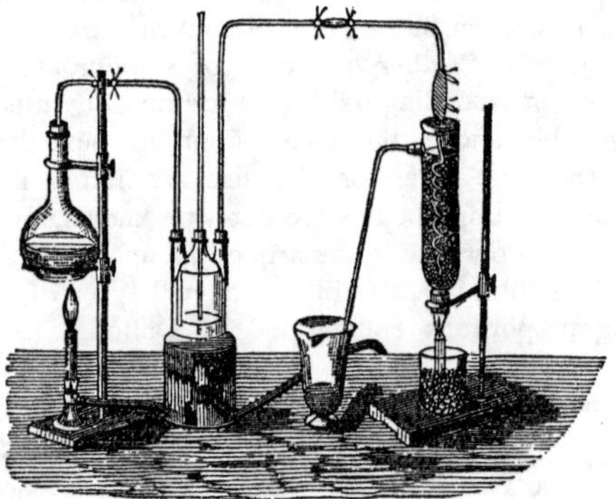

FIG. 32.

blowpipe at the narrow part, and the liquefied sulphuric dioxide preserved. This liquid evaporates very quickly when brought into the air, and the heat which thus becomes latent is so considerable, that the temperature may in this way be reduced to $-60°$; this is easily shown by pouring some of the liquid upon the bulb of an alcohol thermometer which has been wrapped in cotton wool.

Sulphur dioxide, like all other gases which are easily condensed, exhibits considerable deviation from Boyle's law of pressures, occupying for equal increments of pressure less space than air under the same conditions; this difference becoming larger the lower the temperature. The volume of this gas formed by the combustion of sulphur is found to be exactly the same as that of the oxygen employed. Hence, as the density of sulphur dioxide is 32, it contains equal weights of its constituent elements, 1 volume of sulphur uniting with 2 volumes of oxygen to give 2 volumes of the dioxide.

Sulphur dioxide is very soluble in water, 1 volume of water at 10° dissolving 51·38 volumes, and at 20° 36·22 volumes of this gas. The solution of the gas in water consists (like that of carbon dioxide, p. 71) of *hydrogen sulphite* or true sulphurous acid, H_2SO_3: but this substance is decomposed on boiling the liquid, water and sulphurous dioxide being reproduced, this latter escaping as a gas. If the solution of this gas in water be cooled below 5°, a crystalline hydrate of sulphurous acid separates out, having the composition $H_2SO_3 + 14H_2O$.

Sulphurous acid is really the hydrogen salt in a series of compounds called sulphites; these compounds, however, are easily decomposed by the stronger acids, sulphur dioxide being liberated as a gas. This substance is largely used as a bleaching agent, especially for silk and woollen goods which cannot be bleached by chlorine; it is also employed as an *antichlor* for the purpose of getting rid of the excess of chlorine present in the bleached rags from which paper is made. The great value of sulphur dioxide in the arts is in the manufacture of sulphuric acid or oil of vitriol, and for this purpose enormous quantities of the dioxide are used.

Sulphurous acid, $H_2\ SO_3$, like carbonic acid, $H_2\ CO_3$, is a dibasic acid, that is, it contains two atoms of hydrogen, both of them being capable of being replaced by metals : thus two classes of salts are derived ; the so-called *acid salts*, where only one atom of hydrogen has been replaced, and the *neutral salts*, where both atoms have been replaced by a metal. Thus —Hydrogen Potassium Sulphite, $HK\ SO_3$, is an acid salt, and Dipotassium Sulphite, $K_2\ SO_3$, is a neutral salt. Similarly, we have Hydrogen Potassium Carbonate, $HK\ CO_3$, and Potassium Carbonate, $K_2\ CO_3$.

In its bleaching action, sulphur di-oxide acts in a manner exactly opposite to that in which chlorine acts, inasmuch as it unites with the oxygen of the water or coloring matter present, forming sulphuric acid and liberating the hydrogen ; so that sulphurous acid bleaches by acting as a reducing or deoxidizing agent, whereas chlorine bleaches by oxidation : similarly, its action as an antichlor depends on the formation of sulphuric and hydrochloric acids, thus :

$$SO_2+2\ H_2O+2Cl=H_2\ SO_4+2HCl.$$

LESSON XIII.

Sulphur Tri-oxide, or *Sulphuric Anhydride*.* Symbol SO_3. Molecular Weight 80. Density 40.

Sulphur di-oxide does not, under ordinary circumstances, combine directly with oxygen to form SO_3, but if these two dry gases be passed together over heated and finely divided metallic platinum, union takes place, and dense white fumes of the sulphur tri-oxide are evolved, condensing to white silky needles. These crystals melt at 29° and boil at 46°, yielding a colorless vapor, which when passed through a red-hot tube

* By anhydride is meant an oxide which forms an acid on treatment with water.

is decomposed into two volumes of sulphuric di-oxide and one volume of oxygen. Sulphur tri-oxide does not redden litmus paper, and may be moulded by the fingers without charring the skin: when brought into contact with water, the two substances combine with great force, forming sulphuric acid, H_2SO_4, hissing as a red-hot iron would do. The combination thus formed cannot be separated again into sulphur tri-oxide and water by boiling. The tri-oxide may likewise be prepared by distilling Nordhausen fuming sulphuric acid (see below).

Sulphuric Acid, or *Hydrogen Sulphate.* Symbol $H_2 SO_4$ Molecular Weight 98.

This substance is the most important and useful acid known, as by its means nearly all the other acids are prepared, and also because it is very largely used in the arts and manufactures, for a great variety of purposes. So valuable are the applications of this acid, that the quantity now manufactured in the South Lancashire district alone exceeds 3,000 tons per week. Indeed, it has been truly said, that the commercial prosperity of a country may be judged of with great accuracy by the amount of sulphuric acid which it consumes.

Sulphuric acid was first prepared by distilling a compound of iron, oxygen, sulphur and water, called ferrous sulphate or green vitriol. The acid thus obtained is known as fuming or Nordhausen acid, and consists of a mixture of hydrogen sulphate and sulphur tri-oxide, $H_2SO_4+SO_3$. This plan of preparation has, however, long been superseded by the following more convenient method, which depends upon the fact, that although sulphur di-oxide does not combine with free oxygen and water to form sulphuric acid, it is capable of taking up the oxygen when the latter is united with nitrogen in the form of nitrogen tri-oxide, N_2O_3, thus:

$$SO_2+H_2O+N_2O_3=H_2SO_4+N_2O_2.$$

Sulphur di-oxide, water, and nitrogen trioxide, yield sulphuric acid and nitric oxide.

10*

The nitric oxide formed in this decomposition takes up another atom of oxygen from the air, becoming N_2O_3, and this is again able to convert a second molecule of SO_2 with H_2O into H_2SO_4,* being a second time reduced to N_2O_2, and ready again to take up another atom of oxygen from the air. Hence it is clear that the N_2O_2 acts simply as a carrier of oxygen between the air and the SO_2 ; an indefinitely small quantity of this nitrogen trioxide being, therefore, theoretically

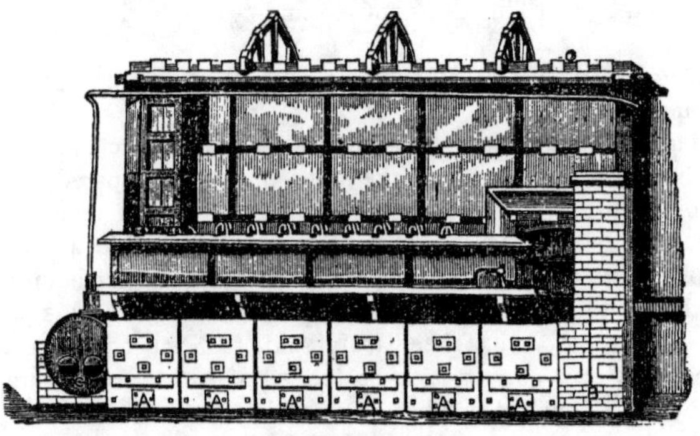

FIG. 33.

able to convert an indefinitely large quantity of sulphur dioxide, water, and oxygen into sulphuric acid.†

This process is carried on, on the large scale, in chambers made of sheets of lead (often of a capacity of 50,000 or 100,000

* A *molecule* is a group of atoms forming the smallest portion of a chemical substance, either simple or compound, that can be isolated, or that can exist alone; it is the smallest amount of substance that can enter into any reaction or be generated by it; an *atom* being the smallest portion of an element that can exist in a compound body as a mass indivisible by chemical forces; thus the *molecule* of water H_2O contains 2 *atoms* of hydrogen.

† There is some doubt respecting the actual decomposition occurring in the leaden chamber, and the above must be taken only as a general explanation of what goes on. If the supply of steam be insufficient, a solid compound of nitric tetroxide, sulphuric dioxide, and water is formed, which is decomposed, on addition of water, into sulphuric acid and nitric oxide. This substance, which has been called the *crystals of the leaden chamber*, is supposed by some to play an important part in the formation of sulphuric acid.

cubic feet), into which the above-mentioned materials are brought. Fig. 33 exhibits the construction of such a *sulphuric acid chamber.* The sulphur dioxide is procured either by burning sulphur in a current of air, or by roasting a mineral called iron pyrites (a compound of sulphur and iron, FeS_2) in a suitable furnace, AA, Fig. 34. The sulphur of the pyrites burns away, and the gaseous product is led, together with atmospheric air, into the chamber, whilst ferric oxide, Fe_2O_3, remains behind in the furnace. A small stove (B), containing nitre, is placed in the hinder part of the furnace, where this salt is decomposed by the action of the sulphur di-oxide, an alkaline sulphate being formed whilst nitrous fumes pass with the other gases into the chamber. Jets of steam are also blown into the chamber from a boiler (C) at various points, and a thorough draught is maintained by connecting the end of the chamber with a high chimney. The sulphuric acid, as it forms, falls on to the floor of the chamber, and when the process is working properly, it is continually drawn off, attaining a specific gravity of about 1·60, whilst the waste gases passing out of the chamber should contain nothing but nitrogen and small quantities of nitric oxide. In order to obtain from this weak chamber-acid the pure sulphuric acid, H_2SO_4, the excess of water must be removed by evaporation; this is conducted, on the large scale, first, by heating the chamber-acid in open leaden pans until the specific gravity rises to 1·72, when the acid is known as the *brown oil of vitriol* of commerce, and then further concentrated in vessels of glass, or of platinum (as lead is attacked by the strong acid), until its maximum strength and specific gravity is attained. The hydrogen sulphate thus obtained is a thick oily liquid, boiling at about 338°,* and freezing at 10°·8; its specific gravity at 0° is 1·854. It combines with water with great force, absorbing moisture rapidly from the air;

* When boiled, hydrogen sulphate undergoes a slight decomposition, sulphur tri-oxide being evolved, and an acid remaining behind which contains only 98·5 of H_2SO_4, and which at 338° C. may be boiled down without further decomposition.

hence it is used in the laboratory as a drying agent. Great heat is evolved when this acid is mixed with water, and care must be taken to bring these two liquids together gradually, otherwise an explosive combination may ensue. Many organic bodies, such as woody fibre and sugar, are completely decomposed and charred by strong sulphuric acid, whilst others, such as alcohol, oxalic and formic acid, are split up into other compounds by the withdrawal of the elements of water by this acid.

One molecule of hydrogen sulphate unites with one of water to form a compound, H_2SO_4, + H_2O, which can be obtained pure by cooling a mixture of acid and water, having a specific gravity of 1·78 down to 7°C., at which temperature rhombic crystals of the hydrated acid are formed. The sulphuric acid of commerce frequently contains large quantities of impurities, especially lead sulphate from the chamber, and frequently arsenic from the pyrites, and nitric acid, as well as the lower oxides of nitrogen. In order to free the acid from these impurities, it must be distilled and subjected to other treatment, for a description of which the reader is referred to the larger treatises. At high temperatures sulphuric acid decomposes into sulphur dioxide, SO_2, oxygen O, and water, H_2O; thus, if a current of the acid be allowed to flow on to red-hot bricks, and the gases resulting from the decomposition passed through water, the sulphur dioxide will be completely absorbed, and a supply of pure oxygen obtained. Hydrogen sulphate is, as its name implies, a dibasic acid, *i.e.* it contains two atoms of hydrogen, either or both of which can be replaced by an equivalent quantity of a metal. As with sulphurous acid, in the case of the alkaline metals, we have two salts, thus—$KHSO_4$ and K_2SO_4, potassium being *mon-atomic;* that is, one atom of potassium is capable of replacing one of hydrogen. As, however, most of the other metals are *di-atomic* —*i.e.* one atom of metal is capable of replacing two atoms of hydrogen—we have only one sulphate of these metals; thus barium sulphate is Ba SO_4, where Ba = H_2: so copper sulphate is $CuSO_4$, where Cu = H_2. Barium and lead sulphates are insoluble in water; hence soluble salts of these metals are used as tests of the presence of a sulphate; calcium, strontium, and potassium sulphates are but slightly soluble in water, while the other sulphates are easily soluble.

Some sulphates crystallize as anhydrous salts, such as K_2SO_4, potassium sulphate; $Ba\ SO_4$, barium sulphate; and Ag_2SO_4, silver sulphate; while others require water to retain their crystalline form, and this water is termed *water of crystallization.* The crystals of iron sulphate, or green vitriol, and of zinc sulphate, or white vitriol, contain seven molecules of water in the solid form; whilst copper sulphate, or blue vitriol, requires but five molecules to preserve its crystalline form. Thus—

$$\begin{matrix} Fe\ SO_4 + 7H_2O \\ Zn\ SO_4 + 7H_2O \end{matrix} \text{ and } Cu\ SO_4 + 5\ H_2O.$$

Hyposulphurous Acid, or *Hydrogen Hyposulphite,* Symbol $H_2S_2O_3$, is not known in the free state. The formula of a metallic hyposulphite, such as that of sodium, is $Na_2\ S_2\ O_3$; this also contains five molecules of water of crystallization; it is largely used in photography for the purpose of fixing the image, the salt possessing the property of dissolving the silver salts which have been unacted on by the light. This useful salt is prepared by passing a current of sulphur dioxide into a solution of sodium sulphide, and purifying by crystallization the sodium hyposulphite obtained.

Sulphur tri-oxide combines directly with hydrochloric acid to form a compound, which is interesting in a theoretical point of view; viz., *chlor-hydrosulphuric* acid, $HClSO_3$. Sulphur dioxide unites with chlorine to form *sulphuryl chloride,* $Cl_2\ SO_2$. The first compound is really sulphurous acid, with one atom of hydrogen replaced by chlorine, whilst the second is sulphur tri-oxide, in which an atom of oxygen is replaced by two of chlorine.

COMPOUNDS OF SULPHUR AND HYDROGEN.

Two of these are known, viz., hydrogen sulphide, H_2S, and hydrogen disulphide, H_2S_2.

Hydrogen Sulphide, or *Sulphuretted Hydrogen.* Symbol H_2S. Molecular Weight 34. Density 17.

This gas is best prepared by the action of dilute sulphuric acid upon iron sulphide, FeS, iron sulphate being also formed thus—

$$FeS + H_2SO_4 = FeSO_4 + H_2S,$$

where two atoms of hydrogen change place with one of divalent iron. Fig. 34 represents a convenient form of apparatus for the production and purification of this gas. It may be collected over warm water, and is a colorless gas, possessing the peculiar oder of rotten eggs ; it burns on application of a light with a bluish flame, forming water and sulphur dioxide. When inhaled, it acts as a poison on the animal economy, even if diluted with large quantities of air. Sulphuretted hydrogen gas dissolves in water to a considerable extent,

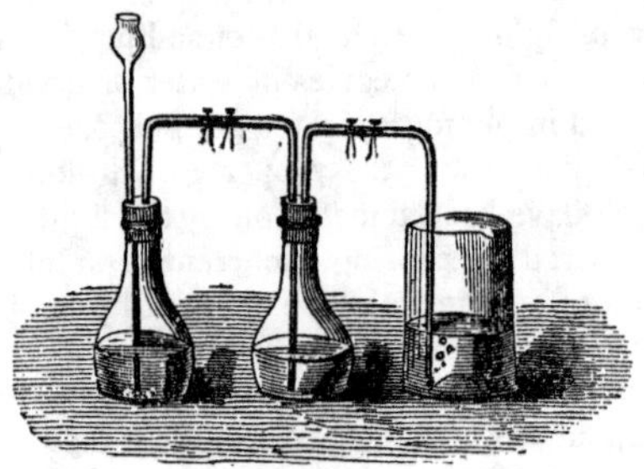

FIG. 34.

imparting its peculiar smell and a slightly acid reaction to the water. One volume of water at 0° dissolves 4·37 volumes of the gas, whilst at 15° 3·23 volumes are soluble. Exposed to a temperature of –74°, this gas condenses to a colorless, mobile liquid, which, when further cooled to –85°, freezes to a transparent, ice-like solid. Under a pressure of about seventeen atmospheres, this gas liquifies at the ordinary temperature of the air. Sulphuretted hydrogen occurs free in nature in volcanic gases, as well as in the water of certain springs; thus Harrogate waters owe their peculiar odor and medicinal power to the presence of this gas. It is likewise generated by the putrefaction of animal matters, such as albumen, or the white of eggs, which contains sulphur; also by the deoxidation of sulphates in presence of decaying organic matter.

The composition of sulphuretted hydrogen may be ascertained either by heating a small piece of metallic tin in a known volume of the gas, when tin sulphide will be formed, and hydrogen liberated; or by decomposing the gas by means of a red-hot platinum wire, when the whole of the sulphur is deposited, and hydrogen set free. In both cases, the volume of hydrogen obtained is found to be equal to that of the gas employed; and hence 2 volumes of sulphuretted hydrogen, weighing 34, consists of 2 volumes of hydrogen, weighing 2, and 1 volume of sulphur vapor, weighing 32.

Sulphuretted hydrogen is an invaluable re-agent in the laboratory, as by its means we are enabled to separate the metals into groups. If we pass a current of this gas through a solution of a copper salt, to which a small quantity of acid has been added, we obtain an immediate precipitate of copper sulphide; if we do the same with a solution of an iron salt we get no such precipitate, because iron sulphide is soluble in an acid; but on the addition of an alkali, iron sulphide is at once precipitated. We may thus divide the metals into groups; first, those which, like copper, are precipitated by sulphuretted hydrogen from an acid solution, or the *copper group;* second, those which are not precipitated by sulphuretted hydrogen in an acid solution, but which are so precipitated in an alkaline one, or the *iron group;* and third, those which are in no case precipitated by this re-agent, as their sulphides are soluble either in water, acids, or alkalies: to this group belong the metals of the *alkalies and alkaline earths.*

Hydrogen Disulphide, H_2S_2. This substance is obtained by pouring a solution of calcium disulphide into hydrochloric acid; an oily liquid falls to the bottom of the vessel, which is the body in question. Hydrogen disulphide closely resembles hydrogen di-oxide in many of its properties. It possesses a peculiar smell, bleaches, and readily decomposes into sulphur and sulphuretted hydrogen.

Carbonic Disulphide. Symbol CS_2. Molecular Weight 76. Density 38.

If the vapor of sulphur be passed over red-hot charcoal, a volatile compound, CS_2, is formed, which may be condensed to a heavy, colorless liquid, possessing a peculiarly disagreeable smell, boiling at 43°·3, and having a specific gravity of 1·272. Carbonic disulphide is very inflammable, its vapor igniting at 149° when mixed with air, forming carbon di-oxide, and sulphur di-oxide (sulphurous acid), SO_2. It is insoluble in water, but acts as a solvent upon gums, caoutchouc, sulphur, and phosphorus; its vapor is, however, very poisonous, and it must be employed with caution.

A remarkable analogy is presented by the foregoing sulphur compounds and the corresponding bodies in the oxygen series: thus water, H_2O, and sulphuretted hydrogen, H_2S; hydrogen di-oxide, H_2O_2, and hydrogen disulphide, H_2S_2; carbon di-oxide, CO_2, and carbon disulphide, CS_2, possess not only an analogous composition, but similar chemical properties, whilst similar relations are seen in many of the compounds of these bodies.

Chlorine and *Sulphur* unite directly to form two compounds, S_2Cl_2 and SCl_2; they are formed by leading a current of chlorine gas over melted sulphur, and are volatile liquids, the first boiling at 138°, and the second at 64°.

LESSON XIV.

SELENIUM.* *Symbol* Se. *Combining Weight* 79·5. *Density* 79·5.

Selenium is an element which closely resembles sulphur in its properties, but it occurs in very small quantities; it was discovered by Berzelius, who found it accompanying sulphur in certain varieties of Swedish pyrites. Selenium also occurs free in nature, and is found in combination with metals in certain rare minerals. Like sulphur, it is capable of existing in various allotropic modifications, one of which is crystalline, the other vitreous; the crystalline form is obtained when selenium is deposited from solution in carbonic disulphide; the vitreous modification results from the cooling of melted selenium. The specific gravity of the former variety is 4·5; that of the latter, 4·7. Crystalline selenium melts at 217°, and boils at a temperature below a red heat, giving off a deep yellow vapor; vitreous selenium softens at a temperature a little above the boiling point of water, and remains in a plastic condition for some time. In a finely divided state, and when seen by transmitted light, selenium has a red color. It burns in the air with a bright blue flame, which, when examined by means of the spectroscope (*p.* 222), exhibits a series of magnificent and characteristic bands. The smell of burning selenium is very peculiar, resembling that of rotten cabbages, and is due to the formation of an oxide, the composition and properties of which are, however, as yet unknown. Selenium forms two well-defined oxides, *selenium di-oxide*, SeO_2, and *selenium tri-oxide*, SeO_3; this latter, however, has not as yet been isolated, but the acid and salts corresponding with it, and those corresponding with the di-oxide, are well known, and correspond exactly with the analogous sulphites and sulphates: they are hence called selenites and selenates.

* From Σελήνη, the moon.

Selenium Di-oxide. Symbol SeO_2. Molecular Weight 111·5. This compound is formed when selenium is burnt in the air or in pure oxygen. It may be prepared, too, by oxidizing selenium in nitric acid or aqua regia. Selenium dioxide is a white crystalline mass, capable of dissolving in water, and thus forming *selenious acid,* H_2SO_3. From this solution selenium is at once deposited on addition of sulphurous acid, sulphuric acid being formed, thus :

$$H_2SeO_3 + 2SO_2 + H_2O = 2H_2SO_4 + Se.$$

The metallic selenites correspond closely with the sulphites.

Selenic Acid, or *Hydrogen Selenate.* Symbol H_2SeO_4.

This is best prepared by fusing a selenite with nitre ; on addition of a lead salt to the solution of the mass thus obtained, insoluble lead selenate is precipitated ; this salt is decomposed by sulphuretted hydrogen yielding selenic acid and lead sulphide, thus :

$$PbSeO_4 + H_2S = H_2SeO_4 + PbS.$$

On evaporating the liquid obtained on filtration selenic acid is left.

Selenic acid decomposes on heating into selenium di-oxide, oxygen, and water ; the metallic selenates correspond to the analogous sulphates, and are *isomorphous* with them, that is, they crystallize in the same forms and have an analogous composition.

Seleniuretted Hydrogen, or *Hydrogen Selenide.* Symbol H_2Se. Molecular Weight 81·5. Density 40·75.

This gas is obtained by the action of an acid upon a selenide, exactly as sulphuretted hydrogen is prepared from a sulphide. It is a colorless inflammable gas, possessing a nauseous smell, and exhibiting properties in every respect analogous to those of its sulphur representative. The most important difference between the two elements, sulphur and selenium, is that the former is oxidized to its highest point by the action of nitric acid, whereas the latter requires to be fused with nitre in order to reach the corresponding degree of oxidation.

TELLURIUM. *Symbol* Te. *Combining Weight* 129. *Density* 129.

Tellurium (from *Tellus*, the earth) is a very rare substance, which, although resembling a metal in its physical properties, bears so strong an analogy to sulphur and selenium in its chemical relations, that its compounds are best considered in this place. It occurs combined with gold and other metals in Transylvania and Hungary. The specific gravity of tellurium is 6·25, and it exhibits a bright white metallic lustre. It melts at about 500°, and may be volatilized at a white heat in a current of hydrogen gas. When heated in the air it burns with a bluish-green flame, forming white fumes of *tellurium dioxide;* this compound is also formed when tellurium is oxidized by nitric acid, and the solution evaporated to dryness. With water the dioxide forms *tellurous acid*, H_2TeO_3, and with metals in place of hydrogen, tellurites of the general form, M_2TeO_3. When tellurium or a tellurite is fused with nitre, potassium tellurate, K_2TeO_4, is formed, from which telluric acid, $H_2TeO_4 + 2H_2O$, and the trioxide TeO_3 can be obtained. With hydrogen, tellurium forms a colorless gas, H_2Te, which cannot be distinguished by its smell from sulphuretted hydrogen.

Oxygen, sulphur, selenium, and tellurium, appear to form a natural group of elements, each uniting with two atoms of hydrogen to form a series of bodies possessing analogous properties, viz. H_2O, H_2S, H_2Se, H_2Te. The last three members of the group exhibit the same kind of striking gradation of properties as was noticed in the case of chlorine, bromine, and iodine. Thus the mean of the combining weights of the two extremes is nearly the combining weight of the mean $\frac{32+129}{2} = 80·5$; whilst their specific gravities, 2·0, 4·5, and 6·25, and their melting and boiling points show a similar gradation.

SILICON. *Symbol* Si. *Combining Weight* = 28.

Silicon, next to oxygen, is the most abundant element known. It does not occur, however, in the free state, but always combined with oxygen to form silicon dioxide (silicic acid, or silica). Silicon dioxide exists nearly pure in quartz or rock crystal, in flint, sand, and in a variety of minerals. Silicon also occurs combined with metals and oxygen, forming metallic silicates; and of these the greater part of almost all known rocks, especially the primary rocks, is composed.

In order to obtain silicon in the free state, a compound of this substance with fluorine and potassium is heated with metallic potassium; a violent reaction occurs, and when the contents of the tube in which the decomposition was effected are put into water, silicon is left undissolved in the form of a brown amorphous powder. Silicon can be obtained in three different modifications—amorphous, crystalline, and graphitoidal, corresponding to the modifications of carbon. The graphite form of silicon is prepared by heating the brown amorphous powder to a high temperature, when the mass contracts, and becomes much more dense. Crystalline silicon is best obtained by fusing the mixture which gives brown silicon with zinc: on cooling the mass, crystals of silicon are found to be deposited on the zinc, which latter can easily be removed by solution in an acid. Silicon thus obtained is hard enough to scratch glass; it has a specific gravity of 2·49, and may be fused at a temperature between the melting points of cast iron and steel.

Silicon Dioxide, or *Silica*, Symbol $Si\,O_2$, Molecular Weight 60, is the only known oxide of silicon; it occurs in the pure state crystallized in six-sided prisms or pyramids, as quartz, and exists in a less pure condition in sandstone, chalcedony, flint, and agate, &c. The aluminium-, potassium-, calcium-, and iron-silicates, mixed together in different proportions, constitute a large number of minerals.

Crystallized silica, in the form of white transparent quartz, has a specific gravity of 2·6, and is hard enough to scratch glass; it is unattacked and undissolved by all acids with the exception of hydrofluoric acid, by the action of which silicon tetrafluoride and water are produced, thus:

$$SiO_2 + 4HF = SiF_4 + 2H_2O.$$

Silica is infusible except at the highest temperature of the oxyhydrogen blowpipe, when it melts to a colorless globule. Silica in an amorphous condition can also be prepared, and then exhibits peculiar properties. For this purpose one part of finely divided quartz or white sand is heated with four parts of sodium carbonate; as soon as the latter begins to fuse, the silica combines with the sodium and oxygen contained in the carbonate, carbonic acid, CO_2, being evolved with effervescence, owing to the formation of a sodium silicate. If the fused mass be boiled with water, it will dissolve, and on the addition of hydrochloric acid, silicic acid, H_4SiO_4, partly separates as a gelatinous mass, partly remains dissolved in the liquid. If this solution be evaporated to dryness and heated a little, and hydrochloric acid then added, silicic dioxide is left as a white powder insoluble in acids; this amorphous silica possesses a specific gravity of 2·2 to 2·3, and can only be obtained again in solution by repeating the process of fusion with an alkali, &c. A pure aqueous solution of hydric silicate can be obtained by allowing the solution of this substance in hydrochloric acid to *dialyse*, or diffuse through a membrane, for some days. All the hydrochloric acid and soluble chlorides in this solution being crystalloids pass through the dialyser, while the colloidal hydric silicate remains behind in solution: the limpid solution thus obtained may be concentrated by boiling until the quantity of hydric silicate in solution reaches 14 per cent.: but this solution is apt to gelatinize on standing, forming a transparent jelly-like mass.

Potassium and sodium silicates are largely used for various purposes in the arts, whilst a mixture of these with calcium, iron, and lead silicates, forms the several descriptions of glass (p. 175). Some remarkable compounds of silicon with oxygen and hydrogen have lately been discovered, but their composition has not yet been accurately determined. A compound of silicon and hydrogen termed *Siliciuretted Hydrogen* is also known: it has, however, not been obtained in the pure state; it is colorless, and takes fire on coming into contact with the air, forming silica and water.

Silicon Tetra-chloride. Symbol $SiCl_4$. Molecular Weight 170. Density 85.

This compound is formed when silicon is heated in chlorine, but may be prepared by passing dry chlorine over a red-hot mixture of finely-divided silica and carbon. Chlorine alone is not able to decompose silica, but in presence of carbon a change is effected, carbon mon-oxide being at the same time formed.

$$SiO_2 + Cl_4 + C_2 = SiCl_4 + 2CO.$$

Silica, chlorine, and carbon yield silicon tetrachloride and carbon mon-oxide. Fig. 35 shows the arrangement employed

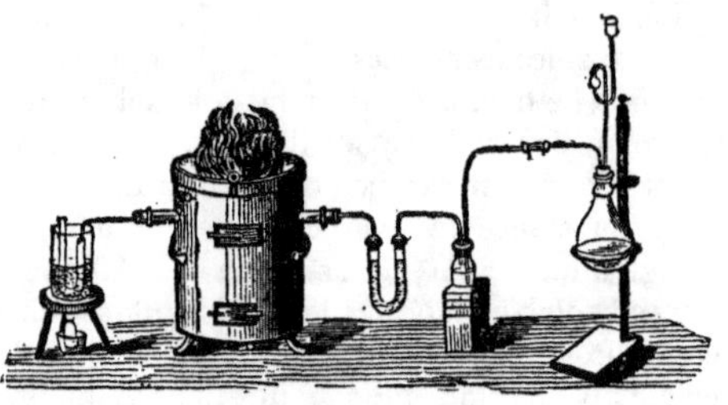

FIG. 35.

for preparing this body; the mixture of silica and carbon is placed in a porcelain tube,. which can be strongly heated by the furnace; dry chlorine gas is passed through the tube, and the volatile silicon chloride collects in the cooled tube, dropping into the bottle placed below to receive it. Silicon chloride is a volatile colorless liquid, boiling at 59° C., and having a specific gravity of 1.52. It is at once decomposed by water, silicic and hydrochloric acids being formed; hence we may see that this body in the chlorine series corresponds to silicon di-oxide in the oxygen series, and that in the formation of the chloride four atoms of chlorine simply replace its equivalent quantity, two atoms of oxygen, in the silica: SiO_2 becomes $SiCl_4$, as one atom of oxygen is equivalent to two of chlorine.

Silicon Tetra-fluoride. Symbol SiF_4. Combining Weight 104. Density 52.

This is one of the most singular compounds of silicon; it is formed whenever free hydrofluoric acid comes in contact with either free or combined silica; this is the cause of the etching which hydrofluoric acid exerts upon glass. Silicon tetra-fluoride is best prepared by heating in a flask equal parts by weight of finely-powdered fluor spar, and white sand, with about eight parts of sulphuric acid: the decomposition first occurring is the one by which hydrofluoric acid is generated, and this then attacks the silica, thus:

$$1)\ CaF_2 + H_2SO_4 = CaSO_4 + 2HF.$$
$$2)\ 4HF + SiO_2 = 2H_2O + SiF_4.$$

Silicon tetra-fluoride is a colorless gas which fumes strongly in the air; it does not burn nor support combustion, and may be condensed by great pressure, or exposure to a very low temperature, to a colorless liquid; it is decomposed by water, but may be collected over mercury, or by displacement; when led into water this gas yields silicic acid, which is deposited in a state of fine division, and a new acid called hydrofluo-silicic acid, or hydric silico-fluoride, having the composition $H_2 Si F_6$, which remains in solution. This substance has an acid reaction; the corresponding potassium- and barium silico-fluorides (K_2SiF_6 and $Ba SiF_6$) are insoluble in water and alcohol.

BORON. *Symbol* B. *Combining Weight* 11·0.

BORON combined with oxygen and sodium is found as *borax*

in nature; it is also found combined with oxygen alone as *boron tri-oxide* (*boracic acid*). It exists in two forms—crystalline and amorphous. Boron is easily obtained as a gray amorphous powder, by heating fused boron tri-oxide, B_2O_3, with sodium. Crystallized boron is prepared by heating the amorphous form strongly with aluminium—this metal in the fused state having the property of dissolving boron, which separates out in nearly colorless crystals when the metal cools, just as the graphitoidal form of carbon does from its solution in iron on cooling (p. 68). Crystallized boron has a specific gravity of 2·68, and occurs in the form of octahedra, which are hard enough to scratch the ruby. In one specimen of these colorless crystals which were analyzed, some quantity of carbon was found to be present; hence carbon may be said to have been prepared artificially in the diamond modification. Boron burns when strongly heated in oxygen or in chlorine, forming the oxide of chloride; it is remarkable as uniting with nitrogen by direct combination—in this respect resembling certain metals, such as titanium.

Boron Tri-oxide (commonly called *Boric* or *Boracic acid*). Symbol B_2O_3. Molecular Weight 70·0.

In certain old volcanic districts in Tuscany constant jets of steam and gas escape from the earth. These steam jets, which are known as *fumerolles* or *soffioni*, contain small quantities of boracic acid, $HBO_2 + H_2O$, which collect in the lagoons formed at the mouth of the jet. By means of the heat of natural steam jets, the solution of boracic acid is concentrated, and the acid obtained by crystallization; about 2,000 tons of crude acid thus prepared are imported every year from Tuscany. Boron likewise occurs in tinkal or borax in Thibet, and as boron tri-oxide on the coast of California.

Boric acid is obtained by decomposing a hot solution of borax, $Na_2\ B_4O_7$, with sulphuric acid; crystals separate out on cooling, having the composition $H\ BO_2 + H_2O$. These crystals on heating lose water and pass into a fused glassy mass, consisting of boracic trioxide, B_2O_3. Boracic acid is slightly soluble in cold, and rather more soluble in hot water; it imparts a peculiar green tint to the blowpipe flame, which exhibits a characteristic series of bands when examined by means of the spectroscope. Metallic Borates are known, and likewise several combinations of these borates with boron trioxide. Thus sodium borate, or boracic acid in which the atom of hydrogen is replaced by sodium, is $Na\,BO_2 + 4H_2O$; whilst fused borax is this salt combined with one molecule of boron trioxide, thus: $2\,Na\,BO_2 + B_2O_3$, or $Na_2\,B_4O_7$. Compounds similar to this latter salt are known amongst the sulphates. Thus Nordhausen sulphuric acid is $H_2SO_4 + SO_3$, and a sodium compound, $Na_2\,SO_4 + SO_3$, is known. Many of the metallic oxides are soluble in fused borax, giving colored glasses. Hence this compound is largely used in the arts as a flux, and in the laboratory as a blowpipe re-agent.

Boron combines with chlorine to form a *trichloride*, BCl_3, and with fluorine to form a corresponding *trifluoride*, BF_3; both these compounds are prepared by a method similar to that adopted for the corresponding silicon compounds, to which, notwithstanding their slightly different constitution, they bear a strong resemblance. Like silicon also, boron forms a borofluoride: hydrofluoboric acid (or hydrogen borofluoride), is HBF_4, and potassium borofluoride, KBF_4.

LESSON XV.

PHOSPHORUS. *Symbol* P. *Combining Weight* 31. *Vapor Density* 62.*

Phosphorus does not occur free in nature, but is found in combination with oxygen and calcium in large quantities in

* The volume occupied by the atom of phosphorus weighing 31 is only half as large as that occupied by the atoms of each of the preceding elements: hence the atomic volume of phosphorus is ½, that of the preceding elements being 1.

the bodies, and especially the bones, of animals, and in the seeds of plants. When bones are burnt, a white solid mass is left behind; this is called Calcium Phosphate (phosphate of lime). Animals obtain the phosphate necessary for the formation of their tissues, &c., from plants. Plants, again, draw their supply from the soil, whilst soils derive their phosphate from small quantities existing in the oldest granite rocks, by the disintegration of which the fertile soils have been produced. Phosphorus appears also to be a very necessary ingredient in the brain and other centres of the nervous action. It was accidentally discovered by Brand of Hamburg in 1669; but Scheele, in 1769, pointed out the existence of phosphorus in the bones, and examined its properties carefully.

Phosphorus is prepared from powdered bone-ash, by mixing it with two-thirds of its weight of sulphuric acid and 15 to 20 parts of water. The sulphuric acid decomposes the bone-ash, forming calcium sulphate, or gypsum, which separates out as a white insoluble powder; whilst the greater part of the phosphorus in the bones comes into solution in combination with calcium, oxygen, and hydrogen, forming a salt, commonly known as *superphosphate of lime.* The liquid is drawn off clear, evaporated down to a syrup, and then mixed with powdered charcoal, dried, and heated to redness in an earthenware retort, the neck of which dips under water. The superphosphate may be considered to contain phosphorus pentoxide, P_2O_5, from which the carbon takes away the oxygen, forming carbon monoxide and liberating phosphorus, which distils over and condenses under water in yellow drops.

Phosphorus is an exceedingly inflammable and oxidizable substance, and requires great care in its preparation; it is manufactured on a very large scale for making the composition for the tips of lucifer matches. In order to purify the phosphorus thus prepared, it may again be distilled, or pressed when melted under hot water through leather; it is then cast into sticks and kept under cold water. Phosphorus is a slightly yellow semitransparent solid, resembling white wax both in appearance and consistency; but at low temperature it becomes brittle. Its specific gravity is 1·83, and it melts at 44°, forming a transparent liquid; it boils at 290°, giving rise to a colorless gas. In the air it gives off white fumes, emitting a pale *phosphorescent* light in the dark—whence its name (φῶς light, and φέρω, I bear; *lucifer*, from *lux*, light, and *fero*, I bear, is its literal Latin equivalent); it is then undergoing a slow combustion, the white fumes consisting of trioxide. At a temperature very little above its fusing point phosphorus takes fire in the air, entering into active combustion, and forming *phosphorus pentoxide*, P_2O_5 (phosphoric anhydride). The ignition of phosphorus takes place by slight friction, or by a blow, and even the heat of the hand may cause this substance to ignite; hence great care must be taken in handling phosphorus, and it should always be cut under water. Phosphorus does not dissolve in water, alcohol, or ether, but it is slightly soluble in oils, and very readily soluble in carbon disulphide, crystallizing from its solution in this liquid in rhombic dodecahedra.

If yellow phosphorus be exposed to a temperature of about 240° for some hours in an atmosphere incapable of acting chemically on it (such as hydrogen or carbonic acid), it is found to have undergone a very remarkable change, being wholly converted into a dark red opaque substance altogether insoluble in carbon disulphide. The weight of red substance produced is exactly equal to that of yellow phosphorus used. This is called *Red, or Amorphous Phosphorus*, and differs much in its properties from the yellow modification, especially in its inflammability, as it does not take fire in the air until heated to above 260°, when it becomes reconverted into the ordinary form and burns with the formation of phosphoric pentoxide. The specific gravity of amorphous phosphorus is 2·14. The sudden conversion of yellow into red phosphorus can be shown by heating a small piece of ordinary phosphorus in a dry tube with a mere trace of iodine; combination at once occurs, a small trace of volatile phosphoric iodine is formed, and the remainder of the phosphorus is converted into the red modification. The red or amorphous modification of phosphorus can also be obtained in a *crystallized* form by heating red phosphorus in a tube with metallic lead. The phosphorus dissolves in the melted lead, and on cooling separates out in crystals, which possess a bright black metallic lustre, and have a specific gravity of 2·34.

Phosphorus forms two oxides, phosphorus trioxide, P_2O_3, and phosphorus pentoxide, P_2O_5.

Phosphorus Trioxide, or *Phosphorous Anhydride.* Symbol P_2O_3. Molecular Weight 110.

This oxide is formed when phosphorus is burnt in a limited current of dry air when it undergoes slow combustion. It forms a white non-crystalline powder which combines with great energy with water, forming thereby *phosphorous acid*, or *hydrogen phosphite* H_3PO_3. This acid is likewise formed when phosphorus is allowed gradually to oxidize in moist air, and also by the action of phosphorus trichloride on water, thus:

$$PCl_3 + 3H_2O = H_3PO_3 + 3HCl.$$

By boiling this solution the hydrochloric acid is driven off, and, on cooling, crystals of phosphorous acid are deposited. There are two classes of metallic phosphites—the one contains those which correspond to phosphorous acid in which two atoms of hydrogen have been replaced by metal; and the second containing those in which one atom only of hydrogen has been thus replaced: the general forms of the two will therefore be $M_2H\ PO_3$ and $MH_2\ PO_3$, the letter M denoting an atom of a monatomic metal.

Phosphorus Pentoxide (Phosphoric Anhydride). Symbol P_2O_5. Molecular Weight 142.

This substance is formed when phosphorus burns brightly in excess of air or oxygen; it is a white amorphous light powder, which absorbs moisture with the utmost avidity, forming hydrogen phosphate or phosphoric acid, $H_3\ PO_4$; it is for this reason frequently used in the laboratory for the purpose of drying gases. Phosphorus pentoxide is volatile, and may be sublimed unchanged by heating in a test tube. It can be best prepared by burning small pieces of phosphorus placed one by one in a cup hung in the centre of a large dry glass globe, and blowing in a sufficient supply of dry air by means of a bellows or aspirator. The white powder falls down, and may be shaken out of the globe when the operation is completed.

Trihydrogen Phosphate (Tribasic Phosphoric Acid). Symbol $H_3 PO_4$. Molecular Weight 98.

When the preceding compound is brought into contact with water, great heat is evolved, and combination takes place with a hissing noise. If the solution be boiled, trihydrogen phosphate is found in solution, being formed thus:

$$P_2O_5 + 3\ H_2O = 2(H_3\ PO_4).$$

The corresponding calcium salt, $Ca_3\ 2PO_4$, occurring in bone-ash and in many minerals, constitutes the main source of all the phosphorus compounds. Trihydrogen phosphate is also formed when phosphorus is heated with nitric acid: the lower oxides of nitrogen are given off as red fumes, and the phosphorus gradually disappears; by evaporating, and boiling the colorless liquid, trihydrogen phosphate may be obtained. The same substance is procured directly from bone-ash by frequent addition of sulphuric acid and continued evaporation: gypsum gradually separates out, leaving a solution from which hydrogen phosphate can be obtained by neutralizing with ammonium carbonate, filtering, evaporating to dryness the clear liquid thus obtained, and igniting the residue. If sodium carbonate be added to a solution of trihydrogen phosphate, effervescence will at once ensue from the liberation of carbonic acid, and if the carbonate be added until the solution ceases to redden litmus paper, a salt will be obtained on evaporation which crystallizes in large transparent prisms. This is *rhombic* or *common neutral sodium phosphate;* its composition is represented by the symbol

$Na_2H\ PO_4$, with twelve atoms of water of crystallization. If caustic soda be added to a solution of this common phosphate, a salt termed the *subphosphate* crystallizes out in small needles on evaporation: the composition of this salt is $Na_3\ PO_4$ with twelve atoms of water of crystallization; and if phosphoric acid be added to a solution of common phosphate, the so-called *Sodium superphosphate* is formed, NaH_2PO_4. We have therefore the following tribasic hydrogen and sodium phosphates:

Trihydrogen Phosphate	$H_3\ PO_4$.
Dihydrogen Sodium Phosphate .	$H_2Na\ PO_4+12H_2O$.
Hydrogen di-Sodium Phosphate .	$H\ Na_2\ PO_4 + H_2O$.
Tri-Sodium Phosphate	$Na_3\ PO_4+12H_2O$.

All these substances are distinguished by giving a *yellow* precipitate with solution of silver nitrate consisting of tri-silver phosphate, Ag_3PO_4; and by producing with ammonia and magnesium sulphate a white crystalline precipitate of ammonium magnesium phosphate, NH_4, $Mg\ PO_4 + 6H_2O$. Small traces of phosphates can readily be detected by the yellow precipitate which forms in nitric acid solution of ammonium molybdate. The three atoms of hydrogen in trihydrogen phosphate may be replaced by three different metals; thus, Microcosmic Salt is Hydrogen sodium-ammonium Phosphate, $H\ Na\ NH_4\ PO_4 + 4H_2O$.

If common sodium phosphate be heated to redness, water is driven off, and a compound called *Sodium pyrophosphate* remains, having the composition $Na_4\ P_2O_7$, two molecules of neutral phosphate yielding one of pyrophosphate, thus:

$$2\ Na_2\ H\ PO_4 = H_2O + Na_4\ P_2O_7.$$

When this salt is dissolved in water, it can be recrystallized, and does not take up water again so as to pass back to the state of common phosphate (except on long-continued boiling of its solution). This substance gives with silver nitrate a *white* precipitate of silver pyrophosphate, $Ag_4\ P_2O_7$, and thus this class of phosphates may be distinguished from the preceding or tribasic phosphates. A so-called acid sodium pyrophosphate, having the composition $Na_2H_2P_2O_7$, is also known.

If microcosmic salt, $Na(NH_4)HPO_4$, is heated, water and ammonia are driven off, and a substance called *Sodium metaphosphate*, $NaPO_3$, is left; this dissolves unaltered in water, forming one of a third class of phosphates termed *monobasic phosphates*, or *metaphosphates*. The solutions of these salts may be distinguished from those of the two preceding classes of salts by their producing gelatinous precipitates with solutions of calcium and silver salts consisting of the metaphosphates of these metals. Monohydrogen phosphate (or metaphosphoric acid), HPO_3, is obtained in the form of a transparent icelike mass by evaporating the solution of trihydrogen phosphate and igniting the residue. On dissolving this glacial acid in cold water, a solution of monohydrogen phosphate is obtained, but this, on boiling, changes to the trihydrogen phosphate.

From the above it is seen that three modifications of phosphoric acid are known, or rather, three different acids, each giving rise to a class of metallic salts. Thus we have—

(1) Trihydrogen phosphate, or phosphoric acid, H_3PO_4, and tri-sodium phosphate, Na_3PO_4.

(2) Tetrahydrogen pyrophosphate, or pyrophosphoric acid, $H_4P_2O_7$, and sodium pyrophosphate, $Na_4P_2O_7$.

(3) Monohydrogen phosphate, or metaphosphoric acid, HPO_3, and sodium metaphosphate, $NaPO_3$.

Each of the above hydrogen phosphates can be prepared by passing sulphuretted hydrogen through water containing in suspension the corresponding silver salts; thus:

$$(1)\ 2(Ag_3PO_4 + 3(H_2S) = 2(H_3PO_4) + 3(Ag_2S).$$
$$(2)\ Ag_4P_2O_7 + 2(H_2S) = H_4P_2O_7 + 2(Ag_2S).$$
$$(3)\ 2(AgPO_3) + (H_2S) = 2(HPO_3) + Ag_2S.$$

In addition to the phosphates and phosphites, a class of salts termed Hypophosphites is also known: the composition of *hydrogen hypophosphite* is represented by the formula HPH_2O_2, and that of sodium hypophosphite by Na PH_2O_2; and these salts may be supposed to be hydrogen or sodium metaphosphate, HPO_3, and Na PO_3, in which one atom of oxygen has been replaced by its equivalent, or two atoms of hydrogen. Sodium hypophosphite is obtained by acting with caustic soda on phosphorus, when phosphuretted hydrogen gas is evolved, and a solution of hypophosphite remains behind.

PHOSPHORUS AND HYDROGEN.

Three compounds of phosphorus and hydrogen are known— PH_3, a gas; P_2H_4, a liquid; and P_4H_2, a solid substance.

Phosphuretted Hydrogen. Symbol PH_3. Combining Weight 34. Density 17.

This gas is obtained in the pure state by the decomposition of hydrogen phosphite, or hydrogen hypophosphite, but it is generally prepared by the action of caustic potash on phosphorus.

$$3\,KHO + P_4 + 3H_2O = 3\,KPH_2O_2 + PH_3.$$

Each bubble of the gas thus prepared takes fire spontaneously on coming into contact with the air, forming singular rings of phosphoric pentoxide, which expand as they rise. This self-inflammability of the gas depends upon the presence of small quantities of the liquid hydride P_2H_4, which may be condensed to a volatile and very inflammable liquid by passing the gaseous hydride through a tube cooled by a freezing mixture. Pure phosphuretted hydrogen prepared from hydrogen phosphite by heat, does not undergo spontaneous combustion in the air. This mode of formation may be thus represented.

$$4(H_3\,PO_3) = 3(H_3\,PO_4) + PH_3.$$

There is some doubt about the exact formulæ of the other two hydrides; the formulæ P_2H_4 and P_4H_2 are in accordance with their physical states, and are, on certain other theoretical grounds, to be preferred to PH_2 and P_2H.

PHOSPHORUS AND CHLORIDE.

Two chlorides of phosphorus are known—*Phosphorus trichloride*, PCl_3 and *Phosphorus pentachloride*, PCl_5. The first of these is a colorless strongly fuming liquid, which is easily formed by passing a current of chlorine gas over phosphorus contained in a retort: when thrown into water it sinks down as a heavy oil, but is gradually decomposed, hydrogen phosphite and hydrochloric acid being formed (*see* p. 128). The specific gravity of the trichloride is 1·45, and its boiling point is 73°·8. Phosphorus trichloride rapidly absorbs chlorine gas, and is converted into the pentachloride, a yellowish solid substance, which is also formed when phosphorus is burnt in an excess of chlorine. Phosphorus pentachloride is decomposed in presence of excess of water, forming trihydrogen phosphate and hydrochloric acid; but when only a limited quantity of water is present, a liquid termed *phosphorus oxichloride* is formed, having the composition PCl_3O, and boiling at 110°. Corresponding compounds with bromine are likewise known. With sulphur, phosphorus forms several compounds, and it is an interesting fact that two of these compounds, P_2S_3 and P_2S_5, correspond in composition with the oxides P_2O_3 and P_2O_5. The oxide corresponding to P_2S, however, is as yet unknown.

LESSON XVI.

ARSENIC. *Symbol* As. *Combining Weight* 75. *Density of Vapor* 150.*

ARSENIC closely resembles phosphorus in its chemical properties and in those of its compounds, although in physical

* The volume occupied by an atom of (gaseous) arsenic weighing 75 is only half of that occupied by the other elements generally; in this respect arsenic resembles phosphorus.

characters, such as specific gravity, lustre, &c., it bears a greater analogy to the metals: indeed, it may be considered the connecting link between these two divisions of the elements, antimony and bismuth being closely connected with it on the one hand, and phosphorus and nitrogen on the other. Arsenic is sometimes found in the free state, but more frequently combined, chiefly with iron, nickel, cobalt, and sulphur. It is also contained in very small quantities in many mineral springs. In order to separate arsenic from any of the metallic ores in which it occurs, the ore is roasted, or exposed to a current of heated air in a reverberatory furnace; the arsenic combines with the atmospheric oxygen, forming arsenic trioxide or arsenious oxide, As_2O_3, which is carried in the state of vapor from the furnace into long chambers or flues in which the trioxide (commonly known as arsenious acid, or white arsenic) is deposited. Metallic arsenic may be prepared from this oxide by mixing it with charcoal and sodium carbonate, and heating in a closed crucible, the upper part of which is kept cool: arsenic condenses in the cool part of this apparatus as a solid with a brilliant grayish lustre. It tarnishes in the air from oxidation; it has a specific gravity of 5·7 to 5·9, and when heated to dull redness, it volatilizes as a colorless vapor without undergoing fusion, and this vapor possesses a remarkable garlic-like smell. Arsenic when heated in the air takes fire, and burns with a bluish flame, forming arsenious oxide, $As_2 O_3$; when thrown into chlorine it instantly takes fire, forming arsenic trichloride, $AsCl_3$.

OXIDES OF ARSENIC.

Two compounds of arsenic and oxygen are known. (1) *Arsenic Trioxide*, As_2O_3. (2) *Arsenic Pentoxide*, As_2O_5.

Arsenic Trioxide. Symbol As_2O_3. Molecular Weight 198. Density (of vapor) 198.

This substance is formed when arsenic is burnt in the air or in oxygen, but it is generally prepared by roasting arsenical pyrites, FeSAs; its specific gravity is 3·6. It exists in two distinct forms, the crystalline and the vitreous; it occurs in the first modification crystallized in brilliant octahedra, and in the second as a semi-transparent glass-like solid, devoid of crystalline structure: this form of the substance, on standing, becomes opaque like porcelain, diminishing in specific gravity. Arsenic oxide is feebly soluble in water; the solution (which may be considered to contain true arsenious acid, or trihydrogen arsenite, H_3AsO_3, analogous to phosphorous acid) has a feebly acid reaction: it dissolves more readily in hydrochloric acid, and is freely soluble in solutions of the alkalies, arsenites being formed of the general form M_3AsO_3; thus, trisilver arsenite is Ag_3AsO_3. The alkaline arsenites are soluble in water; those of the metals of the alkaline earths and heavy metals are insoluble in water. Sodium arsenite is used largely in calico printing; Scheele's green and Emerald green are compounds containing arsenic trioxide and copper, both of which are made in large quantities for employment as a pigment. All the soluble arsenites are dreadfully poisonous; the best antidote is freshly prepared ferric hydrate (hydrated ferric oxide), or magnesia, which form insoluble arsenites, and thus prevent the poison from entering into the system. When heated to about 220°C., arsenic trioxide volatilizes without melting, forming an inodorous and colorless vapor. It is occasionally met with crystallized in long needles of the same form as the crystals of the corresponding oxide in the antimony series (*see* p. 200).

Arsenic Pentoxide or *Arsenic Oxide* (arsenic anhydride). Symbol $As_2 O_5$. Molecular Weight 230.

This oxide is obtained by acting upon the trioxide with nitric acid, evaporating to dryness, and heating to a temperature of 270°. It forms a non-crystalline white powder, which, when strongly heated, decomposes into As_2O_3 and O_2. This powder is readily dissolved by water, and the solution yields crystals of arsenic acid, or trihydrogen arsenate, H_3AsO_4: the metallic compounds corresponding to this are called arsenates, and resemble the corresponding tribasic phosphates (p. 130) in composition, whilst they are identical with them in crystalline form. Thus we have,

Tri-Sodium Arsenate	$Na_3AsO_4 + 12\ H_2O$.
Hydrogen Di-Sodium Arsenate	$HNa_2AsO_4 + 12\ H_2O$.
Di-hydrogen Sodium Arsenate .	$H_2NaAsO_4 + H_2O$.
Tri-hydrogen Arsenate . . .	H_3AsO_4.

With solutions of magnesium and ammonium together, soluble arsenates, like phosphates, form an insoluble precipitate, having the composition $NH_4MgAsO_4 + 6H_2O$ (ammonium magnesium arsenate). The arsenates of the alkaline metals are soluble, those of the other metals insoluble in water. Trisilver arsen*ate* is a characteristic salt of a brownish-red color, whereas trisilver arsen*ite* has a bright yellow tint. Arsenic acid acts as a poison, but is less powerful than arsenious acid.

No arsenates corresponding to the pyro- and metaphosphates have been, as yet, obtained; compounds having the composition $Na_4As_2O_7$, and $NaAsO_3$, have been indeed prepared by heating a tribasic salt; but on solution in water they combine again with it and present only the characteristics of the tribasic acid.

ARSENIC AND HYDROGEN. *Arseniuretted Hydrogen.* Symbol AsH_3. Combining weight 78. Density 39.

This compound, which corresponds to phosphuretted hydrogen, and to ammonia, is formed by decomposing an alloy of arsenic and zinc with sulphuric acid. It is a colorless gas, possessing a fetid odor of garlic, and acts as a most deadly poison, a single bubble of the pure gas having been known to act fatally: when cooled to $-40°$ it condenses to a colorless liquid. Arseniuretted hydrogen burns with a bluish flame and deposits arsenic upon a cold body held in the flame: below a red heat it is decomposed into arsenic and hydrogen.

Arsenic unites with chlorine, bromine, and iodine, to form *arsenic trichloride*, *tribromide*, and *triiodide.* The trichloride is a colorless volatile liquid, boiling at 132°, which decomposes in contact with water, yielding arsenious and hydrochloric acids. *Three sulphides of arsenic* are known—

Arsenic Disulphide, As_2S_2, which occurs naturally as Realgar; Trisulphide, As_2S_3, also occurring in nature as Orpiment; and Pentasulphide, As_2S_5. Orpiment may be obtained by passing a stream of sulphuretted hydrogen gas through the acid solution of the corresponding oxide, when it is precipitated as a yellow powder. The arsenic sulphides form with the sulphides of the alkaline metals compounds bearing the same analogy to the trisulphide and pentasulphide that arsenites and arsenates do to the trioxide and pentoxide: in short, these compounds are sulphur and salts, the arsenites and arsenates being oxysalts; hence they are called *sulph*arsenites and *sulph*arsenates.

Arsenic possesses characters of so peculiar a kind, that its presence even in very minute traces can be detected with certainty. From its solutions it can be precipitated as sulphide, by the aid of sulphuretted hydrogen: and this sulphide, when dried and fused in a small test tube with a mixture of potassium cyanide and sodium carbonate, yields a ring of metallic arsenic: on heating, the metal is oxidized to the trioxide, which deposits in minute octahedral crystals. These, when boiled with water, yield a solution giving a bright green precipitate with neutral copper solutions, and a bright yellow one with neutral silver salts. Arsenic in solution may be also detected by the evolution of arseniuretted hydrogen, on adding zinc and sulphuric acid to the solution to be tested: on burning the gas, arsenic is deposited in the metallic state upon a piece of cold porcelain held in the flame. This mirror dissolves in solution of sodium hypochlorite; and if treated with nitric acid, and the solution neutralized, yields with silver nitrate solution a red precipitate of trisilver arsenate. Many compounds of arsenic heated on charcoal in the inner blowpipe flame give a garlic odor of arsenic. Solutions containing arsenic boiled with hydrochloric acid and clean copper, deposit a coating of arsenic upon the copper: this coating, on drying and heating in a test tube, gives a ring or mirror of arsenic, which may be oxidized to trioxide and tested as before. By these, and other reactions, the pres ence of the minutest portion of arsenic may be detected with certainty.

The general chemical analogy between nitrogen, phosphorus, and arsenic, is well seen when their corresponding compounds are examined; thus the hydrides, oxides, and chlorides, have an analogous composition.

N_2O_3	N_2O_5	NH_3	NCl_3 (?)
P_2O_3	P_2O_5	PH_3	PCl_3
As_2O_3	As_2O_5	AsH_3	$AsCl_3$

These three elements are all *Trivalent*, that is, one atom of each of these bodies is equivalent to, and capable of, replacing three atoms of hydrogen. Antimony and bismuth (*see* pp. 199, 201) exhibit in their chemical relations a striking resemblance to the foregoing group.

Quantivalence of the Elements.

If we compare together the compounds of the preceding elements with hydrogen, we find that the molecule of certain of these compounds contains one atom of hydrogen; that is to say, the combining weight (or two volumes) of the compound contains the combining weight (or one volume) of hydrogen; whilst the molecule of others contains two atoms of hydrogen; that is, two volumes of the compound contains two volumes of hydrogen: in others, the molecule contains three or four atoms of hydrogen, united in each case to one atom of some other element. Thus one atom, or one volume of chlorine, bromine, iodine, or fluorine, combines with one atom or *one* volume of hydrogen to form *one molecule*, or two volumes of the several hydracids, hydrochloric, and hydrobromic, &c.: whilst one volume of oxygen, sulphur, and selenium, on the other hand, unites with *two* volumes of hydrogen to form two volumes or *one molecule* of water gas, or sulphuretted hydrogen. One volume of nitrogen (or fourteen parts by weight), and half a volume of arsenic and phosphorus (or, respectively, seventy-five and thirty-one parts by weight) unite, on the other hand, with *three* volumes of hydrogen (or three parts by weight) to form two volumes or *one molecule* of ammonia, phosphuretted and arseniuretted hydrogen; whilst twelve parts by weight of carbon unite with *four* volumes of hydrogen, to form the molecule of the typical compound, marsh gas.

Hence we come to distinguish the elements into groups, according to their power of combining with, or replacing, different quantities of hydrogen : thus we call chlorine, bromine, iodine, and fluorine, *monovalent elements*, or *monads;* oxygen, sulphur, selenium (and tellurium), *divalent elements*, or *dyads;* nitrogen, phosphorus, arsenic (antimony, bismuth, and boron), *trivalent elements*, or *triads;* carbon (and silicon), *tetravalent elements*, or *tetrads*. The class of monads possess only one combining power, or *quantivalence*, while the dyads possess two, the triads three, and the tetrads four such combining powers. In like manner the *metallic* elements can be divided into classes according to their quantivalence, their power of combining with chlorine and other monads being regarded as the measure of their quantivalence, few compounds of metals with hydrogen being known.

Not only can the *elements* thus be considered as possessing varying quantivalence, but also those groups of elementary atoms, which act collectively as elements, and to which the name of *compound radicals* is given. Thus, nitric acid may, as we have seen, be considered as water, in which one atom of hydrogen is replaced by the *monad radical*, NO_2; thus, water $\left.\begin{matrix}H\\H\end{matrix}\right\}O$, nitric acid, $\left.\begin{matrix}H\\NO_2\end{matrix}\right\}O$. Nitrogen pentoxide is again nitric acid with the second atom of hydrogen replaced by the same radical $\left.\begin{matrix}NO_2\\NO_2\end{matrix}\right\}O=N_2O_5$. Sulphuric acid, on the other hand, may be considered as built up on the type of two molecules of water, in which two atoms of hydrogen are replaced by the *dyad radical* SO_2.

$$\text{Water} \left.\begin{matrix}H_2\\H_2\end{matrix}\right\}O_2. \quad \text{Sulphuric Acid} \left.\begin{matrix}H_2\\ \text{II}\\SO_2\end{matrix}\right\}O_2$$

Again, phosphoric acid may be regarded as three molecules of water, in which three atoms of hydrogen have been replaced by the *triad radical* PO; thus—

$$\left.\begin{matrix} H_3 \\ H_3 \end{matrix}\right\} O_3 \qquad \left.\begin{matrix} \overset{III}{PO} \\ H_3 \end{matrix}\right\} O_3$$

The quantivalence of an element, or of a compound radical, may be conveniently expressed by placing the Roman numerals above the symbol for those which are not monads, thus:

$$H, \overset{II}{O}, \overset{III}{N}, \overset{IV}{C}, NO_2, \overset{II}{SO_2}, \overset{III}{PO}, \&c.$$

LESSON XVII.

The Metallic Elements.

The metals are much more numerous than the non-metallic elements; there are forty-nine of the former, and only fourteen of the latter: very many metals are, however, found in small quantities, and the properties of these and their compounds are but little known, so that in this work we shall only consider the most important and commonly occurring metals.

It has already been stated that the division of the elements into these two classes is one of convenience only, and is not founded on any essential difference; thus arsenic and antimony may, in some respects, be considered as metals, and in others as non-metals.

All metals, with the single exception of mercury (a liquid), are solid at the ordinary temperature; they possess a high power of reflecting light, causing the bright, glittering appearance known as the metallic lustre; they are better conductors of heat and electricity than the non-metals, and, as a rule, they have a higher specific gravity than these. The metals differ widely from each other, both in their physical and chemical properties, and are, accordingly, adapted for different uses; those metals which are lightest exhibit the greatest power of union with oxygen, whilst the heavier metals undergo oxidation with difficulty.

Physical Properties of the Metals.

Specific Gravity.—The following table, giving the specific gravities of the most important metals (water at 0° C. = 1·00), shows the great variation which they exhibit in this respect:

Table of Specific Gravities.

Iridium	21·8	Iron	7·8
Platinum	21·5	Tin	7·3
Gold	19·3	Zinc	7·1
Mercury	13·596	Antimony	6·7
Thallium	11·9	Arsenic	5·9
Palladium	11·8	Chromium	5·9
Lead	11·3	Aluminium	2·56
Silver	10·5	Strontium	2·54
Bismuth	9·8	Magnesium	1·75
Copper	8·9	Calcium	1·58
Nickel	8·8	Rubidium	1·52
Cadmium	8·6	Sodium	0·972
Cobalt	8·5	Potassium	0·865
Manganese	8·0	Lithium	0·593

Fusibility.—The melting points of the metals differ even more widely than their densities, mercury fusing at 40° below zero, and platinum only melting at the highest temperature of the oxyhydrogen blowpipe.

Table of Melting Points.

Mercury	— 40°	Bismuth	270°
Tin	+235°	Cadmium	315°

Lead	334°	White Cast Iron .	1,050°
Zinc	423°	Gray ditto .	1,200°
Antimony	425°	Steel. . .	1,300° to 1,400°
Silver	1,000°	Wrought Iron	1,500° to 1,600°
Copper	1,090°		

Some of the metals can be easily converted into vapor, or volatilized; thus mercury boils at 350°, arsenic passes into vapor even before it assumes the liquid form, whilst potassium, sodium, magnesium, zinc, and cadmium, can be distilled at a red heat. Even the more infusible of the metals, such as copper and gold, are not absolutely fixed, but gige off small quantities of vapor when strongly heated in a furnace.

The color of most of the metals is nearly uniform, varying from the bright white of silver to the bluish-gray of lead; copper is the only red-colored metal known, whilst gold, strontium, and calcium are yellow. In ductility, or the power of being drawn out into wire, and malleability, or the power of being hammered out into thin sheets, the metals differ considerably. Gold is the most malleable of all the metals, being capable of being beaten out to the thickness of the $\frac{1}{200000}$th part of an inch; it is, likewise, the most ductile metal. Hardness, brittleness, and tenacity, are physical properties of great importance, in which the metals differ widely.

Specific Heat and Atomic Heat.

When equal weights of different bodies are raised through the same number of degrees of temperature, they take up different amounts of heat; or different bodies possess different *capacities for heat.* Thus the amount of heat needed to raise a kilogram of water through 100° C. is 31 times as large as that required to raise the same weight of platinum through the same interval of temperature; or, in other words, the same amount of heat which raises 1 kilo. of water through 100° will raise 31 kilos. of platinum through the same temperature. Hence the *specific heat* of platinum is said to be $\frac{1}{31}$, or 0·032; that of water being taken as the unit. The specific heat of the same substance is different according as the substance is solid, liquid, or gaseous; but the specific heats of the metals in the solid state exhibit a remarkable relation to their atomic weights. It has been found if, instead of calculating the specific heats for *equal* weights, we take the *atomic* weights of the metals, that the numbers expressing the capacity for heat of the atoms are all equal; *or the metals all possess the same atomic heat.* This is clearly seen if we multiply the specific heats of the metals by the corresponding atomic weights; thus:

	Specific Heat.		Atomic Weight.		Atomic Heat.
Lead	0·031	×	207	=	6·41
Platinum	0·032	×	197·5	=	6·33
Silver	0·059	×	108	=	6·37
Tin	0·054	×	118	=	6·37
Zinc	0·095	×	65·2	=	6·39

Hence we have in the determination of specific heat a means of checking the atomic weight of a metal or of ascertaining it in a doubtful case. Thus, in the case of the newly discovered metal thallium, chemists were in doubt whether it most resembled lead or the alkali metals: if it was to be classed as a dyad with lead, its atomic weight must be 408; if it was placed with the monad alkali metals, its atomic weight would be $\frac{408}{2} = 204$. The specific heat of thallium was, however, found to be 0·033; and if we divide this into 6·4, the common atomic heat of the metals, we get the number 194—a number much nearer to 204 than to 408. The difference here noticed between 194 and 204 is due to the great difficulty of accurately determining the specific heat of bodies, and the errors which arise from the variation of physical condition.

The following non-metals have the same atomic heat as the metals:

NITROGEN.	SELENIUM.
CHLORINE.	TELLURIUM.
BROMINE.	ARSENIC.
IODINE.	

Nitrogen and chlorine are, it is true, not known in the solid state, but their atomic heats can be calculated from the molecular heats of their solid compounds, for *the elements in the solid state possess the same atomic heats as in their compounds;* and hence the molecular heat is the sum of the atomic heats of the combined elements, as is shown in the following list:

	Specific Heat.		Molecular Weight.		
Silver chloride, Ag Cl . . .	0·089	×	143·5	=	2 × 6·4
Sodium chloride, Na Cl . .	0·219	×	58·5	=	2 × 6·4
Potassium bromide, K Br .	0·107	×	119·1	=	2 × 6·4
Tin di-chloride, Sn Cl_2 . .	0·102	×	189	=	3 × 6·4
Mercuric iodide, Hg I_2 . .	0·0423	×	454	=	3 × 6·4
Platinum potassium chloride, K_2PtCl_6 . .	0·118	×	488·6	=	9 × 6·4

The remaining elements have all an atomic heat smaller than 6·4; thus the atomic heats of sulphur and phosphorus are 5·4; of fluorine, 5; of oxygen, 4; of silicon, 3·8; of boron, 2·7; of hydrogen, 2·3; and of carbon, 1·8. In the case of these elements the atomic heats are calculated from the molecular heats of their compounds in accordance with the above-mentioned law, as the following examples show:

	Spec. Heat.		Molec. Weight.		Molec. Heat.	
Ice, H_2O . .	0·478	×	18	=	8·6	$= 4 + 2 \times 2·3.$
Mercuric oxide, HgO	0·048	×	216	=	10·4	$= 6·4 + 4.$
Arsenic trioxide, As_2O_3	0·125	×	198	=	24·8	$= 2 \times 6·4 + 3 \times 4.$
Calcium Carbonate, $CaCO_3$	0·202	×	100	=	20·2	$= 6·4 + 1·8 + 3 \times 4.$
Potassium sulphate, K_2SO_2	0·196	×	174·2	=	34·2	$= 2 \times 6·4 + 5·4 + 3 \times 4.$
Carbon hexachloride, C_2Cl_6	0·177	×	237	=	42·0	$= 2 \times 1·8 + 6 \times 6·4.$

Occurrence and Distribution of the Metals.

Only a few of the metals occur in the free or uncombined state in nature; in general they are found combined with oxygen, sulphur, or some other metalloid. These metallic compounds exist most variously distributed throughout the earth's crust; some are known to occur in only one or two localities, and even then only in minute quantity, whereas others are found widely distributed in enormous masses. As is seen by reference to the table on p. 8, the metals aluminium, iron, calcium, magnesium, and sodium occur in very large quantities, forming, when united with oxygen and silicon, the whole mass of granitic rocks composing our globe; but it is not from these sources that the metals in question can be obtained for the purposes of the arts. For this object, we employ other combinations, found in smaller quantity, from which the metals can be more easily extracted than from the silicates, and these compounds are termed the *metallic ores.*

The heavy metals and their ores generally occur interspersed throughout the old granitic or early sedimentary rocks in the form of veins or lodes, which are cracks or fissures,

13*

running through the rock in a particular direction, and filled up with a metallic ore. Other ores, such as ironstone, are found amongst the more recent sedimentary formations, having been deposited in large masses, probably from aqueous solution.

The consideration of the occurrence and distribution of the various metallic ores belongs to the science of geology; the study of the modes of procuring the ores is the province of the miner and engineer; whilst the processes by means of which the metal itself is obtained from the ore, although mainly dependent upon chemical principles, are generally classed as belonging to the branch-science of metallurgy.

Classification of the Metals.

The metals can be conveniently grouped into classes, in which the several members possess certain properties and general characters in common.

Class I. Metals of the alkalies.—1, Potassium. 2, Sodium. 3, Cæsium. 4, Rubidium. 5, Lithium. (6, Ammonium.)

General Characteristics.—The metals of this class are monovalent; they are soft, easily fusible, volatile at higher temperatures; they combine with great force with oxygen, decompose water at all temperatures, and form basic oxides, which are very soluble in water, yielding powerfully caustic and alkaline bodies, from which water cannot be expelled by heat. Their carbonates are soluble in water, and each metal forms only one chloride. Ammonium, NH_4, is added to the list of alkaline metals proper, from the general similarity of the ammoniacal salts to those of potash and soda.

These metals and their compounds are closely analogous in their properties, and they exhibit a remarkable relation as regards their atomic weights; thus, sodium, which stands between potassium and lithium in properties, has a combining weight, which is the arithmetic mean of the other two, $\frac{39 + 7}{2} = 23$; so, too, rubidium, standing half-way between cæsium and potassium, has a mean atomic weight, $\frac{113 + 39}{2} = 86$.

Class II. Metals of the alkaline earths.—1, Calcium. 2, Strontium. 3, Barium.

General Characteristics.—The metals of this class are divalent; they cannot be reduced by hydrogen or carbon alone; they decompose water at all temperatures, producing oxides, which combine with water to form hydroxides, from some of which the water can be driven off by heat. Their carbonates are insoluble in water, but soluble in water containing carbonic acid in solution.

Class III. Metals of the Earths.—1, Aluminium. 2, Glucinium. 3, Zirconium. 4, Thorium. 5, Yttrium. 6, Erbium. 7, Cerium. 8, Lanthanum. 9, Didymium.

With the exception of aluminium, these metals are hardly known in the free state, as their compounds occur so rarely that they are not employed for any useful purpose, and their properties cannot be considered in an elementary work. The oxides of this group are insoluble in water; and they cannot be reduced to the metallic state by hydrogen or carbon. Aluminium decomposes water at a high temperature.

Class IV. Zinc Class.—1, Magnesium. 2, Zinc. 3, Cadmium. These metals are divalent; they are all volatile at high temperatures, and burn when heated in the air; they decompose water at a high temperature, or in presence of an acid, and form only one oxide and one chloride.

Class V. Iron Class.—1, Manganese. 2, Iron. 3, Cobalt. 4, Nickel. 5, Chromium. 6, Uranium. These are not volatile at the temperature of our furnaces; they decompose water like the preceding class, and they form several oxides, chlorides, and sulphides.

Class VI. Tin Class.—1, Tin. 2, Titanium. 3, Zirconium. 4, Thorium. 5, Niobium. 6, Tantalum. Tin is the only one of this class employed in the arts. These metals decompose water at high temperatures and in presence of alkalies; the first four form dioxides and volatile tetrachlorides, and are tetravalent and closely connected with silicon.

Class VII. Tungsten Class.—1, Molybdenum. 2, Tungsten. These metals are of rare occurrence; they decompose water at a high temperature, and form trioxides and volatile hexachlorides.

Class VIII. Arsenic Class.—1, arsenic. 2, Antimony. 3, Bismuth. 4, Vanadium. The metals of this class are trivalent; they form the junction between the metals and metalloids, and they closely resemble phosphorus in their properties.

Class IX. Lead Class.—1, Lead. 2, Thallium: heavy metals, allied in their general properties to the two first classes. Lead is divalent, but thallium is monovalent.

Class X. Silver Class.—1, Copper. 2, Mercury. 3, Silver. These metals do not decompose water under any circumstances; they are oxidized by nitric and strong sulphuric acids; copper and mercury form two basic oxides which, except in the case of copper, are decomposed by heat alone. Copper and mercury are divalent; silver is monovalent.

Class XI. Gold Class.—1, Gold. 2, Platinum. 3, Palladium. 4, Rhodium. 5, Ruthenium. 6, Iridium. 7, Osmium. These metals are not acted upon by nitric acid, but only by chlorine or aqua regia, and the oxides are reduced by heat alone; and they with silver and mercury constitute the noble metals. Gold is trivalent, and platinum is tetratomic.

Chemical Properties of the Metals.

The metals combine (1) with each other to form alloys;

(2) with the metalloids to form oxides, sulphides, chlorides, &c. In the alloys the metallic appearance and properties are preserved, whereas in the compounds with the metalloids the physical properties of the metals as a rule disappear.

Alloys. The compounds formed by the metals amongst themselves are not so stable as those which are formed by union with a metalloid; nevertheless the alloys are largely used in the arts, as they possess many valuable properties not exhibited by the metals separately. Thus gold and silver are too soft to be used alone as a medium of currency, but the addition of 7·5 per cent. of copper gives an alloy of the requisite hardness; then copper is too soft and tough to be wrought in the lathe, but when alloyed with half its weight of zinc it forms a hard and most useful substance known as brass. Gun-metal, or bronze, is a hard and tenacious alloy of 90 parts of copper and 10 of tin. Bell-metal, a still harder alloy, contains the same metals in the proportion of 80 of the former to 20 of the latter, whilst an alloy of 33 parts of tin to 67 of copper possesses a white color, takes a high polish, and is known as speculum-metal, and employed for the reflectors of telescopes. For making printing type a peculiar alloy is employed, containing 80 parts of lead to 20 of antimony; this possesses many properties necessary for type metals, which are found to belong to no single metal or other alloy.

The chemical composition of the alloys is not so definite or so well marked as that of the other metallic compounds, but they may frequently be obtained in crystals in which the constituents are contained in atomic proportions. The melting point of an alloy is often much lower than the melting points of its constituent metals. Thus lead melts at 334°, bismuth at 270°, tin at 235°, and cadmium at 315°; whereas an alloy of 2 parts bismuth, 1 of tin, and 1 of lead, melts at 95° to 98° C., and one containing 8 of lead, 15 bismuth, 4 of tin, and 3 of cadmium, softens at as low a temperature as 60°, and is perfectly fluid at 65° C. The alloys of metals with mercury are termed *Amalgams.*

Compounds of the Metals with Non-Metals.

1. *Metallic Oxides.* Oxygen acts very differently on the different metals. Some metals, such as zinc, magnesium, and calcium take fire when heated, and burn with the evolution of intense light; whilst others, such as gold and silver, do not combine directly with oxygen, and are only obtained in combination with it by indirect means and with difficulty.

The oxides differ widely in properties and composition; they may, however, all be represented as water in which the hydrogen has been replaced by metal. Thus the monoxides may be considered to be water in which either each atom of hydrogen is replaced by a monad, as K_2O, Ag_2O, or the two atoms of hydrogen are replaced by a dyad, as $Ba\,O$, $Zn\,O$; whilst the higher oxides are regarded as two or more molecules of water, in which the hydrogen is in like manner replaced by its equivalent of metal. The most important of these higher oxides are the sesquioxides, such as alumina, $Al_2\,O_3$, and ferric oxide, $Fe_2\,O_3$; the dioxides, such as black oxide of manganese, $Mn\,O_2$; the trioxides, as chromium trioxide, $Cr\,O_3$.

2. *Metallic Sulphides.* Metals combine directly with sulphur to form sulphides, and these occur frequently in nature, forming many of the metallic ores. These compounds resemble in composition the corresponding oxides, and may be represented as sulphuretted hydrogen, H_2S, in which the hydrogen is replaced by its equivalent of metal. The sulphides of the first and second class of metals are soluble in water; those of the remaining almost all insoluble in water, but some of them soluble, and others insoluble, in acids and alkalies. In the laboratory this difference in

the solubility of the sulphides is employed as a means of separating the different metals in the processes of chemical analysis.

3. *Metallic Chlorides.* The metals unite directly with chlorine to form an important class of compounds, which may be considered as one or more molecules of hydrochloric acid, in which the hydrogen is replaced by its equivalent of metal; thus $K\ Cl$ is the chloride of a monad; $Ba\ Cl_2$ that of a dyad, and this may be considered as two molecules of hydrochloric acid, in which two atoms of hydrogen are replaced by the dyad barium; whereas antimony trichloride, $SbCl_3$, may be represented as three molecules of hydrochloric acid, in which the three atoms of hydrogen are replaced by one atom of triatomic antimony.

4. Metals unite with bromine, iodine, fluorine, and cyanogen to form bromides, iodides, fluorides, and cyanides, all of which bear a strong resemblance to the corresponding chlorides, and may be considered to possess an analogous constitution.

5. Metals also unite with nitrogen, phosphorus, boron, silicon, carbon, and hydrogen; but the compounds thus formed are in general of slight importance.

6. *Metallic Salts.*—The replacement of the hydrogen in acids (the hydrogen salt), by its equivalent of metal, gives rise to the formation of a very large class of bodies to which the name of *metallic salts* has been given; thus, if we replace both atoms of hydrogen in hydrogen sulphate, $H_2\ SO_4$, by potassium, we get potassium-sulphate, $K_2\ SO_4$; if only one is replaced, we have hydrogen-potassium sulphate, $KH\ SO_4$; if we replace both the atoms of hydrogen with a dyad, we have $Ba\ SO_4$, or $Zn\ SO_4$. The same holds good for all the other acids; thus, $H_3\ PO_4$ (trihydric phosphate) yields $Ag_3\ PO_4$ (trisilver phosphate); $H\ NO_3$ yields $K\ NO_3$, or $Ba\ N_2\ O_6$. The metals forming sesquioxides also give rise to salts which may be considered as being composed of several molecules of the acid with the hydrogen replaced by the metal; thus aluminium sulphate is three molecules of hydric sulphate,

$3 (H_2 SO_4)$, in which the 6 of hydrogen are replaced by 2 atoms of aluminium; thus, $Al_2 3 SO_4$. The constitution of the other classes of salts will be best understood from the special descriptions. Many of the metallic salts when crystallized contain a definite number of atoms of water, and this is termed *water of crystallization.*

LESSON XVIII.

Crystallography.

Most chemical substances, when they pass from the liquid or gaseous into the solid state, assume some definite geometric form, or are said to *crystallize.* Crystals are produced when a substance, such as nitre, is dissolved in water and the solution allowed gradually to evaporate; or when a body, such as sulphur, is melted and allowed to solidify by cooling; or when a volatile substance, such as iodine or arsenic trioxide, is vaporized and the vapor condensed on a cool surface. Many naturally occurring minerals exhibit very perfect crystalline forms; we are ignorant of the mode in which such crystals are in most cases produced, but we know that the process of their formation has been a very slow one, and we find that, in general, a crystal is larger and more perfect the more gradually it has been formed. Crystalline bodies exhibit, in addition to their regular form, a peculiar power of splitting in certain directions more readily than in others, called *cleavage;* as well as in many cases the property of allowing the rays of light and heat to pass more readily in one direction than another, giving rise to the well-known phenomena of double refraction.

Inorganic bodies which do not exhibit these peculiarities, or assume crystalline structure, are said to be *amorphous;* but certain highly complicated structures found in the vegetable and animal world exhibit a structure which, although it is non-crystalline, is not devoid of arrangement, and to which the name *organized* or *cellular structure* has been given. As a rule, every particular substance possesses a definite form in which it always crystallizes, and by which it can be distinguished; when a crystal is formed, from aqueous solution, for example, the smallest visible particle possesses the complete form of the largest crystal, and simply increases in size without undergoing any change of form.

It has been found possible to arrange the many thousand different known crystals in *six systems*, to each of which belongs a number of forms having some property in common. In order to classify these different crystals, the existence of certain lines within the crystal called *axes* is supposed, round which the form can be symmetrically built up. These axes are assumed to intersect in the centre of the crystal, and pass through from one side to the other.

1*st*, or *Regular System.*—Three axes, all equal and at right angles. The simple forms of this system are (1) the cube (Fig. 36); (2) the regular octohedron (Fig. 37); (3) the rhombic dodecahedron (Fig. 38); and (4) the regular tetrahedron (Fig. 39). The following are a few of the substances crystallizing in this system—alum, common salt, diamond, fluor-spar, iron pyrites, and garnet.

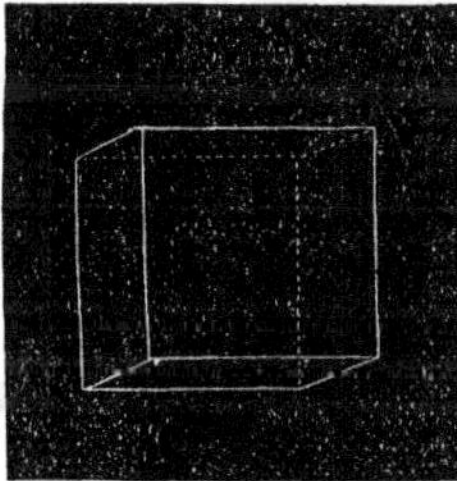

FIG. 36.

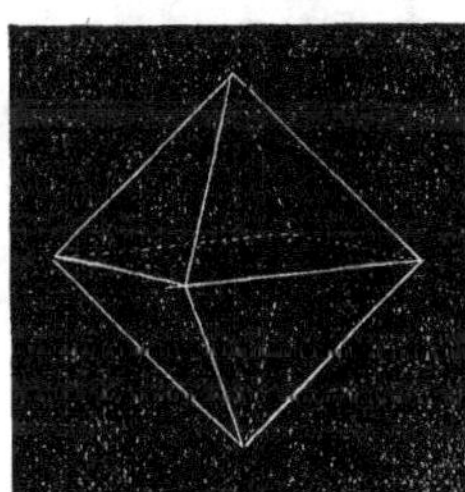

FIG. 37.

2d, or *Quadratric System.*—Three axes, all at right angles one shorter or longer than the other two. The simple forms of this system are the first and second right square prisms

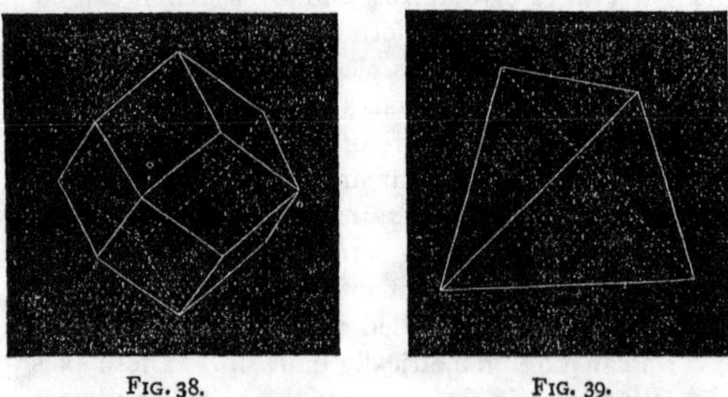

FIG. 38. FIG. 39.

(Fig. 40 *a* & 41), and the first and second right square octohedra (Fig. 41 *a* and *b*). In the first square prism the axes

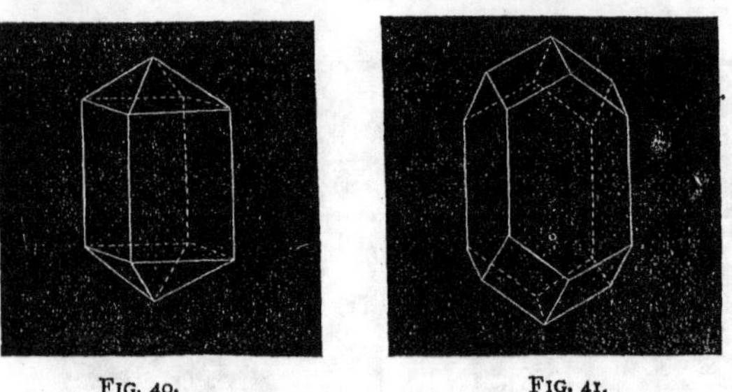

FIG. 40. FIG. 41.

terminate in the centre of each of the sides, and in the second the axes terminate at the intersection of the sides, and this is reversed with regard to the octohedra. Some of the common substances which crystallize in this system are—yellow prussiate of potash, zircon, and tin dioxide.

The 3d, or *Hexagonal System.*—Four axes, three equal

and in one plane, making angles of 60°, and one longer or shorter, at right angles to the plane of the other three. The

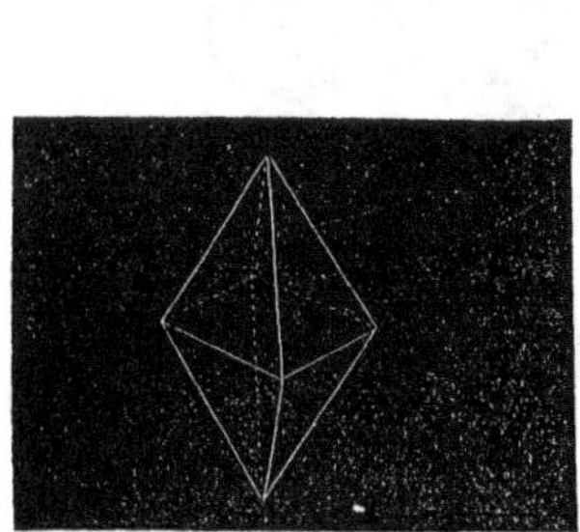

FIG. 42 *a*.

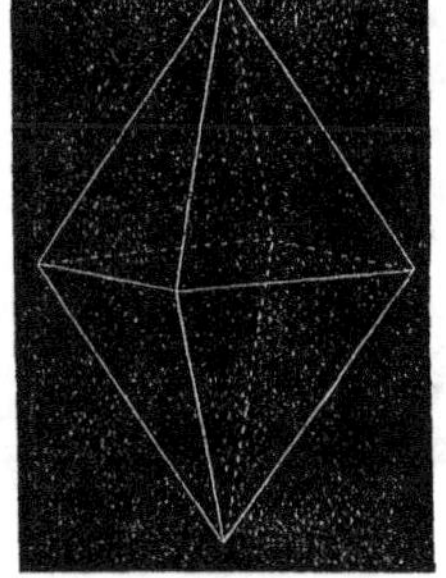

FIG. 42 *b*.

regular six-sided prism (Fig 43), the regular six-sided pyramid (Fig. 44), and the rhombohedron (Fig 45), are the common

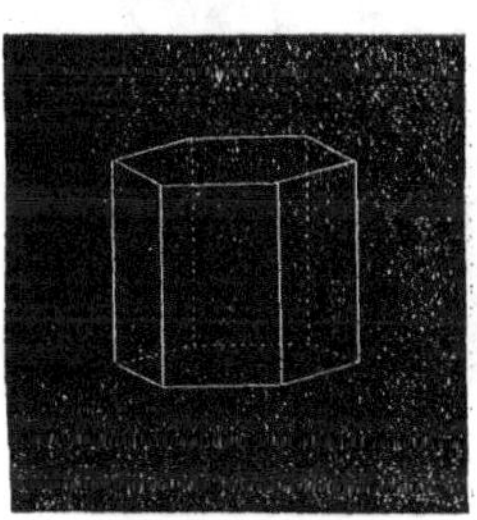

FIG. 43.

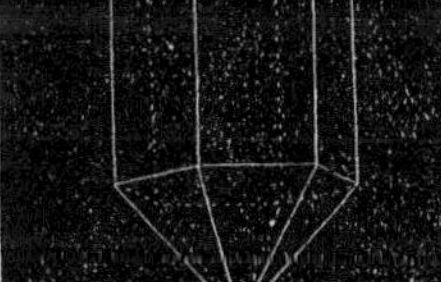

FIG. 44.

forms of this system. Quartz, calc-spar, beryl, corundum, graphite, ice, &c., crystallize in the hexagonal system.

4*th* or *Rhombic System.*—Three axes, all unequal, and all at right angles. The chief forms of the crystals in this system are the right octohedron with rhombic base (Figs. 46 and 47), and the right rhombic prism (Fig. 48). In this system the following substances are found—nitre, barium sulphate, arragonite, topaz, and native sulphur.

5th or *Monoclinic System.*—Three axes, all unequal, two

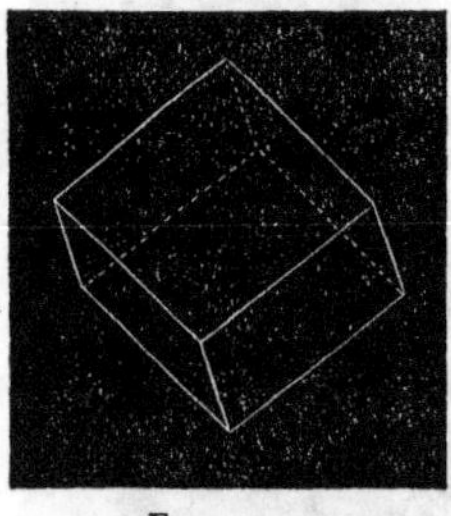

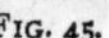

FIG. 45.

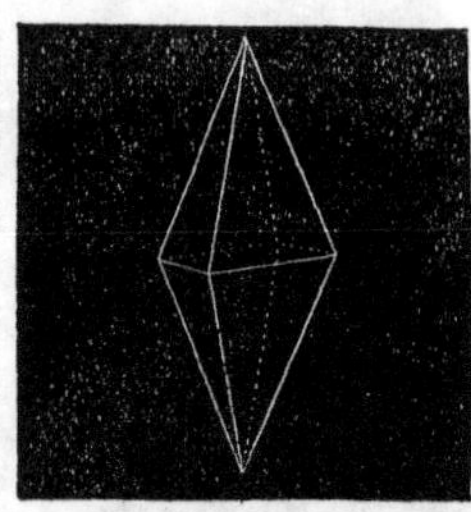

FIG. 46.

cut one another obliquely, and one is at right angles to the plane of the other two. The oblique rhombic octohedron (Fig. 49) belongs to this system. Many substances crystal-

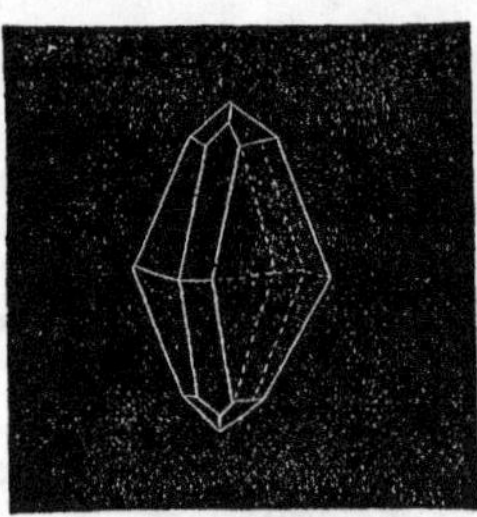

FIG. 47.

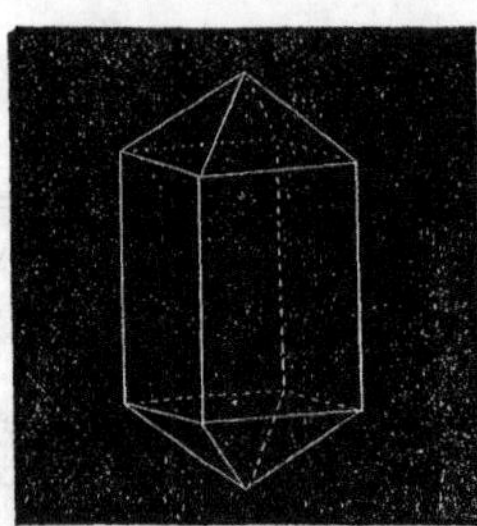

FIG. 48.

lize in this system; amongst the most common are—sulphur deposited from fusion, sodium carbonate and phosphate, ferrous sulphate, borax, and cane sugar.

6th, or *Triclinic System.*—Three axes, all unequal, and all oblique. The doubly-oblique octohedron, and the doubly-oblique prism (Fig. 50), are the leading forms in this system.

Copper sulphate, boric acid, the mineral albite, and a few other substances are found to crystallize in this system, the forms of which are in general very complicated. The crystalline form of copper sulphate is shown in Fig. 51.

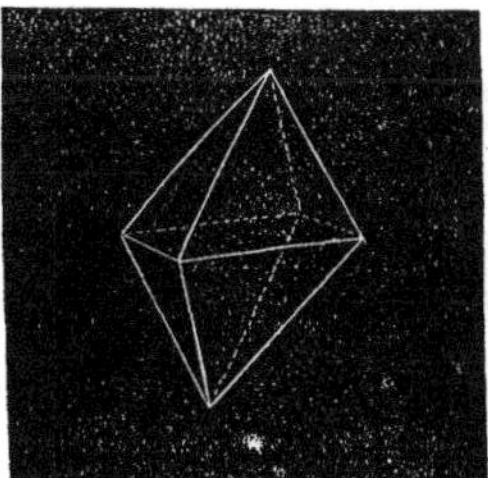

FIG. 49.

Under one or other of these six divisions all the known forms of crystals can be classed. In every distinct crystal belonging to any one of these systems, in which the axes are not all equal, or all at right angles, certain relations exist between the lengths of the axes, and these have certain mutual inclinations to one another. These relations and inclinations vary with different substances, but are constant for the same; so that different bodies all crystallizing in the same system,

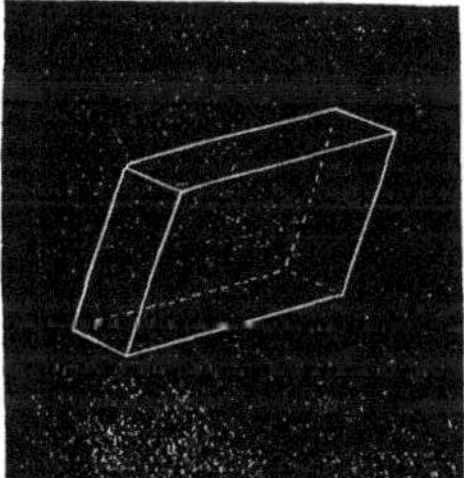

FIG. 50.

as a rule, have different relations between the lengths of the axes, and these generally have different inclinations to one another.

Certain substances exhibiting a similarity in their chemical

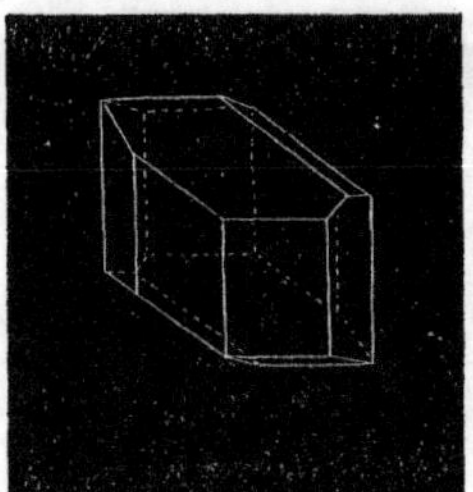

FIG. 51.

constitution are found to crystallize in the same forms: these are said to be *isomorphous;* whilst, when the same body occurs crystallized in two different systems, it is said to be *dimorphous*. Examples of these peculiar relations between chemical composition and crystalline form will be given later on.

LESSON XIX.

CLASS I. METALS OF THE ALKALIES.—1, POTASSIUM. 2, SODIUM. 3, CÆSIUM. 4, RUBIDIUM. 5, LITHIUM. (6, AMMONIUM.)

POTASSIUM. *Symbol* K. *Combining Weight* 39·1. *Specific Gravity* 0·865.

The metal potassium was discovered in the year 1807, by Sir Humphrey Davy, who decomposed the alkali potash into the metal, hydrogen, and oxygen, by means of a powerful galvanic current. Before this time the alkalies and alkaline earths were supposed to be elementary bodies. The metal is now prepared by heating together potash and carbon to a high temperature in an iron retort. The carbon, at the high temperature, is able to take the oxygen from the potash, forming carbon monoxide, which escapes as a gas, whilst the metal potassium, which is volatile at a red heat, distils over. The preparation of this metal is attended with many difficulties, and requires special precautions, as the vapor of potassium not only takes fire when brought in contact with the air, but decomposes water, combining with the oxygen and liberating hydrogen ; hence the metallic vapor must be cooled by rock oil or naphtha, which contains no oxygen. The metal thus prepared must be distilled a second time, in order to purify it and free it from a black, explosive compound, which invariably forms in the original preparation, and has caused several fatal accidents.

Potassium, thus prepared, is a bright, silver-white metal, which can be easily cut with a knife at the ordinary atmospheric temperature ; it is brittle at 0°, and melts at 62°·5, and does not become pasty before melting; when heated to a temperature below red heat, potassium sublimes, yielding a fine green-colored vapor. This metal rapidly absorbs oxygen when exposed to the air, and gradually becomes converted into a white oxide. Thrown into water, one atom of potassium displaces one of hydrogen from the water, forming potassium hydroxide, KHO ; this takes place with such force that the heat developed is sufficient to ignite the hydrogen thus set free, and the flame becomes tinged with the peculiar purple tint characteristic of the potassium compounds, whilst the water attains an alkaline reaction from the potash which is formed. Potassium combines also with chlorine and sulphur, and many other metalloids, evolving, when heated with these substances, heat and light.

Sources of the Potassium Compounds.—The original source of potassium compounds is in the felspar of the granitic rocks, of which the earth is composed, as these contain from two to three per cent. of this metal. Up to the present time, this source has not been available for the manufacture of the potassium salts, as no cheap and easy mode has yet been made available for separating the potash from the silicic acid, with which it is combined in felspar. Plants, however, are able slowly to separate out and assimilate the potash from these rocks and soils ; so that, by burning the plant and extracting the ashes with water, soluble potash-salt is obtained. This is the crude potassium carbonate, called, when purified by recrystallization, *pearl-ash*, and it is from this substance that a large number of the potassium compounds are obtained. Some of the other potassium salts, such as the nitrate and chloride, are found in large quantities in various localities as deposits on the surface, or in the interior, of the earth. Potassium chloride occurs in beds together with rock salt, in Stassfurt in Germany. Another inexhaustible source of potassium compounds, which, however, has only just begun to be utilized, is sea-water : a plan has lately been proposed by which those compounds can be obtained from the sea.

Potassium Oxides.—Potassium combines with oxygen in three proportions, forming three well-defined oxides of the formulæ—

(1) Potassium monoxide . . . . K_2O
(2) Potassium dioxide . . . . K_2O_2
(3) Potassium tetroxide . . . . K_2O_4

These oxides are unimportant bodies ; but *potash*, the compound of the first with water, is a substance of great importance.

Potassium monoxide, K_2O, is obtained by allowing thin pieces of the metal to oxidize in dry air ; it is a grayish-white, brittle substance, which melts a little above red heat, and volatilizes only at a very high temperature. This oxide combines with water with evolution of great heat, producing potassium hydroxide, or potash, from which water cannot again be separated by heat. The reaction may be represented as an exchange of hydrogen for potassium, thus :

$$K_2O + H_2O = 2(KHO).$$

Potassium Hydroxide, or Caustic Potash, HKO, is obtained as above, or more conveniently prepared by boiling one part of potassium carbonate with twelve parts of water, and adding slacked lime prepared from two-thirds part of quick lime. In this reaction calcium carbonate is formed, which falls to the bottom as a heavy powder, caustic potash remaining in solution.' The clear liquid, which should not effervesce on addition of an acid, is evaporated in a silver basin to dryness, fused by exposure to a stronger heat, and cast into sticks in a metallic mould. Thus prepared, caustic potash is a white substance, soluble in half its weight of water, acting as a powerful cautery, destroying the skin. It is largely used in the arts and manufactures, and is employed in the laboratory for various purposes.

Potassium Carbonate. Symbol K_2CO_3. This salt receives the commercial name of potashes, or pearl-ashes, and is imported in large quantities from Russia and America. The crude substance is prepared by boiling out the ashes of plants with water, and evaporating the solution to dryness ; a pure salt may be afterwards obtained by separating the impurities by crystallization. The leaves and small twigs of plants contain more potash than the stems and large branches. Potassium carbonate can be obtained perfectly pure by heating pure potassium tartrate to redness, and separating the carbonate formed by dissolving in water. This salt absorbs water from the air, or is deliquescent, and is, therefore, very soluble in water ; it also turns red litmus blue, or possesses a strongly alkaline reaction.

Hydrogen Potassium Carbonate, $HKCO_3$. This substance is formed when a current of carbonic acid gas, CO_2, is passed through a strong solution of the preceding salt. It may be considered as dibasic carbonic acid, H_2CO_3, in which one atom of hydrogen is replaced by one of potassium. It is a white salt, not so soluble as potassium carbonate ; the solution is nearly neutral to test paper.

Potassium Nitrate, Nitre, or *Saltpetre*, KNO_3. This important salt occurs as an efflorescence on the soil of several dry tropical countries, especially that of India. It may be artificially prepared by the process of nitrification, in which animal matter (containing nitrogen) is exposed in heaps, mixed together with wood-ashes and lime, to the action of the air ; the organic matter gradually undergoes oxidation, nitric acid being formed, and this unites, first with the lime and then with the potash, to form nitre. The salt is obtained from both of these sources by boiling out the soil or deposit with water, and allowing the nitre to crystallize out. Nitre crystallizes in rhombic prisms ; it dissolves in seven parts of water, at 15°, and in its own weight of hot water. It contains nearly half its weight of oxygen, with which it parts on heating with carbon or other combustible matter. For this reason, nitre is largely used in the manufacture of gunpowder and fireworks. Gunpowder consists of an intimate mixture of nitre, charcoal, and sulphur. The general decomposition which occurs when gunpowder is fired may be expressed by saying that the oxygen of the nitre combines with the charcoal, forming carbonic acid and carbonic oxide, whilst the nitrogen is liberated, and the sulphur combines with the potassium. Hence gunpowder can burn under water or in a closed space, as it contains the oxygen needed for the combustion in itself, and the great explosive power of the substance is due to the violent evolution of large quantities of gas, and a rapid rise of temperature causing an increase of bulk sudden and great enough to produce what is termed an explosion. It has been found by practice that the best gunpowder is that which contains nearly two atoms of nitre, one of sulphur, and three of carbon ; but the decomposition which actually occurs in the explosion is a more complicated one than has been expressed above, and cannot be represented in an equation. The following table (see p. 160) gives the composition of musketry powder, as manufactured by different nations.

Potassium combines with sulphur to form several compounds, of which the best known are K_2S, K_2S_2, K_2S_3, and K_2S_5. They are soluble substances, which evolve sulphuretted hydrogen when heated with an acid, and are not used in the arts.

	English and Austrian.	Prussian.	Chinese.	French.
Nitre	75	75	75·7	75·0
Charcoal	15	13·5	14·4	12·5
Sulphur	10	11·5	9·9	12·5
	100	100·0	100·0	100·0

The remaining compounds of potassium of interest and importance are—potassium chloride, KCl, occurring in sea-water and also in certain deposits ; potassium iodide, KI, mentioned under the head of Iodine ; potassium bromide, K Br (*see ante*, p. 95) ; potassium chlorate, $KCl\ O_3$, the preparation of which was described under Hydric Chlorate, (p. 93) ; potassium sulphate, K_2SO_4, one of the most insoluble of the salts of this metal ; and hydric potassium sulphate, $HKSO_4$.

General Characteristics of the Potassium Compounds.

All the potassium compounds impart a violet color to the flame, and the spectrum of this flame (*see* p. 220 spectrum analysis) is distinguished by the presence of two bright lines, one in the red, and another in the violet. Almost all the potassium salts are soluble in water ; the three which are least soluble are—(1) potassium perchlorate ; (2) hydrogen potassium tartrate, which is precipitated in the form of a white crystalline powder, when a solution of a potassium salt is mixed with an excess of tartaric acid ; and (3) potassium-platinum chloride, $2(KCl) + Pt\ Cl_4$, which precipitates in small yellow cubical crystals, when platinum chloride solution is added to a soluble potassium salt. These reactions serve to distinguish the potassium salts.

2. SODIUM. *Symbol* Na (*natrium*). *Combining Weight* 23. *Specific Gravity* 0·97.

This metal was discovered by Sir H. Davy immediately after the isolation of potassium, by the decomposition of soda with the galvanic current. It can be procured more easily than potassium by reducing the carbonate in presence of carbon, and is now manufactured in large quantities for the preparation of other metals, especially magnesium and aluminium. The apparatus employed for the preparation of this metal is the same as that used for potassium; the metal distils over when condensed, and drops into rock oil. Sodium is a silver-white metal, soft at ordinary temperatures, and melting at 95°·6; it volatilizes below a red heat, yielding a colorless vapor. When thrown upon water it floats, and rapidly decomposes the water with disengagement of hydrogen, soda being formed. If the water be hot or be thickened with starch, the globule of metal becomes so much heated as to enable the hydrogen to take fire. The compounds of sodium are very widely diffused, being contained in every speck of dust (*see* spectrum analysis, p. 221); they exist in enormous quantities in the primitive granitic rocks (*see* p. 8), but they are most readily obtained from sea-water, which contains nearly three per cent. of sodium chloride (common or sea-salt), or from the large deposits of this substance which occur in Cheshire, Galicia, &c. Sodium carbonate was formerly obtained from the ashes of sea-plants or kelp, as potassium carbonate is still prepared from the ashes of land plants; but at present the sodium carbonate is altogether manufactured, on an enormously large scale, from sea salt.

Sodium Oxides. There are two compounds of sodium and oxygen known. Sodium oxide, $Na_2 O$; and sodium dioxide, $Na_2 O_2$—*Sodium Oxide*, $Na_2 O$, is formed when sodium is oxidized in dry air or oxygen at a low temperature, a white powder being formed; this takes up moisture with great avidity, forming H Na O, *sodium hydroxide*, or *soda*, from which the water cannot again be separated by heat alone, but which can again be converted into the oxide by heating with sodium; thus, H NaO + Na = Na_2O + H.

Sodium Dioxide, $Na_2 O_2$, is a yellowish-white powder, which is formed when sodium is heated in oxygen to 200° C.; it is soluble in water, but the solution readily decomposes, giving off oxygen and leaving sodium hydroxide. Sodium dioxide is not used in the arts.

Sodium Hydroxide, or *Caustic Soda*, H Na O, is a white solid substance, fusible below a red heat, and less volatile than the corresponding potassium compound. It is very soluble in water, acts as a caustic, is powerfully alkaline, and is largely used in soap-making. The manufacture of solid caustic soda is now carried on on a large scale, by boiling lime and sodium carbonate together with water, and evaporating down the clear solution.

$$Ca\,O + Na_2\,CO_3 + H_2O = Ca\,CO_3 + 2(NaH\,O).$$

Sodium Chloride—Common Salt. Na Cl. It is from this salt that almost all the other sodium compounds are prepared. Sodium chloride occurs in thick beds in various parts of the world, especially in Cheshire, Galicia, Tyrol, Spain, and Transylvania. It is likewise prepared from sea-water by evaporation or by freezing; and from certain brine springs by evaporation. When slowly deposited sodium chloride crystallizes in regular cubes, it is soluble in about two and a half parts of water at 15°, and does not dissolve sensibly more in hot than in cold water.

Sodium Carbonate, $Na_2\,CO_3$. This substance, known in commerce as *soda-ash*, is manufactured in England on an enormous scale, and used for glass-making, soap-making, bleaching, and various other purposes in the arts. Formerly it was prepared from barilla or the ashes of sea-plants, but now it is wholly obtained from sea-salt by a series of chemical decompositions and processes, which may be divided into two stages.

(1) Manufacture of sodium sulphate, or salt-cake, from sodium chloride (common salt); called salt-cake process.

(2) Manufacture of sodium carbonate, or soda ash, from salt cake; called soda-ash process.

(1) *Salt-cake process.* This process consists in the decomposition of salt by means of sulphuric acid: this is effected in a furnace called the *Salt-cake Furnace.* Fig. 52 shows the section, and Fig. 53 the elevation of such a furnace: these are

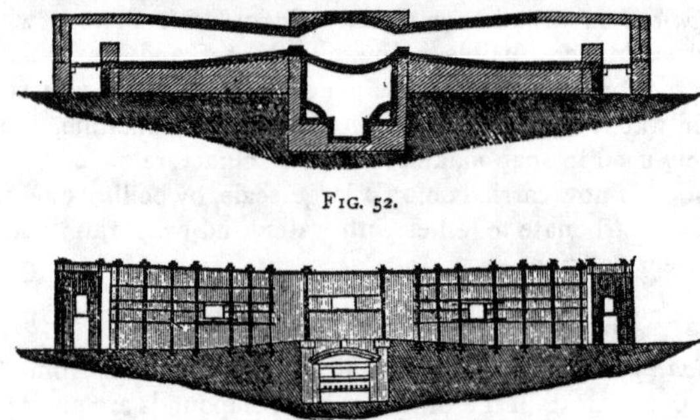

FIG. 52.

FIG. 53.

drawn to scale from one actually in use. It consists of (1) a large covered iron pan placed in the centre of the furnace, and heated by a fire placed underneath; and (2) two roasters or reverberatory furnaces, placed one at each end, and on the hearths of which the salt is completely decomposed. The charge of half a ton of salt is first placed in the iron pan, and then the requisite quantity of sulphuric acid allowed to run in upon it. Hydrochloric acid gas is evolved, and escapes through a flue with the products of combustion into towers, or scrubbers, filled with coke or bricks moistened with a stream of water; the whole of the acid vapors are thus condensed, and the smoke and heated air pass up the chimney. After the mixture of salt and acid has been heated for some time in the iron pan, and has become solid, it is raked by means of the doors seen in Fig. 53, on to the hearths of the furnaces at each side of the decomposing pan, where the flame and heated air of the fire complete the decomposition into sodium sulphate and hydrochloric acid.

(2) *Soda Ash process.* This process consists (1) in the preparation of sodic carbonate, and (2) in the separation and purification of the same. The first chemical change which the salt-cake undergoes in its passage to soda-ash, is its reduction to sulphide by heating it with powdered coal or slack:

$$Na_2\,SO_4 + C_4 = Na_2\,S + 4\,CO.$$

The second decomposition is the conversion of the sodium sulphide into sodium carbonate, by heating it with chalk or limestone (calcium carbonate):

$$Na_2S + Ca\,CO_3 = Na_2\,CO_3 + Ca\,S.$$

These two reactions are in practice carried on at once; a mixture of ten parts of salt cake, ten parts of limestone, and seven and a half parts of coal being heated in a reverberatory furnace called the *Balling Furnace* (shown in section in Fig. 54, and in elevation in Fig. 55), until it fuses and the above

FIG. 54.

FIG. 55.

decomposition is complete, when it is raked out into iron wheelbarrows to cool. This process is generally termed the *black-ash process*, from the color of the fused mass.

The next operation consists in the separation of the sodium carbonate from the insoluble calcium sulphide and other impurities. This is easily accomplished by *lixiviation*, or dissolving the former salt out in water. On evaporating down the solution, for which the waste heat of the black-ash furnace is used, the heated air passes over a leaden pan (shown in Fig. 54) containing the liquid. On calcining the residue, the soda-ash of commerce is obtained. No less than 200,000 tons of common salt are annually consumed in the alkali works of Great Britain, for the preparation of nearly the same weight of soda-ash, of which the value is about two millions sterling. The soda-ash of commerce contains from 48 to 56 per cent. of pure caustic soda, Na_2O, as carbonate and hydrate, the remainder being impurities, consisting generally of sulphate, sulphite, and chloride. If soda-ash be dissolved, and the saturated solution allowed to stand, large transparent crystals (monoclinic) of the hydrated carbonate, of the formula $Na_2CO_3 + 10H_2O$ separate out; this substance is commonly known as *soda-crystals*, and is much used for softening water for washing purposes.

Hydrogen Sodium Carbonate, or *Bicarbonate of Soda*, $HNaCO_3$, is obtained by exposing the crystallized carbonate in an atmosphere of carbonic acid gas. It is a white crystalline powder, which on heating is readily converted into sodium carbonate. The bicarbonate is chiefly used in medicine, and for the production of effervescing drinks.

The other sodium salts of importance are sodium nitrate or Chili saltpetre, $NaNO_3$, found in Northern Chili in large beds, imported in large quantities, and chiefly used as a manure; sodium sulphate, or Glauber salts, $Na_2SO_4 + 10H_2O$ (in the anhydrous state known as salt-cake); hydrogen sodium sulphate (or bisulphate), $HNaSO_4$; sodium hyposulphite, $Na_2S_2H_2O_4 + 4H_2O$, mentioned under the compounds of sulphur and oxygen (p. 113); the sodium phosphates, mentioned under phosphorus (p. 130); borax, $Na_2O\ 2B_2O_3 + 10H_2O$ (*see* p. 125); sodium sulphide, Na_2S, a soluble salt formed by reducing the sulphate with carbon; sodium silicates, or soluble glass (*see* p. 175).

General Characteristics of the Sodium Compounds. All

the sodium salts, with the single exception of the antimoniate, are soluble in water. The presence of sodium compounds can be detected by the peculiar yellow tinge which they impart to the flame. The spectrum of sodium is distinguished by one fine bright double line, identical in position with the dark solar line called D.

3 and 4. CÆSIUM AND RUBIDIUM. Cæ = 133. Rb = 85·4.

These two metals were discovered in 1860–61 by Bunsen and Kirchoff, by means of spectrum analysis (*see* p. 221). They so closely resemble one another and potassium in their chemical properties, that they had previously been mistaken for the latter well-known metal. They occur widely distributed, although generally occurring in small quantities. They were originally discovered in the mineral water of Dürkheim, but since that time they have been found in many other springs, in several kinds of mica and other old plutonic silicates, as well as in the ashes of several plants, viz. beetroot, tobacco, coffee, and grapes. These metals can be separated from potassium by the greater insolubility of the double chloride which they form with platinum; if a mixture of K, Cæ, and Rb salts be completely precipitated by platinic chloride and the precipitate boiled out with water, the insoluble residue will contain the new metals. Cæsium may be separated from rubidium by the greater solubility of the acid tartrate of the former metal. The salts of cæsium and rubidium are isomorphous with the corresponding potassium compounds. The fused chlorides of these metals are easily decomposed by the galvanic current, and the metallic element deposited. The metals can also be prepared by reduction with carbon, as potassium. Rubidium is a white metal which rapidly undergoes oxidation, its specific gravity is 1·52, and it forms a greenish-blue vapor. Cæsium is the most electropositive of the metals.

5. LITHIUM. *Symbol* Li. *Combining Weight* 7. *Specific Gravity* 0·59.

This metal is prepared by decomposing the fused chloride by electricity; it is of a white color, it fuses at 180°, and is the lightest metal known. The lithium salts were formerly supposed to be very rare, only being known to occur in three or four minerals, but spectrum analysis has shown that this is a widely-distributed substance; it occurs in small quantities in almost all waters, in milk, tobacco, and even in human blood. A spring in Cornwall contains large quantities of this metal in the form of chloride. Lithium in its chemical relations stands between the class of alkaline and alkaline-earth metals, the hydrate, carbonate, and phosphate, being only sparingly soluble in water. All the volatile lithium compounds impart a magnificent crimson tinge to the flame, and the spectrum of this flame exhibits the presence of one bright and very characteristic red line, by means of which the presence of the minutest trace of this substance can be detected with certainty and ease.

6. Ammonium and the Salts of Ammonia.

With the class of alkaline metals the ammoniacal salts may conveniently be considered, as in their chemical properties they present a remarkable analogy with the salts of the alkalies proper. In all these salts the existence of a quasi-metal called *Ammonium*, NH_4, is supposed, and if this substance be substituted for an atom of potassium or sodium in the alkaline salts, a corresponding salt of ammonium is formed thus.:

Potassium chloride, KCl. Ammonium chloride, NH_4Cl.
Potassium sulphate, $\left.\begin{matrix}K\\K\end{matrix}\right\}SO_4$. Ammonium sulphate, $\left.\begin{matrix}NH_4\\NH_4\end{matrix}\right\}SO_4$.

The radical ammonium $\left.\begin{matrix}NH_4\\NH_4\end{matrix}\right\}$ has been prepared in the free state; it is a dark blue liquid possessing a metallic lustre, which can only exist under high pressure, and at a low temperature, and very readily decomposes into ammonia and hydrogen. The most important of the ammoniacal salts, which are all volatile, are ammonium-chloride, or sal-ammoniac, NH_4Cl, originally prepared by subliming camel's dung, but now obtained by neutralizing the ammoniacal water of the gas works (*see* p. 80) with hydrochloric acid, evaporating the liquor to dryness, and by subliming the volatile sal-ammoniac. The sulphate, 2 (NH_4) SO_4, is likewise prepared from gas-liquor by neutralization with sulphuric acid. The carbonate, nitrate, and sulphide of ammonium correspond closely to the same potassium salts. The salts of ammonia can easily be recognized by their giving off an alkaline gas possessing a pungent smell of ammonia when they are heated with caustic lime or a caustic alkali. The acid tartrate and the double platinic chloride are both insoluble, and resemble the corresponding potash compounds so closely that the two sets of salts cannot be distinguished by means of these tests. In order to test for potash in presence of ammoniacal salts, all the latter must first be driven off by heating.*

LESSON XX.

CLASS II. METALS OF THE ALKALINE EARTHS.

These are three in number—1. CALCIUM, 2. STRONTIUM, and 3. BARIUM; they chiefly differ from the preceding metals by forming insoluble carbonates, sulphates, phosphates, and oxalates.

1. CALCIUM. *Symbol* Ca. *Combining Weight* 40. *Specific Gravity* 1·58.

Calcium forms a considerable portion (*see* p. 8) of the plutonic rocks of which the earth is composed, and occurs in very large quantities, forming whole mountain chains of limestone, chalk, gypsum, and mountain limestone. The metal

* Ammonia, NH_3, is only the first term of a series of volatile bodies possessing closely-similar properties and forming definite salts; these bodies will be described in the part relating to organic chemistry.

calcium is obtained by the decomposition of the chloride by the electric current, or by heating the iodide with sodium; it is a light yellow metal which easily oxidizes in the air, and when heated in air it burns with a bright light—lime, CaO, the only oxide of calcium, being formed.

Calcium Oxide, or *Lime*, CaO. Pure lime is obtained by heating white or black marble to redness in a vessel exposed to the air. Lime is prepared on a large scale for building and other purposes, by heating limestone (the carbonate) in kilns by means of coal mixed with the stone; the carbonic acid escapes, and *quick-* or *caustic*-lime remains. Pure lime is a white infusible substance, which combines with water very readily, giving off great heat, and falling to a white powder called calcium *hydroxide*, or *slacked lime*, CaO H_2O. The hydrate is slightly soluble in water, 1 part of it dissolving in 730 parts of cold, but only in 1300 parts of boiling water, and forming lime-water, which, like the hydrate, has a great power of absorbing carbonic acid from the air. It is indeed partly owing to this property that the hardening or setting of mortars and cements made from lime is due. Mortar consists of a mixture of slacked lime and sand; a gradual combination of the lime with the silica occurs, and this helps to harden the mixture. Hydraulic mortars, which harden under water, are prepared by carefully heating an impure lime containing clay and silica; a compound silicate of lime and alumina appears to be formed on moistening the powder, which then solidifies, and is unacted upon by water. Lime is largely used in agriculture, its action being, 1st, to destroy the excess of vegetable matter contained in the soil; and, 2dly, to liberate the potash for the use of the plants from heavy clay soils by decomposing the silicate.

Calcium Carbonate, or *Carbonate of Lime*, Ca CO_3. This salt occurs most widely diffused, as chalk, limestone, coral, and marble; many of those enormous deposits being made up of the microscopic remains of minute sea-animals. Calcium carbonate exists crystalline as calc spar, or Iceland spar (rhombohedral, or hexagonal system, Fig. 45), and also in a different form, arragonite (rhombic, Fig. 46), so that this substance is dimorphous. The carbonate is almost insoluble in pure water, but readily dissolves when the water contains carbonic acid, giving rise to what is termed *temporarily hard water.* Such a water deposits a crust of calcium carbonate on boiling, owing to the escape of the carbonic acid. The well-known evil of boiler crust is caused by these deposits: the formation of such a crust may be checked if not avoided by adding a small quantity of sal-ammoniac to the water, soluble calcium chloride and volatile ammonium carbonate being formed. Water hard with dissolved carbonate may be softened by the addition of lime suspended in water in such quantity that the excess of carbonic acid is neutralized.

Calcium Sulphate, $Ca\ SO_4$. This occurs in nature as a mineral termed Anhydrite, and combined with 2 H_2O as selenite, gypsum, or alabaster. It is soluble in 400 parts of water, and is a very common impurity in spring water, giving rise to what is termed permanent hardness, as it cannot be removed by boiling. Gypsum when moderately heated loses its water, and is then called plaster of Paris; this when moistened takes up two atoms of water again and sets to a solid mass, and is therefore much used for making casts and moulds.

Calcium Chloride, $Ca\ Cl_2$. This soluble salt is formed when limestone or marble is dissolved in hydrochloric acid (*see* p. 70); if the solution be then evaporated, colorless needle-shaped crystals of the hydrated chloride, $Ca\ Cl_2 + 6\ H_2O$, are formed. When these are dried the substance still retains 2 H_2O, and forms a porous mass which takes up moisture with great avidity, and is much used for drying gases. When this mass is more strongly heated it fuses and parts with all its water.

Calcium Fluoride, Fluor spar, $Ca\ F_2$. Found crystallized in cubes in Derbyshire and Cumberland. When heated with sulphuric acid, calcium sulphate and hydrofluoric acid (*see* p. 100) are formed. It is sometimes used as a flux in the reduction of metals, whence its name Fluor Spar is derived. The foregoing are the most important salts of calcium.

Among the remaining compounds of calcium may be mentioned bleaching-powder, or chloride of lime, $Ca\ Cl_2\ Ca\ 2\ ClO$, the properties and mode of preparation of which salt are described on p. 92; calcium phosphate, or bone-phosphate, $Ca_3\ 2\ PO_4$ (*see* p. 129); calcium sulphide, Ca S, an insoluble substance formed in the soda-ash process (*see* p. 164); and calcium penta-sulphide, $Ca\ S_5$, a soluble salt. The spectrum of calcium is a very peculiar one, containing a number of distinct bright lines by which the presence of this metal can be easily ascertained.

2. STRONTIUM. *Symbol* Sr. *Combining Weight* 87·5.

This element occurs in much smaller quantities than calcium, or even barium, being found in only a few mineral species, especially *strontianite* the carbonate, and *celestine* the sulphate. Strontium likewise occurs in minute quantities in certain spring waters. The metal has a yellowish-white color, and is prepared by the action of a current of electricity on the fused chloride. It resembles calcium closely in its properties; its specific gravity is 2·54. When heated in the air it burns, forming the monoxide strontia.

Strontium Oxide, or *Strontia*, Sr O. This oxide is best obtained by decomposing the nitrate by heat; it unites with water, evolving great heat, and forming the hydrate $Sr\ O + 9\ H_2\ O$; this is soluble in water, and absorbs carbonic acid with avidity. The native salts of strontium, viz., the carbonate and sulphate, are insoluble, and serve for the preparation of the remaining salts. The nitrate, $Sr\ 2\ NO_3$, and the chloride, $Sr\ Cl_2$, are soluble in water; these are the only salts of this metal which are employed in the arts, and these are used for the preparation of red fires, as the volatile salts of strontium have the power of coloring the flame crimson. The spectrum of strontium is a very characteristic one (*see* Chromo); the minutest trace of this substance can thus be easily and certainly detected, even in presence of calcium and barium salts.

3. Barium. *Symbol* Ba. *Combining Weight* 137.

Barium compounds occur somewhat more widely dispersed than those of strontium, the two most common barium minerals being the sulphate, or *heavy spar*, and the carbonate, or *witherite*. The metal barium has not yet been obtained in the coherent state, but the metallic powder may be prepared in a similar way as the two former metals, which it closely resembles in its properties.

Barium Monoxide, or *Baryta*, Ba O. The best way of forming this oxide is to decompose the nitrate by heat; it is a grayish porous mass, which fuses at a high temperature, and takes up water with evolution of much heat, forming a crystalline hydrate, $H_2 Ba O_2 + 8 H_2 O$. This hydrate is soluble in twenty parts of cold water, and the solution on exposure to the air rapidly absorbs carbonic acid and becomes milky.

Barium Dioxide, $Ba O_2$. When baryta is gently heated in a current of oxygen gas, the two substances combine together to form a dioxide containing twice as much oxygen as baryta; this additional atom of oxygen is, however, evolved at a higher temperature, and it has been proposed to use this decomposition for the manufacture of oxygen from the air. For this purpose, as soon as the dioxide BaO_2 has been reduced to Ba O, the temperature is lowered, and air passed over the baryta; this again takes up oxygen, passing into $Ba O_2$, which again is decomposed by a higher temperature. This interesting process has, however, been found not to work in practice. There are no salts known corresponding to this oxide.

Barium Chloride, $Ba Cl_2$. This soluble salt is one of the most important compounds of barium; it crystallizes in flat scales containing two atoms of water. It may be prepared by dissolving the native carbonate in hydrochloric acid, and it is largely used as a precipitant for sulphuric acid. *Barium Sulphate*, $Ba SO_4$, occurs native and crystalline as heavy spar; specific gravity 4·6 (whence the name Barium, from βαρὺς (heavy). It is one of the most insoluble salts known, and falls as a white crystalline precipitate when any soluble barium salt is brought into a solution of a sulphate. It is used as a paint, and the precipitated salt is termed *blanc fixé*, whilst the native heavy spar, when ground, is largely used to adulterate white lead.

The other more important salts of barium are the nitrate $Ba\ 2\ NO_3$, a soluble salt; the sulphide, a soluble salt, BaS, obtained by heating heavy spar with coal and dissolving in water; the carbonate, $Ba\ CO_3$, an insoluble substance, occurring native as witherite: the silicofluoride and the phosphate are insoluble salts, whilst strontium silicofluoride is soluble in water. The volatile salts of barium have the power of communicating a peculiar green color to the flame, and the spectrum of barium contains a number of characteristic green lines, by means of which the presence of minute traces of this substance can be detected. (*See Chromo.*)

Class III. Metals of the Earths.

1. Aluminium. *Symbol* Al. *Combining Weight* 27·4. *Specific Gravity* 2·6.

This metal occurs in large quantities combined with silicon and oxygen in felspar and all the older rocks, and also in clay, marl, slate, and in many crystalline minerals. Metallic aluminium is obtained by passing the vapor of aluminium chloride over metallic sodium. It has recently been manufactured on a large scale both in England and France, and, from its lightness (specific gravity 2·6) and its bright lustre, it has been used for optical purposes as well as for ornamental work.

Alumina, $Al_2\ O_3$, specific gravity, 3·9. This is the only oxide of aluminium known. It occurs native in a nearly pure and crystalline state as corundum, ruby, sapphire, and emery. Alumina is prepared by adding ammonia to a solution of alum; a white precipitate of the hydroxide falls down, and this on being heated yields a white amorphous powder of pure alumina. This substance is attacked with difficulty by acids, but the hydrate is easily soluble in acids and in the fixed caustic alkalies. Alumina acts as a weak base; the commonest aluminium salts are the alums, and their solutions have an acid re-action. Alumina is largely used in dyeing and calico-printing as a mordant, as it has the power of forming insoluble compounds called lakes with vegetable coloring matter, and thus renders the color permanent by fixing it in the pores of the cloth so that it cannot be washed out; such colors are termed *fast*.

Aluminium chloride, $Al_2 Cl_6$, is a volatile white solid body, obtained by heating a mixture of alumina and charcoal in a current of chlorine gas; it is used in the manufacture of the metal.

The soluble *aluminium sulphate*, $Al_2 3SO_4$, is prepared on a large scale for the use of the dyer by decomposing clay by acting upon it with sulphuric acid; the solid mixture of silica and aluminium sulphate thus obtained goes by the name of alum-cake. The most useful compounds of alumina are, however, the alums, a series of double salts, which aluminium sulphate forms with the alkaline sulphates. *Common potash alum*, or *aluminium potassium sulphate*, has the composition $Al_2 K_2 4SO_4 + 24H_2 O$, and crystallizes in regular octohedra (Fig. 38). It may be prepared by dissolving the two sulphates together, and allowing the compound salt to crystallize, but it is usually obtained from the decomposition of a shale or clay containing iron pyrites, $Fe S_2$; this substance gradually undergoes oxidation when the shale is roasted, absorbs oxygen from the air producing sulphuric acid, which unites with the alumina of the clay, and on the addition of a potassium compound, alum crystallizes out. A salt called ammonium alum, and containing $N H_4$, instead of K, is at present prepared on a large scale, the ammonia liquor of the gasworks, together with sulphuric acid, being added to the burnt shale, instead of a potash salt.

There are a large number of other alums known, in which the isomorphous sesquioxides of iron, chromium, and manganese, are substituted for the alumina in common alum; all these alums occur in regular octohedra, and cannot be separated by crystallization when present in solution together.

Clay is an aluminium-silicate resulting from the disintegration and decomposition of felspar by the action of air and water, the soluble alkali being washed away. Kaolin or porcelain clay is the purest form of disintegrated felspar, containing no iron or other impurities. There are many very beautifully crystalline minerals, consisting of aluminium silicates combined with silicates of the metals of the alkalies and alkaline earths; amongst others, garnet, idocrase, mica, lepidolite, &c. Some silicates, such as stilbite, analcime, &c., retain water of crystallization, and are termed zeolites. Aluminium salts can be detected when in solution by giving with ammonia a white precipitate, insoluble in excess, but soluble in caustic soda; and by assuming a blue color when moistened with cobalt solution and heated before the blowpipe.

GLASS, PORCELAIN, AND EARTHENWARE.

The silicates of the alkalies are, as we have seen, soluble in water and non-crystalline; those of the alkaline earths are soluble in acid and crystalline, whilst compounds of the two are insoluble in water and acids, and do not assume a crystalline form. Such a compound when fused is termed a *glass*. There are four different descriptions of glass used in the arts, differing in their chemical composition and exhibiting corresponding differences in their properties. (1) Crown or window and plate-glass, composed of silicates of sodium and calcium. (2) Bohemian glass, consisting of silicates of potassium and calcium. (3) Flint-glass or crystal, containing silicates of potassium and lead, and (4) Common green bottle-glass, composed of silicates of sodium, lime, iron, and aluminium. The first and third of these kinds of glass are easily fusible, whilst the second or potash glass is much more infusible; the addition of oxide of lead increases the specific gravity, and the lustre of the glass, as well as its fusibility. The common glass articles of household use are generally made of flint glass, whilst for chemical apparatus a soda-lime glass is to be preferred. The potash-lime glass is much employed where a difficultly fusible, or hard, glass is needed, as for instance in the manufacture of combustion tubes for organic analysis (*see* p. 236). The fourth description of glass is an impure mixture of various silicates, employed for purposes in which the colors and fineness of the glass is not of consequence.

In the preparation of all the fine qualities of glass, great care is requisite in the selection of pure materials, as well as in the processes of manufacture; generally the materials are melted together with a quarter to half their weight of "cullet" or broken glass of the same kind. After the glass articles have been blown or cast, they must all be exposed to the process of "annealing," or slow cooling, otherwise they are so brittle as to be perfectly useless, breaking with the slightest touch, owing to the irregular contraction of the different parts brought about by rapid cooling. The following table shows the composition of the chief varieties of glass.

INGREDIENTS FOR VARIOUS GLASSES.

Crown Glass.		*Mirror Plate.*	
Quartz Sand . .	100 parts.	Pure Sand . .	100 parts.
Mild Lime . .	36 „	Soda Ash . .	35 „
Soda Ash . .	24 „	Mild Lime . .	5 „
Sodium Sulphate	12 „	Arsenic Trioxide	$\frac{1}{6}$ „
Arsenic Trioxide	$\frac{1}{8}$ „	Cullet. . . .	100 „
Cullet . . .	100 „		
Bohemian Glass.		*Flint Glass.*	
Pure Sand . .	100 „	Pure Sand . .	100 „
Pure Pearlashes	60 „	Red Lead . .	20 „
Chalk . . .	8 „	Pearlash. . .	40 „
Cullet . . .	40 „	Nitre	2 „
Manganese Dioxide	$\frac{3}{4}$ „	Cullet .	50 to 100 „

Colored Glass. Certain metallic oxides possess the power of coloring glass when they are added in small quantity. Thus ferrous oxide produces a deep green color (bottle glass), whilst the oxides of manganese impart a purple tint to glass; these facts are made use of in the preparation of colorless glass, for as it is difficult to obtain materials perfectly free from iron, which imparts a green color, a small quantity of manganese dioxide is added to the mixture, and the violet color thus produced is complementary to the green, and a nearly colorless glass is the result. The addition of arsenic Trioxide effects the same end by oxidizing the ferrous to ferric-oxide. The colors of precious stones are imitated by adding certain oxides to a brilliant lead glass called "paste;" thus the blue of the sapphire is given by a small quantity of cobalt oxide, whilst cuprous oxide imparts a ruby-red color, and ferric oxide a yellow color resembling topaz.

Porcelain and Earthenware.—The various forms of porcelain and earthenware consist of silicate of aluminium, in fact clay, in a more or less pure state, covered with some substance which fuses at a high temperature, and forms a glaze, giving a smooth surface and binding the material together, and thus counteracting the porous nature of the baked clay. For the manufacture of porcelain the finest white or China clay is used, resulting from the gradual decomposition of felspar, whilst for the common earthenware a colored clay may be employed. The glaze used for porcelain is generally finely powdered felspar, the biscuit or porous ware being dipped into a vessel containing this substance suspended in water, and then strongly fired. The articles thus coated can be used for chemical purposes, as this glaze withstands the action of acids. For earthenware the so-called "salt glaze" is used; the mode of obtaining this glaze consists in throwing some common salt into the furnaces containing the strongly heated ware, when the salt is volatilized and undergoes decomposition on the heated surface, causing a deposit of a fusible silicate upon it, and rendering the ware impervious to moisture.

LESSON XXI.

CLASS IV. (1.) MAGNESIUM. (2.) ZINC. (3.) CADMIUM.

1. MAGNESIUM. *Symbol* Mg. *Combining Weight* 24·0. *Specific Gravity* 1·74.

This metal occurs in large quantities as carbonate, along with calcium carbonate, in *dolomite* or mountain limestone, and also in sea-water and certain mineral springs, as chloride and sulphate. The metal itself has only recently been prepared in quantity; it is best obtained by heating magnesium chloride with metallic sodium, sodium chloride and metallic magnesium being formed. This metal is of a silver-white color and fuses at a low red heat, it is volatile, and may be easily distilled at a bright red heat; when soft it can be pressed into wire, and with care it may be cast like brass, although when strongly heated in the air it takes fire and burns with a dazzling white light, with the formation of its only oxide, magnesia. The light emitted by burning magnesium wire is distinguished for its richness in chemically active rays, and this substance is therefore employed as a substitute for sunlight in photography.

Magnesium does not oxidize in dry air; it is only slowly acted upon by cold water, but more rapidly by hot water; it rapidly dissolves in sulphuric and hydrochloric acids, with evolution of hydrogen.

Magnesium Oxide, or *Magnesia*, Mg O. A light white amorphous infusible powder, obtained by heating the carbonate or nitrate. It unites with acids to form the magnesium salts, but it does not possess a strong alkaline reaction. The most important salts of magnesium are:—*Magnesium Chloride*, $Mg\,Cl_2$, a fusible salt obtained by heating a solution of magnesia in hydrochloric acid with an equal quantity of salammoniac; on fusion, the latter salt volatilizes, and the magnesium chloride remains behind. *Magnesium Sulphate*, $Mg\,SO_4 + 7H_2O$: this is a soluble substance known as Epsom Salts; it occurs in a spring in Surrey, and contains seven atoms of water of crystallization; it is now largely made from dolomite by separating the lime with the sulphuric acid. Magnesium sulphate forms, with the alkaline sulphates, double salts, in which the alkaline sulphate takes the place of one atom of water of crystallization; thus, $MgSO_4K_2SO_4 + 6H_2O$ is the potash double salt.

Magnesium carbonate, Mg CO_3, is an insoluble compound, occurring as a crystallized mineral termed magnesite. The magnesia alba of the shops is a varying mixture of carbonate and hydrate, made by precipitating a hot solution of magnesium sulphate with sodium carbonate. Magnesium sulphide is not formed in the wet way. Magnesium resembles in many respects the metals of the alkaline earths, but it may be distinguished from these by the solubility of the carbonate in ammonium chloride, as well as by the ready solubility of the sulphate in water. Magnesium forms an insoluble double phosphate with ammonia, Mg $NH_4PO_4 + 6H_2O$, and it is in this form that the metal is usually estimated.

2. ZINC. *Symbol* Zn. *Combining Weight* 65·2. *Specific Gravity* 6 8 to 7·2.

Zinc is an abundant and useful metal, closely resembling magnesium in its chemical characters; but it is much more easily extracted from its ores than this latter metal. The chief ores of zinc are the sulphide or *blende*, the carbonate or *calamine*, and the *red oxide*. In order to extract the metal the powdered ore is roasted, or exposed to air at a high temperature, so as to convert the sulphide of carbonate or carbonate into oxide; the roasted ore is then mixed with fine coal or charcoal and strongly heated in crucibles or retorts of peculiar shape; the zinc oxide is reduced by the carbon, carbon monoxide comes off, and the metallic zinc distils over, and is easily condensed.

Zinc is a bluish-white metal, exhibiting crystalline structure; it is brittle at the ordinary temperature, but when heated to about 130°, it may be rolled out or hammered with ease, whilst if more strongly heated to 200°, it is again brittle, and may be broken up in a mortar. Zinc melts at 423°, and at a bright red heat it begins to boil and volatilizes, or if air be present it takes fire and burns with a luminous greenish flame, forming zinc oxide. Zinc is not acted upon by moist or dry air, and hence it is largely used in the form of sheets, and is employed as a protecting covering for iron, which when thus coated, is said to be *galvanized*.

Zinc Oxide, Zn O, is the only known compound of this metal with oxygen, and is obtained by burning the metal, or by precipitating a soluble zinc salt with an alkali, and heating the precipitate. Zinc oxide is an insoluble white amorphous powder, which when heated becomes yellow, but loses this color on cooling; it dissolves easily in acids, giving rise to the zinc salts. The most important salts of zinc are: *Zinc sulphate*, Zn SO_4 + 7 H_2O, a soluble salt, crystallizing in long prisms, and commonly called white vitriol; this salt is isomorphous with magnesium sulphate, and like the latter salt it forms a series of double salts with alkaline sulphates. *The chloride*, Zn Cl_2, a white soluble, deliquescent substance, formed by burning zinc in chlorine; or better, by dissolving the metal in hydrochloric acid. *The sulphide*, Zn S, occurs as a crystalline mineral called blende, generally colored, from presence of iron and other impurities; it is obtained artificially as a white gelatinous precipitate, insoluble in acetic, but soluble in a mineral acid, formed when an alkaline sulphide is added to a soluble zinc salt. *The carbonate*, Zn CO_3, an insoluble substance, occurring native as calamine: it cannot be prepared by precipitating a solution of zinc salt by an alkaline carbonate, as a quantity of oxide is precipitated along with the carbonate. The salts of zinc can be distinguished by the solubility of the oxide in excess of both potash and ammonia; by the white sulphide insoluble in acetic acid, and by the green color which a solution of cobalt chloride imparts to these salts when heated before the blowpipe.

3. CADMIUM. *Symbol* Cd. *Combining Weight* 112. *Specific Gravity* 8·6.

This is a comparatively rare metal, occurring in certain zinc ores. In its chemical relations it closely resembles zinc. It is, however, more volatile than the other metal, and therefore distils over first in the preparation of zinc. Cadmium is a white ductile metal, melting at 315°; it may be easily distinguished and separated from zinc by yielding a bright yellow sulphide insoluble in hydrochloric acid. The metal takes fire when heated in the air, forming a brown oxide, Cd O. The chloride and sulphate are soluble well-crystallizing salts.

INDIUM. *Symbol* In. Combining Weight 74·0. A metal lately discovered by means of spectrum analysis, in certain zinc ores. Its compounds impart a blue color to flame, and its spectrum is characterized by two fine indigo-colored lines (*see Chromo*). The properties of indium and its compounds have as yet not been fully examined; it is, however, a soft white metal resembling cadmium.

CLASS V. IRON CLASS.—1, MANGANESE. 2, IRON. 3, COBALT. 4, NICKEL. 5, CHROMIUM. 6, URANIUM.

1. MANGANESE. *Symbol* Mn. *Combining Weight* 55. *Specific Gravity* 8·0.

Manganese occurs in nature as an oxide, and it can be obtained, though with difficulty, in the metallic state by heating the oxide very strongly with charcoal. The metal is of a reddish-white color; it is brittle, and hard enough to scratch glass. It decomposes water at the ordinary temperature with evolution of hydrogen; it cannot be preserved in the air without undergoing oxidation, and must be kept under naphtha, or in a sealed tube; it is slightly magnetic, and, like iron, combines with carbon and silicon. Metallic manganese is not used in the arts, but an alloy of this metal and iron is now made on a large scale, and used in the manufacture of steel. Some of its oxides are used for the purpose of evolving chlorine from hydrochloric acid, and also for tinting glass a purple color.

Manganese forms several well-characterized oxides: (1) *Manganous oxide*, or *manganese monoxide*, MnO, is a basic body, furnishing the series of well-known manganous salts, in which the oxygen is replaced by its equivalent of another element, or of a salt radical; thus, MnO, $MnCl_2$, $MnSO_4$, $Mn_2\ NO_3$; (2) *manganic oxide*, or *manganese sesquioxide*,

Mn_2O_3, which also forms salts, but of a much less stable character, and occurs in nature as the mineral braunite; (3) *red*, or *mangano-manganic oxide*, Mn_3O_4, a neutral body, corresponding to the magnetic oxide of iron, and occurring in nature as hausmannite; (4) *black oxide*, or *manganese dioxide*, MnO_2, a neutral substance, occurring as the ore of manganese in the minerals pyrolusite and varvacite; (5) *manganese heptoxide*, Mn_2O_7, a dark-green heavy liquid obtained by the action of strong cold sulphuric acid upon potassium permanganate.

Manganous Oxide, MnO, is a greenish powder, obtained by heating the carbonate in absence of air; it forms, with acids, a series of pink-colored salts, and rapidly absorbs oxygen, passing into a higher state of oxidation. The hydrate is precipitated as a white gelatinous mass, when an alkali is added to a solution of manganous salt; this, however, rapidly becomes brown, owing to absorption of oxygen. Of the manganous salts, the chief soluble ones are, the *sulphate*, $MnSO_4 + 5H_2O$, a pink-colored crystalline salt, prepared by acting on the dioxide with sulphuric acid, oxygen gas being evolved,

$$MnO_2 + H_2SO_4 = MnSO_4 + O + H_2O\text{—)}$$

the *chloride*, $MnCl_2 + 4\,H_2O$, is a salt obtained by crystallization from the residues in the manufacture of chlorine from the dioxide and hydrochloric acid,

$$MnO_2 + 4\,HCl = MnCl_2 + 2H_2O + Cl_2.$$

Among the insoluble manganous salts of importance are the *sulphide*, MnS, obtained as a flesh-colored precipitate by the addition of an alkaline sulphide to a soluble manganous salt; and the *carbonate*, $MnCO_3$, which occurs native, crystallizing like calcspar in rhombohedra, and is prepared as a white powder by precipitating a manganous salt by an alkaline carbonate.

Manganese Sesquioxide, $Mn_2\,O_3$, exists in nature as braunite, and may be prepared artificially by exposing manganous oxide to a red heat. It forms a series of somewhat unstable salts, of which the manganese alum is one of the most interesting, being isomorphous with common alum, in which Mn_2O_3 is substituted for Al_2O_3.

Manganese Dioxide, MnO_2, is the common black ore of manganese, and is termed pyrolusite by mineralogists; it can be artificially formed by adding a solution of bleaching powder to a manganous salt. This substance yields one-third of its oxygen when heated to redness (*see* p. 11), forming the red oxide, $3\,MnO_2 = Mn_3O_4 + O_2$, and gives up half its oxygen when heated with sulphuric acid (*see* above.) It is largely used for the manufacture of chlorine.

Manganic, and Per-manganic Acids. When an oxide of manganese is fused with caustic alkali, a bright green mass is formed, which yields a dark green solution; this contains potassium manganate, K_2MnO_4, which may be crystallized, and is isomorphous with the corresponding sulphate and chromate. If this green solution be allowed to stand, it slowly changes to a bright purple color, and hydrated manganic dioxide is deposited, hence its common name of mineral chamelion: it then contains a new salt in solution, viz. a per-manganate, $KMnO_4$, which may be obtained in the crystalline state by evaporation, and is isomorphous with potassium perchlorate. The presence of a few drops of acid at once effects this decomposition on the green solution.

The manganates and per-manganates readily give up a part of their oxygen in presence of organic matter, and they are now largely used as disinfectants, and known as Condy's liquids, as well as being employed in the laboratory for the purpose of volumetric analysis. Manganese is chiefly characterized by the flesh-colored sulphide, and by the formation of the green sodium manganate, a most delicate reaction.

LESSON XXII.

2. IRON. *Symbol* Fe. *Combining Weight* 65. *Specific Gravity* 7·8.

Iron is of all metals the most important to mankind. The uses of iron were long unknown to the human race, the age of iron implements being preceded by those of bronze and stone. Pure metallic iron exists only in very small quantity on the earth's surface, almost entirely occurring in those peculiar structures known as meteoric stones, which possess an extra-terrestrial origin.

The process of obtaining iron from its ores is a somewhat difficult one, and requires an amount of knowledge and skill which the early races of men did not possess. The iron of commerce exists in three different forms, exhibiting very different properties, and possessing different chemical constitutions: 1, *wrought iron;* 2, *cast iron;* 3, *steel.*

The first is nearly pure iron, the second is a compound of iron with varying quantities of carbon and silicon, and the third a compound of iron with less carbon than that needed to form cast iron. The modes of manufacture of these three kinds of iron are essentially different, and will be best understood when the properties of the metal have been described.

Pure iron in the form of powder may be obtained by reducing the oxide, moderately heated in a current of hydrogen; it must, however, be retained in an atmosphere of hydrogen, as finely-divided iron takes fire and burns to oxide when exposed to the air. A button of pure iron may be prepared by exposing fine iron wire mixed with some oxide of iron to a very high temperature in a covered crucible, the oxide retaining the traces of impurity which the wire contained. Iron has a bright white color, and is remarkably tough, though soft, an iron wire two mm. in thickness not breaking until when weighted with 250 kilogs. The pure metal crystallizes in cubes; iron which has been uniformly hammered exhibits, when broken, a granular and crystalline structure: this structure becomes, however, fibrous when the iron is rolled into bars, and the more or less perfect form of the fibre determines to a great extent the value of the metal. This fibrous texture of hammered bar iron undergoes a change when exposed to long continued vibration, the iron returning to its original crystalline condition; and many accidents have occurred in the sudden snapping of railway axles, owing to this change from the fibrous to the granular texture. Wrought iron melts at a very high temperature; but as it becomes soft at a much lower point, it can be easily worked, especially as, when hot, it possesses the peculiar property of "welding;" that is, the power of uniting firmly when two clean surfaces of hot metal are hammered together.

Iron and certain of its compounds are strongly magnetic, but the metal loses this power when red-hot, regaining it upon cooling. A solid mass of iron does not oxidize or tarnish in dry air, at the ordinary temperature, although iron powder takes fire spontaneously; but if heated, it oxidizes with the production of black scales of oxide, and when more strongly heated in the air, or plunged into oxygen gas, it burns with the formation of the same black oxide. In pure water, iron does not lose its brilliancy; but if a trace of carbonic acid is present, and access of air is permitted, the iron begins at once to oxidize at the surface, or to rust, forming a hydrated sesquioxide. Iron decomposes steam at a red heat, liberating hydrogen (*see* p. 15), and forming the black oxide produced by the combustion of iron in oxygen. The oxides of iron are four in number: (1) the monoxide, or ferrous oxide, FeO, from which the green or ferrous salts are derived; (2) the sesqui or ferric-oxide, Fe_2O_3, yielding the yellow ferric salts; (3) the magnetic, or black oxide, Fe_3O_4, which does not form any definite salts; (4) ferric acid, H_2FeO_4, a weak acid, forming colored salts with potassium.

Ferrous Oxide, or *Iron Monoxide*, FeO. This substance has not been prepared in the pure state, owing to the great readiness with which it absorbs oxygen, passing into the higher oxides. Hydrated ferrous oxide is thrown down as a white precipitate, when potash or soda is added to a soluble ferrous salt; this white precipitate can only be obtained in complete absence of oxygen, as it at once absorbs this gas, yielding a greenish-brown precipitate of a higher oxide. This oxide colors glass green (*see* p. 176), and gives the peculiar tint to common bottle-glass. The most important of the ferrous salts are—

Ferrous Sulphate, $Fe\,SO_4 + 7\ H_2O$. This soluble salt, sometimes called green vitriol, is obtained by dissolving (1) metallic iron, or (2) ferrous sulphide, in sulphuric acid, or by the slow oxidation of pyrites, $Fe\,S_2$.

$$(1)\ Fe + H_2SO_4 = Fe\,SO_4 + H_2$$
$$(2)\ FeS + H_2SO_4 = Fe\,SO_4 + H_2S.$$

The solution thus obtained yields on evaporation large green crystals of the salt. It is largely used in the manufacture of several black dyes, and is one of the constituents of writing-ink. Like ferrous oxide, this salt easily takes up oxygen, producing a new salt called ferric sulphate.

Ferrous Chloride, $Fe\,Cl_2$. When dry hydrochloric acid gas is passed over hot metallic iron, ferrous chloride and hydrogen are formed; the hydrated chloride is also produced when iron is dissolved in aqueous hydrochloric acid, green crystals being deposited, having the composition $Fe\,Cl_2 + 4\ H_2O$.

Ferrous Carbonate, $Fe\,CO_3$. This is an insoluble compound, and occurs largely as a mineral, constituting the clay iron-stone; the ore of iron from which a large proportion of our iron is prepared.

Ferrous Sulphide, $Fe\,S$, an invaluable compound, formed by fusing equivalent quantities of sulphur and iron together, is employed in the laboratory for the generation of sulphuretted hydrogen (*see* p. 114). A disulphide, $Fe\,S_2$, called iron pyrites, is found in large quantities, and is much used in the production of sulphuric acid (*see* p. 111).

Ferric Oxide, $Fe_2\,O_3$. This oxide occurs native, as the minerals red hæmatite and specular iron ore, whilst, combined with water, it forms brown hæmatite. It may be readily prepared artificially by heating ferrous sulphate to redness; or by adding solution of ammonia or caustic potash to a solution of a ferric salt, when the hydrated oxide falls down as a bulky brownish-red powder, which dissolves in acids, forming the ferric salts: thus, when acted upon by sulphuric acid, *ferric sulphate*, $Fe_2\,3SO_4$, is produced; and by hydrochloric acid, *ferric chloride*, $Fe_2\,Cl_6$. Of the ferric salts, the chloride is the most important; the anhydrous salt forms in brilliant red crystals when chlorine gas is passed over heated metallic iron. Solutions of the ferric salts can be reduced by various deoxidizing agents to the corresponding ferrous salts, whilst these latter, in contact with an oxidizing agent, pass into the ferric salts. The ferrous salts are distinguished by their light green color, by their solutions giving (1) a white precipitate, with caustic alkalies; (2) a light blue precipitate, with potassium ferrocyanide, which rapidly becomes dark: whilst the ferric salts are yellow-colored, and their solutions yield (1) a deep reddish-brown precipitate, with the caustic alkalies; and (2) a deep blue precipitate, with potassium ferrocyanide. Ferrous oxide and the ferrous salts are magnetic, whilst the ferric oxide and salts are not magnetic.

The Magnetic, or *Black Oxide*, $Fe_3\,O_4$, occurs native, crystallyzed in cubes, and, as the mineral loadstone, it constitutes one of the most valued ores of iron. It is the oxide formed when iron is oxidized at a high temperature in the air, in oxygen, or in aqueous vapor. A corresponding sulphide, Fe_3S_4, is also magnetic.

Ferric Acid. The potassium salt of this acid is prepared by fusing ferric oxide and nitre together; the mass yields, with water, a purple-colored solution, and contains potassium ferrate, $K_2\,FeO_4$. It is an exceedingly unstable substance; neither the acid $H_2\,Fe\,O_4$, nor the oxide $Fe_2\,O_6$ have been prepared.

Manufacture of Iron. The oldest method of manufacturing wrought iron was to reduce it at once from the ore by heating in a wind-furnace with charcoal or coal, and to hammer out the spongy mass of iron thus obtained. This plan can only be economically employed on a small scale, and with the purest forms of iron ore, and has been superseded by a more complicated method, applicable, however, to all kinds of iron ore. This consists in the formation of cast iron as the first product, and the subsequent separation of the carbon and silicon which the cast iron contains. Cast iron is manufactured in England chiefly from clay iron-stone, which generally occurs in masses, situated in the immediate neighborhood of a coal seam. The clay iron-stone (ferrous carbonate, with clay) is first roasted, in which operation the carbonic acid is driven off, and ferric oxide formed, the ore afterwards being thrown, together with coal and limestone, into a blast furnace, the construction of which is seen in Fig. 56. It has the shape of a double cone (A B, Fig. 56), built of strong fire-

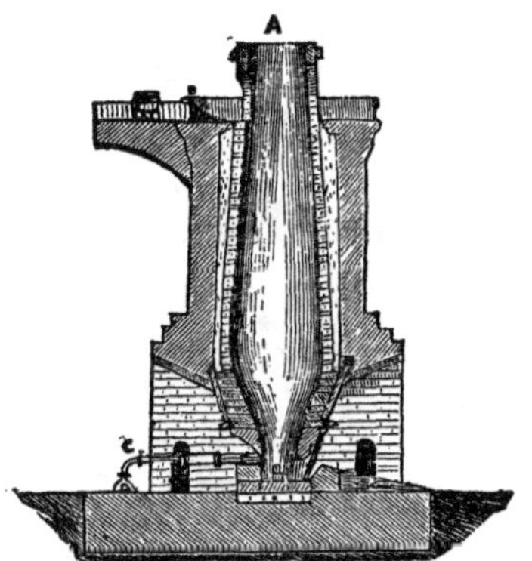

FIG. 56.

brick and masonry, and is about fifty feet in height, and fifteen to eighteen feet in width at the broadest part. The furnace is closed at the bottom, the air necessary for the maintenance of the combustion being supplied in a powerful blast, blown through pipes called *tuyères* (C), whilst the mixture of fuel and ore, being cast in at the top of the furnace (D), is added continually as the burning mass sinks down, and the molten mass is drawn off at the bottom so that one furnace often does not stop working for several years. At the lowest part of the structure is the earth (E), where the melted metal and fused slag collect, the former being occasionally tapped from the bottom of the earth, and cast into pigs in moulds made in the sand, whilst the lighter slag, which swims on the surface of the metal, runs continually out from an opening at the upper part of the hearth.

The first chemical change which the roasted iron ore, or impure ferric oxide, undergoes in its passage from the top to the bottom of the furnace, is its reduction to a porous mass of metallic iron, by the carbonic oxide gas proceeding from the lower layers of burning coal; the temperature of this portion of the furnace is, however, much too low to melt the iron, and it therefore sinks down unchanged, together with the clay and limestone, until it reaches a point at which the heat is greater. Here the second change occurs; viz. the clay, sand, and other impurities of the ore unite with the limestone to form a fusible silicate or slag, whilst the heated metal coming in contact with carbon, unites at once with it to form cast iron, a fusible compound, which runs down to the bottom of the furnace. This, in passing through the hottest portion of the furnace, reduces the silica, with which it meets, to silicon, and, combined with this, it forms cast iron.

The properties and appearance of cast irons vary much with the quantity of carbon and silicon which they contain; for cast iron is not a definite chemical compound of these elements with iron. The carbon is found in cast iron, (1) as scales of graphite, giving rise to mottled cast iron; and (2) in combination, forming white cast iron. Sometimes sulphur and phosphorus are also found in cast iron, but these must be considered as impurities. A great saving of fuel in the working of blast furnaces has lately been effected by employing the heat of combustion of the waste gases—which usually escape and burn at the top of the furnace—to raise the temperature of the blast of air supplying the furnace.

In order to obtain *wrought* from cast iron, the latter must undergo the processes of "refining" and "puddling." These consist essentially in burning out the carbon, silicon, sulphur, and phosphorus, by exposing the heated metal to a current of air in a reverberatory furnace; the melted cast iron becomes first covered with a coat of oxide, and gradually thickens so as to allow of its being rolled into large lumps or balls. During this process the whole of the carbon escapes as carbonic oxide, and the silicon becomes oxidized to silica, which unites with the oxide of iron, and forms a fusible slag; any phosphorus or sulphur contained in the pig iron is also oxidized in this process. The ball is then hammered to give the metal coherence, and to squeeze out the liquid slag, and the mass is afterwards rolled into bars or plates.

Another interesting branch of the iron trade is the manufacture of steel; this useful substance is formed when bars of wrought iron are heated to redness for some time in contact with charcoal: the bar is then found to have become fine-grained instead of fibrous, the substance is more malleable and more easily fusible than the original bar iron, and is found to contain carbon varying in amount from one to two per cent. Steel possesses several important properties, especially the power of becoming very hard and brittle when quickly cooled, which fits it for the preparation of cutting-tools, &c.; these are, however, generally made of bar-steel, which has been previously fused and cast into ingots.

A new and very rapid mode of preparing cast steel, which is both of high scientific interest and industrial importance, is that known as the Bessemer process. This process consists in burning out all the carbon and silicon in cast iron by passing a blast of atmospheric air through the molten metal, and then in adding such a quantity of pure cast iron to the wrought iron thus prepared as is necessary to give carbon enough to convert the whole mass into steel; the melted steel is then at once cast into ingots. In this way six tons of cast iron can at one operation be converted into steel in twenty minutes. The Bessemer steel is now largely manufactured for railway axles and rails, for boiler-plates, and other purposes, for which it is much more fitted than wrought iron, so that this process bids fair to revolutionize the old iron industry.

LESSON XXIII.

3. COBALT. *Symbol* Co. *Combining Weight* 58·7. *Specific Gravity* 8·5.

Cobalt is a reddish-white, very tenacious metal, which is as infusible as iron, and, like the latter metal, is strongly magnetic. It is not found native, but occurs in combination with arsenic and sulphur, as two distinct minerals. The metal dissolves slowly in sulphuric and hydrochloric acids with evolution of hydrogen. The cobalt compounds are distinguished for the brilliancy of their color; they are employed as pigments, and they impart a magnificent blue tint to glass. There are two oxides of cobalt, the monoxide, CoO, and the sesquioxide, $Co_2 O_3$: the former, on solution in acids, forms the protosalts of cobalt, which are pink when hydrated, and blue when anhydrous; whilst the sesquioxide does not form any salts. The *monoxide*, CoO, is obtained as a brown powder by carefully heating the rose-colored hydrate, precipitated by potash in solutions of cobalt; and the *sesquioxide*, Co_2O_3, is prepared by adding a solution of bleaching-powder to a soluble protosalt. The most important salts of cobalt are the *Chloride*, $Co Cl_2$, obtained by acting on the oxide, or on the metallic ore with hydrochloric acid; the solution yields on evaporation pink crystals of the hydrated chloride, or, if further heated, blue crystals of the anhydrous salt. The *Nitrate* and *Sulphate* are also soluble salts; the latter is isomorphous with magnesium sulphate. *Cobalt sulphide*, Co S, is a black powder, insoluble in dilute acids. Cobalt can be easily recognized by the deep blue tint which very minute traces impart to glass, or to a borax bead, made by fusing borax into a colorless mass on the loop of a platinum wire.

4. NICKEL. *Symbol* Ni. *Combining Weight* 58·7. *Specific Gravity* 8·8.

Nickel occurs in large quantities, combined with arsenic, as *kupfernickel;* also together with cobalt in *speiss;* and it is now prepared in considerable quantities for the manufacture of German silver, an alloy of nickel, zinc, and copper. Nickel is a white, malleable and tenacious metal; it melts at a somewhat lower temperature than iron, and is strongly magnetic, but loses this property when heated to 350°. There are two oxides of nickel, the *monoxide*, Ni O, and the *sesquioxide*, Ni_2O_3: the former of these gives rise to the nickel salts, which possess a peculiar apple-green color. The monoxide is obtained by heating the nitrate or carbonate, or by precipitating a soluble nickel salt with caustic potash, and heating the apple-green hydrate, which is thrown down. The sesquioxide is a black powder, prepared by adding a solution of bleaching-powder to a soluble nickel salt. The important soluble nickel salts are the *sulphate*, $NiSO_4$, + $7H_2O$, crystallizing in green prisms; the *nitrate*, Ni 2 (NO_3); and the *chloride*, $NiCl_2$. Like cobalt, nickel forms a black sulphide, Ni S, insoluble in dilute acids. The nickel salts may be distinguished from those of the former metal by imparting a reddish-yellow color to the borax bead, as well as by their green color.

5. CHROMIUM. *Symbol* Cr. *Combining Weight* 52·5. *Specific Gravity* 6·8.

Chromium is a substance whose compounds do not occur very widely distributed, or in large quantities, but they are, nevertheless, much employed in the arts as pigments, many of them possessing a fine bright color. The chief ore of this metal is a compound of the oxides of chromium and iron, called chrome iron-stone, found in America, Sweden, and the Shetlands; a compound of chromium, lead, and oxygen, called lead chromate, is also found in some quantity. Pure chromium appears to be the most infusible of all the metals, as it cannot be melted at a temperature sufficient to fuse and volatilize platinum; it has, however, been obtained by another process in the form of bright crystals belonging to the cubic system. Chromium unites with oxygen in four different proportions: (1) chromium monoxide, CrO; (2) chromium sesquioxide, $Cr_2 O_3$; (3) chromo-chromic oxide, $Cr O Cr_2O_3$; (4) chromium trioxide, CrO_3. The two first of these oxides yield corresponding chlorides and salts; thus, CrO, $CrCl_2$, Cr_2O_3,Cr_2Cl_6; the third oxide is a neutral body, corresponding to the magnetic oxide of iron; and the fourth oxide forms with water an acid.

Chromium Monoxide, CrO, is only known in the hydrated state, as both it and its compounds absorb oxygen with great avidity. The hydrate is prepared as a brown precipitate by adding potash to the solution of chromium dichloride.

Chromium Dichloride, $Cr Cl_2$, is a white crystalline body, which dissolves in water, forming a blue solution. It is obtained by passing hydrogen over heated chromic chloride.

Chromium Sesquioxide, or *Chromic Oxide*, Cr_2O_3, is a dark-green, perfectly stable powder, obtained by igniting the hydro-oxide formed by precipitating any soluble chromic salt with ammonia. It is employed as a green color for painting on porcelain, and produces the green of the emerald. A splendid green color is also obtained by heating potassium bichromate with boron-trioxide: on dissolving in water a grass-green hydroxide remains behind, which is termed Guignet's green ($Cr_4H_6O_9$).

Chromic Chloride, Cr_2Cl_6, the anhydrous chloride, is obtained as a sublimate, in beautiful violet crystals, by passing a current of chlorine gas over a red-hot mixture of chromium sesquioxide and charcoal. These crystals do not dissolve easily in water, but are readily soluble if a trace of chromium dichloride is present. The most ready way of preparing a solution of chromic chloride is to boil a solution of chromic acid on a chromate with hydrochloric acid and alcohol, the red or yellow solution after a few minutes being changed to a deep greenish-blue color. A solution of chromic sulphate, $Cr_2 3SO_4$, may be obtained in the same way, by substituting sulphuric acid for hydrochloric acid. Chromium sulphate forms a series of *alums* with potassium and ammonium sulphates, which have a deep purple tint, and are isomorphous with common alum, $\left.\begin{matrix} K_2 \\ Cr_2 \end{matrix}\right\} 4SO_4 + 24H_2O$. The chromic salts are green, but a violet-colored modification also occurs.

Chromic Acid and Chromates.

If any chromic compound be fused with potassium carbonate, it becomes oxidized, and a soluble yellow chromate is formed, K_2CrO_4: this is the mode in which the chromium compounds are prepared from chrome-iron ore. This yellow chromate is isomorphous with potassium, sulphate and manganate. When sulphuric acid is added to a solution of this yellow salt in sufficient quantity to combine with half the base, large red crystals of the *bichromate* $K_2Cr_2O_7$, separate out. This salt is largely used for the preparation of the chrome pigments. If to the solution of the bichromate a solution of chromium trioxide be added, a third salt, termed *ter-chromate*, $K_2Cr_3O_{10}$, crystallizes out. The constitution of these three salts may be represented as follows:

$$(1)\ \left.\begin{matrix} CrO_2 \\ K_2 \end{matrix}\right\} O_2; \quad (2)\ \left.\begin{matrix} CrO_2 \\ CrO_2 \\ K_2 \end{matrix}\right\} O_3; \quad (3)\ \left.\begin{matrix} CrO_2 \\ CrO_2 \\ CrO_2 \\ K_2 \end{matrix}\right\} O_4.$$

Chromium Trioxide, CrO_3, is obtained in the form of long ruby-red needle-shaped crystals by adding an excess of strong sulphuric acid to a concentrated solution of the bichromate. The crystals are very soluble in water, forming an acid solution of chromic acid, H_2CrO_4. The excess of sulphuric acid may be removed by washing with concentrated nitric acid, and the crystals then dried in a current of air in a glass tube. The crystals of chromium trioxide are very easily reduced to sesquioxide in presence of organic matter: so energetic is this evolution of oxygen, that ignition occurs when alcohol is dropped on the dry crystals.

If a solution of chromium trioxide or of potassium bichromate is heated with hydrochloric acid, chromic chloride is formed and chlorine liberated; whereas, if chromium trioxide is heated with sulphuric acid, a chromic sulphate is formed and oxygen gas is given off:

$$(1)\ 2CrO_3+12HCl=Cr_2Cl_6+6H_2O+3Cl_2.$$
$$(2)\ 2CrO_3+3H_2SO_4=Cr_2(SO_4)_3+3O_2.$$

The chief of the insoluble chromates are *lead chromate*, $PbCrO_4$, or chrome yellow, obtained by precipitating potassium chromate by a soluble lead salt, and largely used for dyeing and other purposes in the arts; *silver chromate*, Ag_2CrO_4, a characteristic, deep-red colored precipitate; and *barium chromate*, $BaCrO_4$, also a yellow insoluble powder.

Chromium Oxychloride, or *Chromyl Chloride*, $CrO_2\left\{\begin{matrix}Cl\\Cl\end{matrix}\right.$.—A compound resembling sulphuryl chloride (p. 135) is obtained by distilling potassium bichromate, sulphuric acid, and common salt. It is a dark red, strongly fuming liquid; it boils at 116·8°, and has a specific gravity of 1·92; and the density of its vapor is 77·7 (H=1). If potassium bichromate is dissolved in warm hydrochloric acid, large red crystals separate out on cooling: these consist of potassium chloro-chromate, $KClCrO_3$, a substance intermediate between chromium oxychloride and potassium chromate. We thus have:

Chromium Oxychloride.	Potassium Chloro-Chromate.	Potassium Chromate.
$CrO_2\left\{\begin{matrix}Cl\\Cl\end{matrix}\right.$;	$CrO_2\left\{\begin{matrix}OK\\Cl\end{matrix}\right.$;	$CrO_2\left\{\begin{matrix}OK\\OK\end{matrix}\right.$.

The presence of chromium and its compounds can be easily detected by the formation of soluble yellow-colored alkaline salts, yielding insoluble yellow lead and silver compounds, and capable of easy reduction to green solutions in presence of organic matter. Chromium sesquioxide imparts to glass or borax a fine deep-green color.

It may also be detected by the splendid, but very fugitive blue coloration which is produced when hydrogen dioxide is added to a very dilute solution of chromic acid: this blue color is due to the formation of a still higher oxide of chromium, which, however, is very readily decomposed.

URANIUM. *Symbol* U, *Combining Weight* 120, *Specific Gravity* 18·4.

Uranium is a metal which occurs but sparingly in nature, existing combined in two somewhat rare minerals, pitchblende and uranite. The metal is of a steel-white color, and it does not oxidize in dry air at ordinary temperatures, but when strongly heated it burns brilliantly. There are two oxides which form salts, viz. *uranous oxide*, U O, and *uranic oxide*, U_2O_3: the uranous salts are green, whilst the uranic compounds are yellow.

CLASS VI. 1, TIN. 2, TITANIUM.

1. TIN. *Symbol* Sn. (*Stannum*). *Combining Weight* 118. *Specific Gravity* 7·3.

The ores of tin—although this metal has been known from very early times—occur in but few localities, and the metallic tin is not found in nature. The chief European sources of tin are the Cornish mines, where it is found as tin dioxide or tin-stone. It is in all probability from these mines that the Phœnicians and Romans obtained all the tin which they employed in the manufacture of bronze. Tin-stone is also met with in Malacca, and Borneo, and Mexico. In order to prepare the metal, the tin-stone is crushed and washed to remove mechanically the lighter portions of rock with which it is mixed, and the purified ore is then placed in a reverberatory furnace with anthracite or charcoal, and a small quantity of lime; the oxide is thus reduced, and the liquid metal, together with the slag, consisting of silicate of lime, falls to the lower part of the furnace. The blocks of tin, still impure, are then refined by gradually melting out the pure tin, leaving an impure alloy behind. English tin generally contains traces of arsenic, copper, and other metals; that imported from Banca is nearly chemically pure.

Tin possesses a white color resembling that of silver; it is soft, malleable and ductile, but possesses little tenacity, a wire two mms. in diameter breaking with a weight of sixteen kilos. When bent, pure tin emits a peculiar crackling sound; tin melts at 235°, and is not sensibly volatile. Tin does not lose its lustre on exposure to the air, whether dry or moist, at ordinary temperature, but if strongly heated it takes fire, and a white powder of stannic oxide (sometimes termed putty powder) is formed. Hydrochloric acid dissolves tin with the evolution of hydrogen and the formation of stannous chloride; nitric acid also attacks the metal with great energy, nitrous fumes being given off and stannic oxide being left as a white powder. There are two well-marked oxides of tin.

Stannous oxide, $Sn\,O$. This is a black powder prepared by heating the stannous hydrate in an atmosphere of carbonic acid; it rapidly absorbs oxygen from the air, passing into stannic oxide. The hydrate falls as a white powder when a solution of a stannous salt is added to an alkaline carbonate. The stannous chloride, $SnCl_2$, is obtained by dissolving tin in hydrochloric acid, and separates out in needle-shaped crystals, $SnCl_2 + 2H_2O$, when the solution is concentrated. Stannous chloride is termed "tin salts" in commerce; it is largely manufactured for the calico printer and dyer, who use it as a "mordant."

Stannic oxide, $Sn\,O_2$, occurs native as tin-stone, and it can be prepared artificially in two conditions, possessing totally different properties. If tin be oxidized by nitric acid, stannic oxide is produced as a white powder insoluble in acids; if, on the other hand, to a solution of stannic chloride an alkali be added, a white precipitate is formed of stannic oxide, which is readily soluble in acids. Both of these varieties of hydrated stannic oxide form salts, the insoluble compound having been termed metastannic, and the soluble compound stannic acid. Sodium stannate, $Na_2\,Sn\,O_3$, $4\,H_2\,O$, formed by boiling stannic oxide with soda, is largely used in calico printing as a mordant, and then termed "*tin prepare liquor.*" *Stannic chloride*, $Sn\,Cl_4$, is obtained by passing chlorine gas over metallic tin; it is a colorless liquid, boiling at 120° C., and having a vapor density of 9·2 It fumes strongly in the air, and forms a crystalline hydrate, when a small quantity of water is added, which easily dissolves in an excess. Stannic chloride is also used by dyers, and is prepared for this purpose by dissolving tin in cold nitro-hydrochloric acid. Of the sulphides of tin, SnS, stannous sulphide, and $Sn\,S_2$, stannic sulphide, are the most important; the former is blackish-gray, and the latter a bright yellow crystalline powder known as mosaic gold, soluble in alkaline sulphides.

Tin can easily be distinguished in solution by the formation of a splendid purple color called purple of cassius, formed when gold chloride is added to a dilute solution of stannous chloride. Tin is also easily reduced before the blowpipe in the form of white malleable beads. Tin is largely used in the arts for covering and thus protecting iron plates, or for "tin-plating," and also for preparing several valuable alloys, as pewter, Britannia metal, plumbers' solder, bronze, bell-metal, &c.

2. TITANIUM. *Symbol* Ti. *Combining Weight* 50.

Titanium is a rare metal, resembling tin in its chemical properties. It is found in combination with iron in the mineral rutile. The oxides of titanium correspond to those of tin; viz., titanous and titanic oxides, $Ti\,O$, and $Ti\,O_2$. Titanium and its compounds are not used in the arts.

NIOBIUM and TANTALUM are two extremely rare metals, whose properties are as yet but imperfectly known.

CLASS VII.

1. MOLYBDENUM. *Symbol* Mo. *Combining Weight* 96.

The chief ore of this metal is molybdenum disulphide, a mineral in appearance resembling graphite. The metal is a gray substance, which oxidizes on heating in the air to molybdenum trioxide, $Mo\,O_3$, a yellow powder which acts as an acid, forming with bases salts called molybdates. The compounds of molybdenum do not occur frequently, and are not used in the arts.

2. TUNGSTEN. *Symbol* W. (*Wolfram*). *Combining Weight* 184.

This metal occurs in tolerably large quantities combined with ferrous oxide in the mineral wolfram, and also with lime as scheelite. The metal has only been obtained as a grayish-black powder, having a specific gravity of 17·4. Tungsten is employed occasionally in the arts—the addition of a small quantity imparts a great degree of hardness and other valuable qualities to steel. Two oxides of tungsten are known,—*Tungsten Dioxide*, $W\,O_2$, and *Tungsten Trioxide*, $W\,O_3$. The former of these is obtained as a brown powder by heating the trioxide in an atmosphere of hydrogen; the latter, sometimes called tungstic acid, is obtained as an insoluble yellow powder by heating the native calcium tungstate with nitric acid. Tungsten trioxide forms a variety of somewhat complicated salts. The sodium compound is soluble, and has been used to add to the starch employed to stiffen light fabrics, the tungstate rendering the fabric uninflammable.

LESSON XXIV.

CLASS VIII.—1, ARSENIC. 2, ANTIMONY. 3, BISMUTH.

1. ARSENIC. The properties of this element and its compounds have been already considered (*see* p. 133).

2. ANTIMONY. *Symbol* Sb (*Stibium*). *Combining Weight* 122. *Specific Gravity* 6·71.

Metallic antimony occurs native, but its chief ore is the trisulphide. The metal is easily reduced by heating the sulphide with about half its weight of metallic iron, when ferrous sulphide and metallic antimony are formed. Antimony may also be reduced by mixing the ore with coal and heating in a reverberatory furnace. Antimony is a bright bluish-white colored metal, crystallizing in rhombohedra, isomorphous with arsenic. It is very brittle, and can be powdered in a mortar; it melts at 450°, and may be distilled at a white heat in an atmosphere of hydrogen. Antimony undergoes no alteration in the air at ordinary temperatures, but rapidly oxidizes if exposed to air when melted, and, if heated more strongly, it takes fire and burns with a white flame, giving off dense white fumes of antimony trioxide. Antimony is not attacked either by dilute hydrochloric or sulphuric acids; nitric acid attacks the metal, converting it into white insoluble antimony pentoxide. Nitro-hydrochloric acid dissolves antimony easily

The alloys of antimony are largely used in the arts; of these, type metal (an alloy of lead and antimony) is the most important: it contains 17 to 20 per cent. of the latter metal. The two important oxides of antimony, (1) antimony trioxide, $Sb_2 O_3$, (2) antimony pentoxide, $Sb_2 O_5$ (sometimes called antimonic acid), correspond to those of arsenic (*see* p. 133). A third oxide exists unknown in the arsenic series: this is an intermediate gray tetroxide, having the composition $Sb_2 O_4$.

Antimony Trioxide, $Sb_2 O_3$. This oxide gives rise to the important series of salts of antimony used in medicine; it is obtained in crystalline needles, which are isomorphous with the rare form of arsenic trioxide (*see* p. 134). Antimony trioxide has also been observed to crystallize in octohedra; hence these two oxides are said to be isodimorphous. The best mode of preparing the pure oxide is by decomposing antimony trichloride with an alkaline carbonate, when the oxide is precipitated as a white powder: thus, $2\ Sb\ Cl_3 + 3\ Na_2 CO_3 = Sb_2 O_3 + 6\ Na\ Cl + 3\ CO_2$. Antimony trioxide dissolves, when boiled with a solution of cream of tartar (hydrogen potassium tartrate), and on concentration the solution deposits crystals of tartar emetic (potassium antimony tartrate); antimony trioxide also dissolves in hydrochloric acid, yielding a solution of the trichloride, but this is rendered turbid by addition of water.

Antimony Pentoxide, $Sb_2 O_5$ (sometimes called Antimonic Acid), obtained by acting on antimony with strong nitric acid, or by decomposing the penta-chloride of antimony with water. It is a light straw-colored powder, which loses oxygen at a red heat, and is converted into the intermediate oxide $Sb_2 O_3\ Sb_2 O_5$. Antimonic oxide forms salts with the alkalies called Antimoniates, from which antimonic acid, $Sb_2 O_6 H_2$, can be separated as a white powder. The oxides prepared by the two methods above given are found to possess different properties as regards their power of uniting with bases. That prepared with nitric acid yields monobasic salts, whilst that obtained from the penta-chloride yields dibasic salts; to the first class of salts the name Antimoniates, and to the second that of Metantimoniates has been given. The gray intermediate tetroxide, $Sb_2 O_4$, is obtained by heating the metal in the air until no further change occurs.

Finely-powdered metallic antimony takes fire spontaneously when thrown into chlorine gas with formation of the chlorides. There are two chlorides of antimony; *antimony tri-chloride*, $Sb Cl_3$, is obtained as a buttery mass by passing chlorine gas over an excess of metallic antimony, or by dissolving the metal or sulphide in hydrochloric acid to which a little nitric has been added: on distilling the liquid thus obtained the trichloride volatilizes, and, on cooling, solidifies to a mass of white crystals. The *penta-chloride*, $Sb Cl_5$, is a mobile strongly-fuming liquid, obtained by passing an excess of chlorine over the trichoride or the metal. The *sulphides of antimony*, $Sb_2 S_3$, and $Sb_2 S_5$, correspond to the oxides, and, like the oxides, are capable of uniting with the alkaline sulphides to form a class of soluble salts.

Like arsenic, antimony unites with hydrogen to form a gaseous compound, *antimoniuretted hydrogen*, $Sb H_3$, analogous to $As H_3$, *arseniuretted hydrogen.* This gas is evolved, together with hydrogen, when an antimony salt is brought in contact with zinc and dilute acid. Like the corresponding arsenic compound, it burns with a bluish flame, evolving a white-colored antimony trioxide, and is decomposed at a red heat with deposition of metallic antimony. The detection and separation of arsenic and antimony is a subject of much importance in medical jurisprudence, as both substances exhibit poisonous characters, and closely resemble one another in their reactions; still, with care it is easy to discriminate between these two metals, and to detect with certainty a very minute quantity of either when present in the body of an animal.

3. BISMUTH. *Symbol* Bi. *Combining Weight* 210. *Specific Gravity* 9·8.

This metal is found in small quantities in the native state, but occurs more often as a sulphide ; it is easily reduced to the metallic state, and then exhibits a pinkish-white color. It crystallizes in large rhombohedra which can scarcely be distinguished from cubes ; it melts at 264°, and is volatilized at a white heat. Bismuth does not oxidize in dry air at the ordinary temperature, but if heated strongly it burns with a blue flame, forming an oxide ; it also takes fire when thrown into chlorine gas. Bismuth dissolves easily in nitric acid. The metal is chiefly used as an ingredient of fusible metal ; its compounds are also used in medicine, and as pigments. Two oxides of bismuth are known, *bismuth trioxide*, Bi_2O_3, and *bismuth pentoxide*, $Bi_2 O_5$. The first of these is a pale yellow powder, formed when the metal is roasted in the air ; the second oxide is obtained by dissolving the first in potash and precipitating the pentoxide by nitric acid and heating: it is a reddish-brown powder. Like the corresponding antimony compound, bismuth pentoxide forms with the alkalies soluble salts. The *nitrate*, $Bi\ 3\ NO_3 + 5H_2\ O$, is the most important soluble salt of bismuth ; the *sulphide*, Bi_2S_3, is a black insoluble compound ; the *trichloride*, $BiCl_3$, is obtained by heating the metal in chlorine. One of the most striking peculiarities of the bismuth compounds is, that solutions of the salts become milky on the addition of water, owing to the formation of insoluble basic compounds. Metallic bismuth is easily reduced from its compounds, before the blowpipe, as a brittle bead.

VANADIUM.

Symbol V, *Combining Weight* 51·3.

This is a very rare metal: its compounds occur in small quantity in certain iron ores, and also in combination as lead vanadate. It forms an interesting oxide, termed *Vanadium pent-oxide*, $V_2\ O_5$, which yields salts called *vanadates*, isomorphous with arsenates and phosphates, and also an oxychloride, $VO\ Cl_3$, corresponding to phosphorus oxychloride, $PO\ Cl_3$.

CLASS IX.—LEAD CLASS. 1, LEAD. 2, THALLIUM.

1. LEAD. *Symbol* Pb (*plumbum*). *Combining Weight* 207. *Specific Gravity* 11·3.

Lead does not occur free in nature; all the lead of commerce is obtained from galena, or lead sulphide. The mode of reducing lead from this ore is a very simple one: the galena is roasted in a reverberatory furnace with the addition of a small quantity of lime to form a fusible slag with any silicious mineral matter present in the ore. By the action of the air a portion of the sulphide is oxidized to sulphate, whilst in another portion the sulphur burns off as sulphur dioxide, and lead oxide is left behind; after the lapse of a certain time the air is excluded and the heat of the furnace raised—the lead sulphate and oxide formed both decompose the remaining sulphide, giving off sulphur dioxide and leaving metallic lead behind, thus:

$$(1)\quad Pb\,SO_4 + Pb\,S = 2\,Pb + 2\,SO_2.$$

$$(2)\quad 2\,Pb\,O + Pb\,S = 3\,Pb + SO_2.$$

Galena almost always contains small quantities of silver, which is extracted by a process described on p. 213. Lead is a bluish-white-colored metal, and so soft that it may be scratched with the nail; it may be drawn out to wire, or hammered into plate, but possesses little tenacity or elasticity, and a wire 2 mms. in diameter breaks with a load of 2 kilos. Lead melts at 334°, and at a higher temperature volatilizes, though not in quantity sufficient to enable it to be distilled.

The bright surface of the metal remains permanently in dry air, but it soon becomes tarnished in moist air, owing to the formation of a film of oxide; and this oxidation proceeds rapidly in presence of a small quantity of weak acid such as carbonic or acetic. The small quantity of certain salts contained in all spring and river waters exerts an important influence on the action of lead; thus waters containing nitrates or chlorides are liable to contamination with lead, whilst those hard waters containing sulphates or carbonates may generally be brought into contact with lead without danger, as a thin deposit of sulphate or carbonate is formed, which preserves the metal from further action. If the water contains much free carbonic acid it should not be allowed to come into contact with lead, as the carbonate dissolves in water containing this substance. The presence of lead in water may easily be demonstrated by passing a current of sulphuretted hydrogen through a deep column of the acidified water, and noticing whether the liquid becomes tinged of a brown color, owing to the formation of lead sulphide.. Three compounds of lead and oxygen are known.

1. *Litharge*, or *Lead Monoxide*, Pb O, a straw-colored powder, obtained by heating lead in a current of air ; it fuses at a red heat, forming scaly crystals termed *Litharge.* Lead oxide is soluble in caustic potash, and is deposited from a hot solution in the form of rhombic prisms. This oxide forms with acids the important series of lead salts, which are generally colorless, and of which the soluble ones act as violent poisons. Lead oxide combines with silica to form an easily fusible silicate, or glass ; thus earthen crucibles in which the oxide is fused are rapidly attacked. A white hydrated oxide is obtained by precipitating a soluble salt of lead by caustic potash, and this if heated yields the oxide. 2. *Lead Dioxide*, or *puce-colored oxide*, PbO_2. This oxide is a brown powder obtained by passing chlorine through the hydrated monoxide, or by digesting red lead with nitric acid. Lead dioxide does not form salts with acids. When heated it yields half its oxygen, and acted upon with warm hydrochloric acid, chlorine is evolved, and lead chloride is formed.

3. *Red Oxide*, or *Red Lead*, a compound of the two last oxides, having the composition $2\,PbO + PbO_2$. It is obtained by exposing *massicot* to the air at a moderate red heat, oxygen being absorbed. Red lead is chiefly used in glass-making (*see* p. 176). When treated with dilute nitric acid the lead oxide dissolves, forming soluble lead nitrate, leaving the puce-colored oxide behind.

Of the soluble salts of lead the *nitrate*, $Pb\,2NO_3$, is the most important. This compound is obtained by dissolving the oxide, the carbonate, or metallic lead in warm nitric acid ; it crystallizes in octohedra, and dissolves in eight parts of cold water, and when heated strongly it yields red fumes of

NO_2 (*see* p. 61). *Lead acetate*, or *sugar of lead*, is also a soluble salt, which will be described under Acetic Acid. Almost all the other lead salts are insoluble in water. *Lead Carbonate*, or *White Lead*, $PbCO_3$, is a substance much used in the arts as a paint, and manufactured on a large scale. The salt may be obtained in the pure state by precipitating a cold solution of the nitrate with an alkaline carbonate, when it falls down as a white powder. For preparing the salt in quantity two plans are employed—the one similar in principle to that described for the pure salt; and the second an old and interesting process, known as the Dutch method. In this process thin sheets of lead are rolled into a coil, and each coil placed in an earthen pot containing a small quantity of crude vinegar (acetic acid); several hundred of these jars and coils are packed on a floor in a bed of stable manure or spent tanbark, and then covered with boards, whilst a second layer of pots similarly charged is placed above, and this is continued until the building is filled. After remaining thus for several weeks, the coils are taken out, when the greater part of the lead is found to be converted into white carbonate. It appears that, to begin with, a lead acetate is formed, and that the acetic acid is gradually driven out from its combination by the carbonic acid evolved from the putrefying organic matter, and thus enabled to unite with another portion of the lead lying underneath that which was first attacked.

Lead sulphide, or *galena*, PbS, is found native, and constitutes the chief ore of the metal. It is prepared as a black precipitate by passing sulphuretted hydrogen gas through a solution of a lead salt. Galena crystallizes in cubes and octohedra, and possesses a bright bluish-white metallic lustre; *lead sulphate*, $PbSO_4$, is a white insoluble salt, which is found native, and is prepared artificially by adding sulphuric acid to a soluble lead salt; *lead chloride*, $PbCl_2$, prepared by adding hydrochloric acid to a strong solution of lead nitrate, when a crystalline precipitate of lead chloride is formed. It dissolves in about thirty parts of boiling water, separating out in shining needles on cooling; *lead iodide* is precipitated in the form of splendid yellow spangles, when hot solutions of potassium iodide and lead nitrate are mixed and allowed to cool; *lead chromate* is a yellow insoluble salt used as a pigment under the name of chrome-yellow. Lead can easily be recognized,—1st, by the black sulphide, soluble in dilute nitric acid; 2dly, by the white insoluble sulphate; 3dly, by the yellow iodide and chromate; and 4thly, by the easy reduction of the metal in the form of a malleable bead when any of the salts are heated before the blowpipe with a reducing agent.

2. THALLIUM. *Symbol* Tl. *Combining Weight* 204. *Specific Gravity* 11·85.

Thallium was discovered in 1861 by Mr. Crookes, by means of spectrum analysis, in the deposit in the flue of a pyrites burner (*see* p. 110). The presence of this new metal is indicated by the occurrence of a splendid green line in the spectrum. Metallic thallium closely resembles lead in its physical properties; the freshly cut surface has a bluish-white lustre, which rapidly tarnishes; it is so soft that it receives impressions of the nail, and can be easily drawn into wire; it melts below a red heat. It is found to occur in many specimens of iron pyrites, and appears to take the place of arsenic, which is a common impurity of this mineral. Metallic thallium undergoes gradual oxidation, so that it is best preserved in water; when strongly heated in oxygen, it takes fire, and burns with a bright green flame. Thallium dissolves easily in nitric and sulphuric acids with evolution of hydrogen, but more slowly in hydrochloric acid, owing to the insolubility of the chloride. Two oxides of this metal are well characterized, *Thallium monoxide*, $Tl_2 O$, and *Thallium trioxide*, $Tl_2 O_3$. Thallium monoxide corresponds in composition, and somewhat resembles in properties, the alkali potash, K_2O, as it is soluble in water, yielding an alkaline caustic solution, which absorbs carbonic acid from the air, and forming a well-defined series of salts. Of these the sulphate, Tl_2SO_4, and the monochloride, Tl Cl, are the most important. The sulphate is a soluble salt, crystallizing in six-sided prisms, and furnishing octohedral crystals of an alum with aluminium sulphate; whilst the chloride is only slightly soluble in water, in this respect more nearly resembling the corresponding lead salt. Thallium carbonate is also a soluble salt, requiring about twenty-five parts of cold water for solution. The sulphide, Tl_2S, is an insoluble black powder, precipitated when an alkaline sulphide is added to any soluble thallium compound. The soluble thallium salts are colorless, and act as strong poisons. The metal is precipitated in a pulverulent form, when a piece of zinc is introduced into its solutions. It will be seen that the properties of thallium are intermediate between those of lead and the alkalies. Thallium is a monad, 204 parts of this metal replacing one part of hydrogen.

LESSON XXV.

CLASS X. 1, COPPER. 2, MERCURY. 3, SILVER.

1. COPPER. *Symbol* Cu. *Combining Weight* 63·5. *Specific Gravity* 8·93.

Copper is an important metal, largely used in the arts, and has been known from very early times, as it occurs native in the metallic state, and is moreover easily reduced from its ores. Metallic copper is found in considerable quantity in North America and other localities, crystallizing in cubic and octohedral forms, but the chief sources of copper are the following ores: (1) A compound of copper, sulphur, and iron, known as copper pyrites, $Cu_2S + Fe_2S_3$. (2) The cuprous sulphide, Cu_2S. (3) The carbonate or malachite; and (4) The red or cuprous oxide, Cu_2O. The Cornish mines yield large quantities of copper, whilst much ore is furnished by Chili and South Australia. Pure metallic copper can be obtained by reducing the oxide in a current of hydrogen gas, or by the electrolytic decomposition of a salt of copper. The process for obtaining copper on a large scale from the carbonate or oxide is a very simple one, viz., merely reducing these ores together with carbon and some silica in a wind furnace. The reduction of the metal is more difficult when the commoner ore, copper pyrites, is employed. In this case the ore is repeatedly roasted, in order partially to convert the cuprous sulphide into oxide, and the roasted ore melted in a reverberatory furnace with the addition of sand or silicious slag; in this operation the cuprous oxide becomes converted into the corresponding sulphide, whilst the iron oxidizes and unites with the silica to form a light and fusible slag. The impure cuprous sulphide fuses and sinks to the lower portion of the furnace, forming the "mat" or coarse metal; and by repeating this operation, a pure cuprous sulphide or "fine metal" is obtained. In order to prepare the metallic copper free from sulphur, this fine metal is roasted, and afterwards fused in contact with the air. During the first part of the operation a portion of the sulphur is burnt off, cupric oxide being formed, and in the later stages of the process this oxide acts upon the remaining quantity of sulphide, forming sulphuric dioxide, and metallic copper, $Cu_2 S + 2 Cu O = SO_2 + 4 Cu$. In order to get rid of the last traces of oxide, the molten copper is "poled" or stirred up with a piece of green wood.

Metallic copper possesses a peculiar deep red color, which is best seen when a ray of light is several times reflected from a bright surface of the metal; it is very malleable and ductile, and possesses great tenacity, a wire of two mms. in diameter supporting a weight of 140 kilos.; it melts at a red heat, and is slightly volatile at a white heat, communicating a green tint to a flame of hydrogen gas, which is passed over it, and it is one of the best conductors of heat and electricity. Copper does not oxidize either in pure dry or moist air at ordinary temperatures, but if heated to redness in the air, it rapidly oxidizes to scales of copper oxide. Steam is not decomposed by metallic copper at a red heat. Finely divided copper dissolves in hydrochloric acid with evolution of hydrogen; when heated with strong sulphuric acid, sulphur dioxide (p. 106) is evolved, and copper sulphate formed; and when acted upon with nitric acid, copper nitrate is produced, and nitric oxide (p. 59) liberated.

Many of the copper alloys are of importance; as brass, yellow, or muntz metal, used for the sheathing of ships, bronze, gun-metal, bell-metal, and speculum-metal.

Copper is a dyad element, and forms two sets of compounds, the *cuprous* and the *cupric* salts: the molecules of the cupric salts contain one atom of copper, whilst the cuprous compounds contain only two atoms of metal. The constitution of the two oxides Cu O and Cu_2O, and the corresponding chlorides $Cu\ Cl_2$ and Cu_2Cl_2, may be represented as follows:

Cupric Oxide, $Cu=O$	Cuprous Oxide $\begin{matrix} Cu \\ \vert \\ Cu \end{matrix}>O$
" Chloride, $Cu\begin{matrix}-Cl \\ -Cl\end{matrix}$	" Chloride $\begin{matrix} Cu-Cl \\ \vert \\ Cu-Cl \end{matrix}$

Cuprous Oxide, or *Red Oxide*, $Cu_2\ O$, occurs native in ruby-red octahedral crystals; it is artificially prepared by heating equivalent quantities of cupric oxide and copper filings, or by boiling a solution of copper sulphate and sugar, to which excess of caustic potash has been added: the sugar reduces the copper salt, and cuprous oxide is precipitated as a bright red powder. Cuprous oxide imparts to glass a splendid ruby-red color; it forms colorless salts with acids which rapidly absorb oxygen from the air, and pass into the corresponding cupric compounds. The most important of these salts is cuprous chloride, $Cu_2\ Cl_2$, a white solid substance obtained by dissolving a mixture of cupric oxide and metallic copper in hydrochloric acid: the solution of cuprous chloride possesses the remarkable property of absorbing carbonic oxide gas.

Cupric, or *Black Oxide*, Cu O. This oxide is formed when copper is heated in the air, or when cupric nitrate is heated to redness; it yields the blue and green cupric salts, and it is largely used in the laboratory as a means of giving oxygen for the combustion of organic substances (*see* p. 236). Hydrated copper oxide is obtained as a light blue precipitate when caustic alkali is added to a cupric salt; when this is heated to 100° it loses its water, and the anhydrous oxide falls as a brown powder. Cupric oxide is soluble in acids, furnishing a series of well crystallizing salts; of these the most important soluble compounds are: (1) *copper sulphate*, $Cu\,SO_4 + 5H_2O$. This salt is sometimes known as blue vitriol, and is largely manufactured by dissolving copper oxide (copper scales) in sulphuric acid. It crystallizes in large blue crystals belonging to the *triclinic* system (Fig. 51); when heated to redness, it loses all its water of crystallization, and forms a white powder, which again at a higher temperature decomposes, leaving copper oxide. Copper sulphate is employed in calico printing, and in the manufacture of Scheele's green, and Brunswick green, and other copper pigments. The sulphate and the other copper salts give, with excess of ammonia, a deep-blue colored solution, forming a remarkable compound, capable of crystallizing; the production of this blue color may be used as a test of the presence of copper. (2) *Copper nitrate*, $Cu\ 2 \left.\begin{matrix}(NO_3)\\(NO_3)\end{matrix}\right\} + 6\,H_2O$, a very soluble salt, crystallizing in large blue prisms obtained by dissolving copper in nitric acid. (3) *Copper chloride*, $Cu\,Cl_2$, formed when copper is brought into chlorine gas, or when copper oxide is dissolved in hydrochloric acid; it forms green needle-shaped crystals soluble in water and alcohol. The alcoholic solution burns with a characteristic green flame. The insoluble copper salts are: (1) the *sulphide*, $Cu\,S$, obtained as a black precipitate, when sulphuretted hydrogen gas is passed through an acidified solution of a copper salt; (2) the *carbonate*, which, however, is not known in the pure state, as the green precipitate obtained by adding a solution of an alkaline carbonate to a copper salt, always contains hydrated oxide; the mineral malachite also possesses a similar composition; (3) *copper arsenite*, or Scheele's green, is a bright green powder used as a pigment, and obtained by mixing solutions of sodium arsenite with copper sulphate. The copper salts act as powerful poisons, and they may be detected—(1) by the black insoluble sulphide ; (2) by the blue hydrate turning black on heating ; (3) by the deep blue coloration with ammonia ; (4) by the deposition of red metallic copper upon a bright surface of iron placed in the solution.

2. MERCURY. *Symbol* Hg (*Hydrargyrum*). *Combining Weight* 200. *Specific Gravity* at 0° 13·596. *Density* 100.*

The chief ore of mercury is the sulphide, or *cinnabar*, which occurs at Almaden in Spain, at Idria in Illyria, in California, and also in China and Japan. The metal is easily obtained by roasting the ore, when the sulphur burns off as the dioxide, and the metal volatilizes, and its vapor is condensed in earthen pipes. Mercury is the only metal liquid at the ordinary temperature ; it freezes at -40°, crystallizing in octohedra ; in the solid state it is malleable and possesses a density of 14·4. It boils at 350° (measured by the air thermometer), and gives off a slight amount of vapor at the ordinary temperature. The specific gravity of its vapor (air = 1) is 6·976. Mercury when pure does not tarnish in moist or dry air, but when heated above 300° it slowly absorbs oxygen, passsing into the red oxide ; and it combines directly with chlorine, bromine, iodine, and sulphur. Hydrochloric acid does not attack mercury ; sulphuric acid, on heating, forms sulphur dioxide (p. 105) and mercuric sulphate ; and nitric acid evolves nitric oxide and forms mercuric nitrate. Mercury is largely used in the process of extracting gold and silver from their ores (p. 213), and in the arts, for silvering mirrors, and other purposes. Mercury is deposited from its solutions upon metallic iron or copper, in the form of a gray powder, which becomes bright on burnishing. Mercury and its salts act as valuable medicines.

Two oxides of mercury are known : the black or *mercurous oxide*, Hg_2O, and the red or *mercuric oxide*, HgO. Mercurous oxide is best obtained by digesting calomel (mercury chloride) with an excess of caustic alkali ; it forms black

* The atom of mercury weighing 200 occupies 2 volumes, and hence its vapor density is *half* its combining weight.

powder, which undergoes decomposition on exposure to light, or, on heating to 100°, into metallic mercury and the red oxide. Mercuric oxide, Hg O, is obtained by moderately heating the nitrate; it may also be prepared by heating the metal to about 300° in the air, or by precipitating the nitrate with caustic potash. The sulphides correspond to the oxides; *mercurous suphide*, Hg_2 S, is an unstable black compound, prepared by passing sulphuretted hydrogen through a solution of mercurous salt. *Mercuric Sulphide*, cinnabar, or vermilion, Hg S, is a more important substance. It occurs native, and may be prepared artificially by heating a mixture of sulphur and mercury. When precipitated from a solution of mercuric salt by sulphuretted hydrogen, the sulphide is black, but on sublimation becomes red. The chlorides of mercury are important compounds. Mercurous chloride, or calomel, Hg_2Cl_2, and mercuric chloride, or corrosive sublimate, $HgCl_2$.

Mercurous Chloride, Calomel, Hg_2 Cl_2, is generally prepared by heating a mixture of three parts of finely-divided metallic mercury with four parts of corrosive sublimate; the metal combines with half the chlorine of the corrosive sublimate, and one atom of calomel is formed, $Hg\ Cl_2 + Hg = Hg_2\ Cl_2$. The calomel sublimes, and is deposited in a solid cake; it must be well washed in order to free it from any soluble mercuric chloride which may remain undecomposed. Calomel, a white powder, is insoluble in water, but is decomposed by potash or ammonia. It is used largely in medicine. *Mercuric Chloride*, or *Corrosive Sublimate*, Hg Cl_2, is prepared on a large scale by heating an intimate mixture of equal parts of mercuric sulphate and common salt; it is also formed when mercury burns in chlorine. It acts as a violent poison; it is soluble in water, crystallizing in rectangular octohedra, fuses at 265°, and boils at 295°. The other important soluble salts of mercury are the mercurous and mercuric nitrates, respectively obtained by acting on mercury with an excess of dilute nitric acid, and by dissolving mercuric oxide in nitric acid. The mercury compounds can be readily recognized: (1) by precipitation of black mercuric sulphide, insoluble in nitric acid; (2) by the reduction of liquid globules of the metal when any compound is strongly heated with sodium carbonate in a small tube; (3) by the deposit of metallic mercury on copper. The mercurous salts are distinguished by precipitating calomel when a chloride is added to a soluble salt; whilst the mercuric salts may be detected by the formation of red mercuric iodide.

3. SILVER. *Symbol* Ag. *Combining Weight* 108. *Speci-Gravity* 10·5.

Silver was known to the ancients. It is found in the native state, as well as combined with sulphur, antimony, chlorine, and bromine. It is also contained in small quantities in galena (p. 202), and it can be extracted with profit from the lead prepared from this ore, even when the lead contains only two or three ounces of silver to the ton. The method thus adopted for the extraction of the silver depends upon the fact that the whole of the silver can be concentrated into a small portion of lead, by crystallization; metallic lead free from silver separates out in crystals, and a rich alloy is left. When this reaches the concentration of 300 oz. silver to the ton, the alloy undergoes the operation of *cupellation*, in which the mixture is melted in a furnace on a porous bed of bone-earth, and a blast of air blown over the surface: the lead oxidizes, and the oxide (litharge) fuses, and partly runs away and partly sinks into the porous bed of the furnace, whilst the silver remains behind in the metallic state.

For the extraction of silver from the other ores, a process termed amalgamation is employed, in which mercury is used to dissolved the metallic silver. The argentiferous ores of Germany, in which the silver occurs in combination with sulphur, are worked in a different manner; the ore is roasted in a furnace with common salt, by which means the silver sulphide is converted into chloride; the mixture is then placed in casks made to revolve, and scrap-iron and water is added. The iron reduces the silver to the metallic state, and when this is fully accomplished, metallic mercury is added; this forms a liquid amalgam with the silver (and gold, if any be present), and by distilling the mercury off, the silver is obtained in an impure state. A somewhat different method is employed in South America, where fuel is expensive. Silver possesses a bright white color and a brilliant lustre, which it does not lose in pure air at any temperature, but when melted in the air it possesses the singular power of absorbing mechanically a large volume (twenty-two times its bulk) of oxygen; this gas it again gives out on solidifying, a phenomenon technically known as the "spitting" of silver.

Silver is probably the best conductor of heat and electricity known, and is extremely ductile; one gramme of metal can be drawn out into a wire of 2,600 metres in length. Sulphur combines at once with silver, forming a black sulphide; silver articles long exposed to the air tarnish from this cause. Silver is easily soluble in nitric acid, the nitrate being formed and nitric oxide gas being evolved.

Alloys of Silver. Silver itself is largely used in the pure state for various purposes in the arts, but it is usually alloyed with a small quantity of copper when employed as coin or for articles of plate.

Silver forms three compounds with oxygen. The first of these is called *silver suboxide*, Ag_4O; it is a black powder which readily undergoes decomposition; the second, termed *silver monoxide*, Ag_2O, is obtained in the form of a brown precipitate, when caustic potash is added to a solution of silver nitrate. From this oxide, which is decomposed into metal and oxygen on heating, the ordinary silver salts may be derived by dissolving in acids; the third is called *silver dioxide*, Ag_2O_2, and is obtained as a black powder by the action of ozone on metallic silver. *Silver nitrate*, $Ag\ NO_3$, is the most important soluble salt of silver. It is obtained in the form of large transparent tabular crystals on evaporating a solution of silver in nitric acid, and is soluble in its own weight of cold and half its weight of hot water, and in four parts of alcohol. Silver nitrate fuses easily on heating, and when cast into sticks goes by the name of lunar caustic. This salt undergoes decomposition when exposed to the sunlight in contact with organic matter, and a black substance, probably consisting of the suboxide, is formed; hence it is employed in the manufacture of an indelible ink for marking linen and other fabrics.

Of the insoluble silver salts, the *chloride*, Ag Cl, is the most important. This salt occurs in nature, and is then known as horn silver, and is precipitated as a white curdy mass when a solution of a chloride and a silver salt are brought together. When exposed to sun- or daylight, the white chloride becomes tinted of a purple color, which increases in shade as the action of light continues; this coloration arises from a partial decomposition of the salt, a small quantity of subchloride and free hydrochloric acid being formed; in presence of organic matter this change takes place much more rapidly, and upon this fact the phenomena of photography depend. Silver chloride fuses at about 260°, and at higher temperatures volatilizes; it is easily reduced to metallic silver in presence of zinc and sulphuric acid. The chloride is perfectly insoluble in pure water, but it dissolves appreciably in strong hydrochloric acid and in a solution of common salt, whilst it dissolves easily in ammonia; it is also easily soluble in a solution of sodium hyposulphite, and it is for this reason that the latter salt is used for "fixing" photographic pictures, that is, dissolving out the unaltered silver salt, and thus rendering the image permanent. *Silver bromide*, Ag Br, falls as a white precipitate when silver nitrate is added to an alkaline bromide; it is also acted upon by the light, and is soluble in ammonia and an alkaline hyposulphite. The *iodide*, AgI, is a yellow powder, insoluble in water and ammonia, but dissolved by an alkaline hyposulphite. *Silver sulphide*, Ag_2S, occurs native in cubic crystals, as silver glance; it is precipitated as a black powder by passing sulphuretted hydrogen through solutions of the salts of silver. Silver can be easily detected when in solution by the precipitation of the white curdy chloride, insoluble in water and nitric acid, and soluble in ammonia: the metal can be easily obtained in malleable globules before the blowpipe, whilst it is reduced from its solutions by iron, copper, and mercury. Silver is always estimated quantitatively either as the chloride or as the metal.

CLASS XI. 1, GOLD. 2, PLATINUM, AND THE RARE PLATINUM-LIKE METALS.

1. GOLD. *Symbol* Au (*Aurum*). *Combining Weight* 197. *Specific Gravity* 19·3.

Gold is always found in the metallic state; it occurs in veins in the older sedimentary or in the plutonic rocks, and in the detritus of such rocks; it occurs in traces in the sand of most rivers, and although found generally in small quantities, it is a widely diffused metal. Previous to the discoveries of the gold-fields of California and Australia, it was obtained from certain iron pyrites. In order to obtain the gold, the detritus or sand which contains the metal is washed in a "cradle" or other arrangement, by means of which the lighter particles of mud or mineral are washed away, whilst the heavier grains of gold sink to the bottom of the vessel. When gold has to be worked in the solid rock, the mineral is crushed to powder and then shaken up with mercury, and the gold thus extracted by amalgamation.

Gold possesses a brilliant yellow color, and, in thin films, transmits green light; it is nearly as soft as lead; it can be drawn out into fine wire, and is the most malleable of all the metals. It does not tarnish at any temperature, in dry or moist air, nor is it affected by sulphur like silver; it is not acted upon by any single acid (except selenic), but dissolves in presence of free chlorine and in nitro-hydrochloric acid. At high temperatures gold is slightly volatile. Pure gold is best prepared by dissolving the ordinary metal in aqua regia, and adding ferrous sulphate, which is oxidized to ferric salt and precipitates the gold as a brown powder. The *standard gold* of our country is an alloy of gold and copper in the proportion of 11 of gold to 1 of copper, or 8·33 per cent. of the latter metal; this alloy is harder and more fusible, but less ductile than pure gold.

Gold unites with oxygen in two proportions, forming *Gold suboxide*, Au_2O, and *Gold trioxide*, Au_2O_3. Neither of these oxides forms salts with acids; but the latter unites with bases to form compounds called Aurates. Gold trioxide is obtained by adding zinc-oxide or magnesia to a solution of gold chloride: the oxide falls as a brown powder, from which the zinc can be separated by nitric acid. Gold trioxide decomposes, in direct sunlight, into metal and oxygen, and is also reduced when heated to a temperature of about 250°. The most important compound of gold trioxide is fulminating gold; this substance is obtained by acting on a solution of gold with excess of ammonia; a yellow-brown powder is precipitated, which, when dry, explodes very easily when heated to 100°, or when struck with a hammer. There are two gold chlorides known: (1) *gold chloride*, $AuCl$, obtained as an insoluble white mass when gold trichloride is heated to the melting point of tin; (2) the *trichloride*, $AuCl_3$, obtained when gold is dissolved in aqua regia. This is the most important compound of gold. On evaporating the solution, crystals of a compound of gold trichloride and hydrochloric acid, are deposited. Gold trichloride also forms crystalline compounds with the alkaline chlorides. Gold salts can be easily recognized by the brown precipitate of metallic gold formed on addition of ferrous salts, which can be reduced to a globule before the blowpipe; and also by the formation of a purple color (purple of Cassius), when gold trichloride is added to a dilute solution of a mixture of the two tin chlorides.

2. Platinum. *Symbol* Pt. *Combining Weight* 197·4. *Specific Gravity* 21·5.

Platinum is a comparatively rare metal, which always occurs in the native state, and generally alloyed with five other metals, viz., palladium, rhodium, iridium, osmium, and ruthenium. This alloy occurs in small grains in detritus and gravel in Siberia and Brazil; it has not been found *in situ* in the original rock, which probably belongs to the old plutonic series.

The original mode of obtaining the metal was to dissolve the ore in aqua regia, and precipitate the platinum (together with several of the accompanying metals) with sal-ammoniac, as the insoluble double chloride of ammonium and platinum, $2NH_4Cl, + PtCl_4$. This precipitate, on heating, yields metallic platinum in a finely divided or spongy state, and this sponge, if forcibly pressed and hammered when hot, gradually assumes a coherent metallic mass, the particles of platinum welding together, when hot, like iron. A new mode of preparing the metal has recently been proposed, the ore being melted in a very powerful furnace heated with the oxyhydrogen blowpipe. In this way a pure alloy of platinum, iridium, and rhodium is formed, the other constituents and impurities of the ore either being volatilized by the intense heat, or absorbed by the lime of which the crucible is composed. This alloy is in many respects more useful than pure platinum, being harder and less easily attacked by acids than the pure metal.

Platinum possesses a bright white color, and does not tarnish under any circumstances in the air; it is extremely infusible, and can only be melted by the heat of the oxyhydrogen blowpipe. It is unacted upon by the ordinary acids, but dissolves in aqua regia, and hence platinum vessels are much used in the laboratory. Caustic alkalies, however, act upon the metal at high temperatures. When finely divided, metallic platinum has the power of condensing gases on to its surface in a remarkable degree; the effect of bringing spongy platinum in contact with a mixture of oxygen and hydrogen has already been mentioned. Platinum and oxygen unite in two proportions to form,—(1) *Platinum monoxide*, $Pt O$; and (2) *Platinum dioxide*, $Pt O_2$. The first of these oxides is a black powder, easily decomposed on heating, and yielding a series of unstable salts; the second is obtained as a brown hydrate, by adding to a solution of platinic nitrate half its equivalent of caustic potash; the hydrate, when heated, first loses its water, forming the anhydrous oxide, and then parts with its oxygen, leaving the metal. *Platinum dichloride* $PtCl_2$, is a green insoluble powder, obtained by heating the higher chloride to 200°; *platinum tetrachloride*, $PtCl_4$, is the most important platinum compound. It is obtained as a yellowish-red solution by dissolving the metal in aqua regia: on evaporation, crystals of a compound of platinum tetrachloride with hydrochloric acid separate out. The platinum tetrachloride combines with many alkaline chlorides to form double salts; these compounds with potassium, rubidium, cæsium, and ammonium, are insoluble in water, and are isomorphous, crystallizing in cubes, whilst the sodium salt is soluble and crystallizes in large prisms.

Platinum dichloride, when acted upon by ammonia, gives rise to several very remarkable compounds, containing platinum, nitrogen, and hydrogen; these substances act as bases, and form a well-defined series of salts. These bases may be considered as molecules of ammonia, in which the hydrogen has been partly replaced by either a diatomic or tetratomic platinum.

For the properties of the rare metals palladium, rhodium, ruthenium, iridium, and osmium, the larger manuals must be consulted.

LESSON XXVI.

Spectrum Analysis.

An entirely new branch of chemical analysis, of great delicacy and importance, has recently been developed, chiefly by the researches of Bunsen and Kirchhoff, the principles of which may here be shortly stated.

It has long been known that certain chemical substances, especially the salts of the alkalies and alkaline earths, when strongly heated in the blowpipe, or other nearly colorless flame, impart to that flame a peculiar color, by the occurrence of which the presence of the substance may be detected; if many of these substances are present together the detection of each by the naked eye becomes impossible, owing to the colors being blended and thus interfering with each other. Thus, for instance, the sodium compounds color the flame an intense yellow, whilst the potassium salts tinge the flame violet; the yellow soda color is, however, so much more intense than the purple potash tint, that a small trace of soda prevents the eye from detecting the purple, even if large quantities of potash salts are present. This difficulty is altogether overcome, and this method of observation rendered extremely sensitive, if, instead of regarding the flame with the naked eye, it is examined through a prism. This consists of a triangular piece of glass, in passing through which the light is refracted, or bent out of its course—each differently colored ray being differently refracted, so that if a source of white light, such as the flame of a candle is thus regarded, a continuous band of differently colored rays is observed—the compound white light being resolved into all its variously colored constituents. This colored band is termed a *Spectrum*, and each source of pure white light gives the same *continuous spectrum*, stretching from red (the least refrangible) to violet (the most refrangible) color, identical in fact with the colors of the rainbow. (*See* No. 1 of the Chromo.)

If these colored flames are examined by means of a prism, the light being allowed to fall through a narrow slit upon the prism, it is at once seen that the light thus refracted differs essentially from white light, inasmuch as it consists of only a particular set of rays, each flame giving a spectrum containing a few *bright bands*. Thus the spectrum of the yellow soda flame contains only one fine bright yellow line, whilst the purple potash flame exhibits a spectrum in which there are two bright lines, one lying at the extreme red, and the other at the extreme violet end. (*See* No. 6 and 2 of Chromo.) These peculiar lines are always produced by the same chemical element, and by no other known substance; and the position of these lines always remains unaltered. When the spectrum of a flame tinted by a mixture of sodium and potassium salts is examined, the yellow ray of sodium is found to be confined to its own position, whilst the potassium red and purple lines are as plainly seen as they would have been had no sodium been present.

The colored flames which are exhibited by the salts of lithium, barium, strontium, and calcium, likewise each give rise to a peculiar spectrum, by means of which the presence or absence of very small quantities of these substances can be ascertained with certainty when mixed together, simply by observing the presence or absence of the peculiar bright bands characteristic of the particular body. (*See* Chromo.)

The advantage which this new method of analysis possesses over the older processes lies in the extreme delicacy, as well as in the great facility, with which the presence of particular elements can be detected with certainty. Thus a portion of sodium salt less than the $\frac{1}{180,000,000}$th part of a grain can be detected; and compounds are found to be most widely disseminated throughout the earth, which were supposed to occur very seldom. The extreme delicacy of the method is seen when we learn that every substance which has even been exposed to the air for a moment gives the soda line, when placed in a colorless flame; and when we find that the lithium compounds, which were formerly supposed to be contained in only four minerals, by aid of spectrum analysis are found to be substances of most common occurrence, being observed in almost all spring waters, in tea, tobacco, milk, and blood, but existing in such minute quantities as to have altogether eluded recognition by the older and less delicate analytical methods. $\frac{1}{6,000,000}$th part of a grain of lithium can thus be detected. A still more striking proof of the value of spectrum analysis lies in the fact of the recent discovery of four new elementary bodies by its means; two new alkaline metals, rubidium and cæsium, having been found, together with potash and soda in certain mineral springs, and two new metals, thallium and indium, having been respectively detected in iron pyrites and zinc ores. The new alkaline metals resemble potassium so closely in their properties, that it would be nearly impossible to have detected them by the ordinary analytical methods, although their spectra exhibit very distinct bright bands not seen in the potassium or any other known spectrum. The metal thallium was discovered by observing a splendid green line which did not belong to any known substance (*see* No. 5 of Chromo; whilst indium was recognized by the presence of a hitherto unobserved fine dark-blue line.

It is not only those bodies which have the power of imparting color to a flame which yield characteristic spectra, for this property belongs to every elementary substance, whether metal or non-metal, solid, liquid, or gas; and it is always observed when such element is heated to the point at which its vapor becomes luminous, for then each element emits the peculiar light given off by it alone, and the characteristic bright lines become apparent when its spectrum is observed. Most metals require a much higher temperature than the common flame, in order that their vapors should become luminous; but they may be easily heated up to the requisite temperature by means of the electric spark, which, in passing between two points of the metal in question, volatilizes a small portion, and heats it so intensely as to enable it to give off its peculiar light. Thus all the metals, among others iron, platinum, silver, and gold, may each be recognized by the peculiar bright lines which their spectra exhibit.

The permanent gases also yield characteristic spectra when they are strongly heated, as by the passage of an electric spark; thus, if the spark be passed through an atmosphere of hydrogen gas, the light emitted is bright red, and its spectrum consists of one bright red, one green, and one blue line; whilst in nitrogen gas the spark has a purple color, and the peculiar and complicated spectrum of nitrogen is observed when this spark is examined with a prism.

The instrument used in these experiments is termed a *spectroscope.* It consists of a prism (A, Fig. 57), fixed upon a firm iron stand, and a tube (B) carrying the slit (D) through which the rays from the colored flame (E) fall upon the prism, and being rendered parallel by passing through a lens. The light, having been refracted, is received by the telescope (F), and the image magnified before reaching the eye. For exact

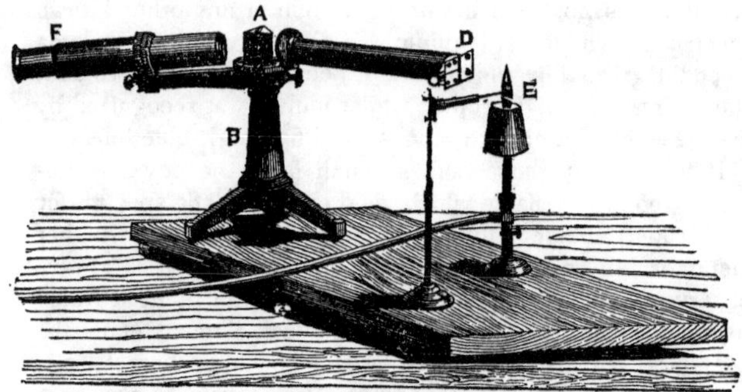

FIG. 57.

experiments, the number of prisms and the magnifying power are increased, and arrangements made for bringing two spectra into the field of view at once, so as to be able to make any wished-for comparison of the lines.

The peculiar appearance of the spectra of the alkalies and alkaline earths, as seen in the above instrument, is well represented by the colored plate placed at the commencement of this volume. On this, No. 1 represents the solar spectrum; No. 2 the spectrum of the potassium compounds; No. 3 that of the new metal rubidium; No. 4 that of the second new alkaline metal, cæsium; No. 5 that of the indium compounds; No. 6 that of the green flame of thallium; No. 7 is the sodium spectrum, the yellow line being identical in position with the dark solar line marked D; No. 8 is the spectrum of lithium; No. 9 that of the calcium compounds; No. 10 that of the strontium compounds; and No. 11 the complicated spectrum of barium salts. It will be at once evident that none of these lines overlie one another, and that if all the nine different substances were present together in a flame, it would be easy to detect the presence of each ingredient by the appearance of all its characteristic lines.

Solar and Stellar Chemistry.

If sunlight be allowed to fall upon the slit of the spectroscope, it is observed that the solar spectrum thus obtained differs essentially from the spectra which we have hitherto considered, inasmuch as it consists of a band of bright light, passing from red to violet, but intersected by a very large number of *fine black lines*, of different degrees of breadth and shade, which are always present, and always occupy exactly the same relative position in the solar spectrum. These lines indicate the absence in sunlight of particular rays, and they may be considered as shadows, or spaces where there is no light; they are called "*Fraunhofer's*" lines, after a German optician, who first satisfactorily mapped and described them.

In the last few years the existence of these lines has become a matter of great importance and interest, as it is by their help that the determination of the chemical constitution of the sun and far-distant fixed stars has become possible. The spectra of the moon and planets (reflected sunlight) are found to exhibit these same lines in unaltered position, whilst in the spectra of the fixed stars, dark lines also occur, but these stellar lines are different from those seen in direct and reflected sunlight. Hence the conclusion has been long drawn that the Fraunhofer's lines are in some way produced in the body of the sun itself; but it is only recently that the cause of their production has been discovered by Kirchhoff, and thus the foundation laid for the science of solar and stellar chemistry.

If the positions of these dark lines in the solar spectrum be carefully compared in a powerful spectroscope with those of the bright lines in the spectra of certain metals, such as sodium, iron, and magnesium, it is seen that each of the *bright* lines of the particular metal coincides not only in position, but also in breadth and intensity, with a *dark* solar line; so that if the apparatus be so arranged that a solar and metallic spectrum be both allowed to fall, one below the other in the field of the telescope, the bright lines of the metal are *all* seen to be continued in dark solar lines. In the case of metallic iron alone, more than sixty such coincidences have been observed, and the higher the magnifying power employed, the more striking and exact does this coincidence appear.

With other metals—such, for instance, as gold, antimony, lithium—no single coincidence can be noticed, whilst all the lines of certain other metals have their dark representatives in the sun. From these facts, it is clear that there must be some kind of connection between the bright lines of these metals and the coincident dark solar lines, as such coincidences cannot be the result of mere chance. Is the coincidence of the *dark* solar lines with the *bright* iron lines caused by the presence of iron in the sun? And if so, how do the lines come to appear *dark* in the solar spectrum?

The explanation of this is given by an experiment, in which the bright metallic lines are *reversed*, or changed into dark lines. Thus the bright yellow soda lines (coincident with Fraunhofer's line D) can be made to appear as a dark line, by allowing the rays from a strong source of white light (such as the oxyhydrogen light) to pass through a flame colored by soda, and then to fall upon the slit of the spectroscope. Instead of then seeing the usual soda spectrum of a *bright* yellow double-line upon a dark ground, a double *dark* line, identical in position and breadth with the soda line, will be seen to intersect the continuous spectrum of the white light. Here then the yellow flame has absorbed the same kind of light as it emits, a consequent diminution of intensity in that part of the spectrum occurred, and a dark line made its appearance. In like manner the spectra of many other substances have been *reversed*, each substance in the state of vapor having the power of absorbing the same rays it emits, or being opaque for such rays.

By observing the coincidences of these dark lines with the bright lines of terrestrial metals, we arrive at a knowledge of the occurrence of such metals in the solar atmosphere with as great a degree of certainty as we are able to attain to in any question of physical science. The metals hitherto detected in the sun's atmosphere are nine in number, viz., iron, sodium, magnesium, calcium, chromium, nickel, barium, copper, and zinc. Hydrogen is also known to exist in the sun.

Stellar Chemistry. The same methods of observation and reasoning apply to the determination of the chemical constitution of the atmospheres of the fixed stars, as these are self-luminous suns; but the experimental difficulties are greater, and the results, therefore, are as yet less complete, though not less conclusive, than is the case with our sun.

The spectra of the stars all contain dark lines, but these are for the most part different from the solar lines, and differ from one another; hence we conclude that the chemical constitution of the solar and stellar atmospheres is different. Many of the substances known on this earth have been detected in the atmosphere of the stars, by Mr. Huggins and Professor W. A. Miller, to whom we owe this most important discovery. Thus the star called Aldebaran contains hydrogen, sodium, magnesium, calcium, iron, tellurium, antimony, bismuth, and mercury; whilst in Sirius only sodium, magnesium, and hydrogen have with certainty been detected.

In examining the spectra of some of the nebulæ, a striking difference is observed: the stellar spectra, it will be remembered, resemble the spectrum of the sun, inasmuch as each consists of a *bright* ground intersected with *dark* lines; the spectra of the nebulæ, on the other hand, consist simply of *bright lines*, like the spectra of hydrogen, nitrogen, or any of the metals. Hence we conclude that the nebulæ are masses of glowing gas, and do not consist, like the sun and stars, of a solid or liquid mass, surrounded by a gaseous atmosphere.

CHEMISTRY OF THE CARBON COMPOUNDS, OR ORGANIC CHEMISTRY.

LESSON XXVII.

ORGANIC CHEMISTRY is defined as the chemistry of the carbon compounds. Many of these compounds exist already formed in the bodies of plants and animals, and hence the name of Organic Chemistry was given to this branch of the science. It is separated from the foregoing, Inorganic portion, not because any real difference exists in the laws regulating the formation of the bodies classed under these two great divisions, but because the number of compounds which have to be studied under organic chemistry is so large, and their constitution frequently so complicated, that they are at present best considered after the more simple inorganic compounds have been described.

Certain organic substances do, indeed, differ fundamentally in constitution and mode of formation from any inorganic compound, inasmuch as these exhibit what is termed an *organized* structure, being the sole and direct product of animal or vegetable life. Such an organized structure is seen in the simple cell, the germ of living organisms. It cannot be artificially prepared from its elementary constituents, whereas any crystalline or liquid organic body may possibly be thus built up from its elements.

The first striking peculiarity which the carbon compounds exhibit, is their extraordinary number, those already known far exceeding all the compounds of the other elements taken together, and new ones being daily brought to light. A second peculiarity of these compounds is, that they are almost all of them formed by the union of carbon in different proportions with one or more of three other elements; viz., hydrogen, oxygen, and nitrogen, whilst the number of *atoms* of these elements contained in the *molecule* of many organic bodies is extremely large ; thus sugar contains 45, and stearine no less than 173 atoms of their constituent elements.

The cause of the multiplicity of the carbon compounds is to be sought in a fundamental and distinctive property of carbon itself. This consists in the power which this element possesses, in a much higher degree than any of the others, of *uniting with itself* to form complicated compounds, containing an aggregation of carbon atoms united with either hydrogen, oxygen, nitrogen, or several of these, bound together to form a distinct chemical whole.

We have already seen (p. 138), that the atoms of different elements possess different powers of combination or *atomicities ;* that is, one atom of an element is capable of replacing either one or more atoms of hydrogen in combinations : thus Chlorine, Potassium, and Silver are *monads*, or can replace only one atom of hydrogen ; whilst Oxygen, Sulphur, and Magnesium are *dyad* elements, capable of replacing two of hydrogen ; and Nitrogen, Phosphorus, and Aluminium are *triads*, or play the part of three atoms of hydrogen. Carbon, on the other hand, is a *tetrad*, or tetravalent element ; and just as the compounds HCl, H_2O, H_3N are the representatives of the compounds of the monad, dyad, and triad elements, so marsh gas, H_4C, is the characteristic and representative compound of tetravalent carbon.

In this compound the *four combining units* of the carbon atom are saturated, or satisfied, by union with the four atoms of hydrogen, and hence this is said to be a *saturated compound.* Four atoms of any other monad would, however, satisfy this condition ; and we find, in fact, that one or more of the four atoms of hydrogen can be substituted, step by step, for chlorine, so that the following *substitution products* are obtained :

$$CH_4. \quad CH_3Cl. \quad CH_2\,Cl_2. \quad CH\,Cl_3. \quad C\,Cl_4.$$

The four combining powers of the carbon atom can be saturated not only by the union of the carbon to four monad atoms, but also by its union to two dyad atoms, or to one triad and one monad, or to one tetrad atom. Thus in carbon dioxide (carbonic acid), $O_2 C$, and carbon disulphide, $S_2 C$, we have a carbon atom saturated with two dyads; in hydrogen cyanide (prussic acid), HNC, we have a carbon atom saturated with a triad (N) and a monad (H) element.

When two atoms of tetravalent carbon unite together, a new radica. or group of atoms is formed; this *duplication* of the carbon element takes place by a combination of one of the four combining units of one atom with one of the four units of the other atom; so that two of the *eight* original combining units are saturated or disposed of, and only *six* remain free to combine. Hence, whilst CH_4 is the type of the monocarbon series, $C_2 H_6$ is that of the dicarbon series, and similarly, $C_3 H_8$ that of the tricarbon series; and no compound of any of these three series is known containing respectively more than four, six, or eight atoms of a monad.

Other groups of bodies exist in which all the combining powers of the carbon are not fully satisfied; such bodies, for instance, as carbonic oxide, CO, and olefiant gas, $C_2 H_4$. These substances are termed *non-saturated compounds*, and possess the peculiar property of uniting directly with other elements in such quantity as to fill up the vacant combining powers. Thus carbonic oxide and olefiant gas both combine directly with Cl_2 to form saturated compounds, which conform to the law above stated; whilst, on the other hand, we find it impossible to obtain a combination of chlorine with CO_2, or with $C_2 H_6$.

The following graphical representation of these three typical compounds (*see* p. 230) may help to render their mode of formation more evident.

From these figures it is seen that an addition of $C H_2$ is necessary to pass up the series. This addition can actually be experimentally made, and the higher and more complicated carbon groups thus obtained from the lowest and simplest one, whilst this, in its turn, can be prepared from its constituent elements. We are well acquainted with no less than fifteen of this series, containing from one to fifteen atoms of carbon, combined with a saturating quantity of hydrogen, and each member of the series forms a starting-point for a number of peculiar derivatives, all containing a common constituent, and exhibiting a family likeness.

The compounds obtained from each of these series of

Monocarbon Series. Dicarbon Series. Tricarbon Series.

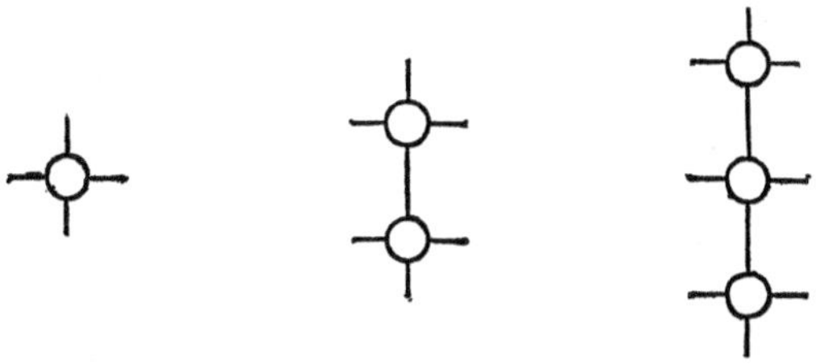

mono- di- tri- and higher carbon groups, may indeed be compared with those of the inorganic metals, and each different carbon series may be supposed to contain a group of atoms of carbon and hydrogen, which plays the same part in these compounds as the metal does in the metallic salts, and to which the name of compound radical has been given. The radical contained in each of the three typical substances just mentioned is found to be a hydrocarbon, containing one atom less hydrogen than the original type, and each of these bodies is therefore termed the *hydride* of a radical, and considered to be a molecule of hydrogen $\left.\begin{matrix}H\\H\end{matrix}\right\}$, in which one atom of the hydrogen is replaced by a radical ; thus we have :

Monocarbon Series.	Dicarbon Series.	Tricarbon Series.
Methyl- CH_3 Hydride H	Ethyl- C_2H_5 Hydride H	Propyl- C_3H_7 Hydride H

By replacing the one of hydrogen out of the radical by chlorine, we obtain the corresponding *chlorides:* viz.—

Monocarbon Series.	Dicarbon Series.	Tricarbon Series.
Methyl- CH_3 Chloride Cl	Ethyl- C_2H_5 Chloride Cl	Propyl- C_3H_7 Chloride Cl

And by replacing the same hydrogen by the monatomic radical H O in each hydride, we obtain an important class of bodies termed the *alcohols.*

Monocarbon Series.	Dicarbon Series.	Tricarbon Series.
Methyl- Alcohol $\left.\begin{matrix}CH_3\\H\end{matrix}\right\}O$	Ethyl- Alcohol $\left.\begin{matrix}C_2H_5\\H\end{matrix}\right\}O$	Propyl- Alcohol $\left.\begin{matrix}C_3H_7\\H\end{matrix}\right\}O$

The molecules of the radicals methyl CH_3, ethyl C_2H_5, and propyl C_3H_7, in these several compounds remain indivisible throughout all the derivatives, and give the peculiar characters to each series.

As in mineral chemistry we have radicals (*see* p. 139), some of which are monads, and some dyads, triads, or tetrads, so amongst the carbon compounds some radicals exist in which more than one combining power remains unsaturated, and which therefore act as *polyatomic radicals:* thus methylene $\overset{II}{CH_2}$, ethylene $\overset{II}{C_2H_4}$, and propylene, $\overset{II}{C_3H_6}$, are dyads; whilst glyceryl, $\overset{III}{C_3H_5}$, is a triad.

These bodies give rise to a large class of derivatives, each containing the radical or group of carbon and hydrogen atoms. All the substances analogous to the foregoing, or derived from them, are classed as the *fatty group* of organic bodies. Other carbon compounds are, however, known, which are saturated, but contain the carbon atoms more intimately united together, than the members of the alcohol group; the principal group of these substances is termed the *aromatic group* of organic bodies. Thus, the formula of Benzol is C_6H_6, and in this body eighteen of the twenty-four combining powers of the six atoms of tetratomic carbon are saturated by combination of carbon with carbon.

We shall first study the properties and mode of formation of some of the most important members of the alcohol group, and then notice the chief properties of the aromatic series of organic bodies.

It follows from the tetratomic character of carbon, that however the carbon atoms may be united together, the combining powers which remain unsaturated must be an even number; hence the sum of the atoms of monad or triad elements united with the carbon must be an even number, whilst the number of dyad elements is not thus restricted.

Empirical and Rational Formulæ.

The simplest mode of expressing the composition of an organic compound is to write down the number of the constituent atoms side by side, thus:

C_2H_6	Ethyl-hydride,
C_2H_6O	Ethyl-Alcohol,
O_2H_7N	Ethylamine,
$C_2H_4O_2$	Acetic Acid.

These represent the molecular weights of the substances, and are called *empirical formulæ.* Amongst the very large number of carbon compounds, it happens not unfrequently that two or more organic bodies possess the same chemical composition (that is, they contain the same number of the same elements), although they differ in their chemical and physical properties. In order to distinguish between these *isomeric* bodies, it is necessary to employ *rational formulæ* for the purpose of giving an idea of the chemical nature of the substances, or representing the decompositions which they undergo. The foregoing compounds can be represented by the following rational formulæ :

Ethyl-hydride, $\left.\begin{matrix} C_2H_5 \\ H \end{matrix}\right\}$; Ethyl-alcohol, $\left.\begin{matrix} C_2H_5 \\ H \end{matrix}\right\} O.$

Ethylamine, $\left.\begin{matrix} C_2H_5 \\ H \\ H \end{matrix}\right\} N$; Acetic Acid, $\left.\begin{matrix} C_2H_3O \\ H \end{matrix}\right\} O.$

This denotes that the monad radical C_2H_5 is contained in the first three compounds; that alcohol may be regarded as water in which one atom of hydrogen has been replaced by ethyl, and that ethylamine stands in the same relation to ammonia. The formula for acetic acid shows that it may be considered to be alcohol in which two atoms of hydrogen are replaced by one atom of oxygen, and that a monobasic acid is thus formed. A single formula cannot possibly represent all the relations of a compound,—hence one body may possess several rational formulæ ; and it is important also to remember that the formula never pretends to point out the actual position of the atoms in the molecule, but simply to show the deportment of the compound. We shall in the future have frequent occasion to employ both empirical and rational formulæ of different kinds for the same substance, according to the nature of the reaction or peculiar property which we desire to explain.

END OF LESSON XXVII.

LESSON XXVIII.

Determination of the Composition of Carbon Compounds.

1. *Organic Analysis. Estimation of Carbon and Hydrogen.*

As all organic compounds contain carbon, and most of them hydrogen, the estimation of these two constituents becomes a matter of importance, and the method of analysis remains nearly the same for all organic substances. It is founded upon the fact, that when any compound of carbon is heated to redness with excess of oxygen, it undergoes complete combustion, the carbon being oxidized to carbon dioxide (carbonic acid), and the hydrogen to water, so that by weighing the quantity of these two products obtained by burning a given weight of the substance, we can ascertain the weight of carbon and of hydrogen which the substance contained.

The combustion of the organic compound can either be made in a current of pure oxygen gas, or by mixing the body with pure copper oxide (Cu O), which readily parts with its oxygen to hydrogen or carbon at a red heat—in either method the products of combustion being carefully collected and weighed. A weighed quantity (generally about 0·3 grm.) of the solid substance about to be analyzed by means of copper oxide is brought into a combustion tube made of hard Bo-

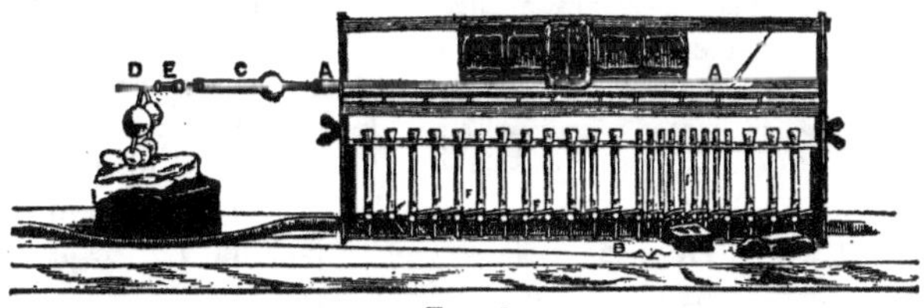

Fig. 58.

hemian glass (AA, Fig. 58), about 50 to 60 centimetres in

length, and drawn out at one end to a fine point, and open at the other. Before the introduction of the substance, a quantity of pure and perfectly dry, freshly-ignited, granulated, copper oxide is brought into the tube sufficient to fill about one quarter of its length, and the substance is well mixed with this oxide by means of a brass wire (B, Fig. 58); fresh oxide is then added, the brass wire being well cleaned from every possible trace of the adhering substance, until the tube is nearly filled.

The apparatus intended to collect the water produced is now attached to the open end of the tube by means of a well-fitting dry cork: it consists of a tube (C), filled with porous pieces of calcium chloride, and accurately weighed; this substance effectually absorbs all the water and aqueous vapor formed in the combustion; the carbon dioxide passes through the tube unabsorbed, and bubbles into a solution of strong caustic potash contained in the bulb apparatus (D), attached to the drying tube by a well-fitting caoutchouc joining (E). The increase in weight of the drying tube and potash bulbs gives respectively the weight of water and carbon dioxide produced.

The combustion tube is placed in a long furnace, so that it can be gradually heated to redness; this is most readily effected by a number of gas-burners arranged in line, so that each part of the tube can be gradually and separately heated (F). A larger number of small burners are placed under the part of the tube containing the substance, in order that the combustion may be more accurately regulated. After the whole arrangement has been shown to be properly air-tight, the part of the tube near the cork, containing only pure copper oxide, is heated; and when a length of 15–20 centimetres of it is red-hot, the part of the tube containing the substance is gently heated, until bubbles of carbonic acid are seen slowly to enter the potash bulbs; and the heat is applied, so that this slow disengagement of carbonic acid continues until the whole of the substance is burnt.

When the gas ceases to come off, the whole length of the

tube is strongly heated for some minutes, and as soon as the potash solution begins to pass back into the bulb nearest the combustion tube (owing to absorption of the carbonic acid), the drawn-out end of the tube is broken, the gas flames extinguished at that end of the furnace, and air drawn for some minutes through the whole apparatus by sucking with the mouth through a tube placed on the end of the potash-bulbs. This operation is necessary, in order to collect in the potash the carbonic acid which still remains in the combustion-tube. As soon as this is finished, the analysis is complete with the exception of weighing the drying tube and potash-bulbs. Many precautions must be taken, and much attention to details must be paid, in order to insure accurate results in organic analysis ; for an enumeration of these the larger manuals must be consulted.

If the substance under examination is a liquid, it is sealed up in a small weighed glass bulb drawn out to a fine point; this is again weighed, the point broken off, and the bulb dropped into the combustion tube, and the operation conducted as above described. When nitrogen is contained in the body about to be analyzed, it is necessary to place some turnings of metallic copper in the front part of the tube to decompose any nitrous fumes which are formed and would be absorbed by the potash, and thus impair the result.

Determination of Nitrogen. Nitrogenous organic bodies, when heated with caustic soda, or potash, yield the whole of the nitrogen which they contain in the form of ammonia. This evolution of ammonia is easily rendered evident by heating a small piece of cheese with solid caustic soda. Upon this reaction a method is based for determining the quantity of nitrogen in organic bodies; it consists simply in heating a given weight of substance with a mixture of caustic soda and quicklime in a tube, and collecting the ammonia formed, in hydrochloric acid, and estimating the weight of ammonium-chloride produced by weighing as double platinum salt. For every 100 parts by weight of this salt obtained, the substance contains 6·35 parts of nitrogen.

In certain cases, viz., when the nitrogen is contained as an oxide in the organic substance, the foregoing method cannot be employed, inasmuch as these oxides are not completely converted into ammonia; it is then necessary to obtain the nitrogen gas in the free state by heating the substance with a mixture of copper and mercury oxides, and passing the gases produced over metallic copper. All the nitrogen comes off in the gaseous form, and may be easily purified by caustic soda from the carbonic acid also evolved, and thus the volume of nitrogen obtained can be accurately measured. From this volume, measured under given circumstances of temperature and pressure, the weight of the nitrogen can, of course, be calculated.

Chlorine, sulphur, and phosphorus exist not unfrequently in organic bodies, and have to be determined: the first is estimated by heating the substance to redness in a tube containing pure quicklime; the chlorine forms calcium chloride, in which, on solution in nitric acid, the chlorine is weighed as silver salt. Sulphur and phosphorus are determined by heating the organic body with a mixture of pure nitre and sodium carbonate, placed in a tube; sulphuric and phosphoric acids are formed and estimated in the usual manner.

Oxygen is usually obtained by difference, that is, by subtracting the sum of the weights of all the constituents which have been directly determined from the weight of the substance taken: several direct methods for the estimation of oxygen have been proposed, but these are not often used.

2. *Determination of the Molecular Weight of an Organic Compound.*

The above methods of analysis give us the percentage composition of the substance, and the *relation* between the number of atoms of carbon, hydrogen, oxygen, &c., contained in the compound, but we need to make a further determination in order to get to know the *formula* and the *molecular weight* of the body. Thus, in an analysis of glacial acetic acid, 0·395 grammes of substance was found to yield 0·580 grammes of carbonic acid, and 0·235 grammes of water; hence 100 parts of glacial acetic acid consist of

Carbon	40·0	
Hydrogen . . .	6·6	
Oxygen	53·4	(By difference.)
	100·0	

If we divide these numbers respectively by the combining weights of carbon, hydrogen, and oxygen, $\frac{40}{12} = 3·3$, $\frac{6·6}{1} = 6·6$, and $\frac{53·4}{16} = 3·3$, we obtain the relation between the combining weights of these constituents present. Thus we see that the number of atoms of carbon and oxygen is equal, whilst that of hydrogen is twice as large; the composition of acetic acid is, therefore, represented by $C_n\,H_{2n}\,O_n$, but we do not know whether the true formula is $C\,H_2\,O$, $C_2\,H_4\,O_2$, or $C_3\,H_6\,O_3$; or whether it contains a still higher number of carbon atoms. In order to decide this point, and, therefore, to determine the *molecular weight* of the substance, we must endeavor to find a compound of it with some well-known element (such as silver), in which one atom of hydrogen in acetic acid is replaced by one atom of silver; that is, we must find the weight of C, H, and O, in the ascertained relative proportion, which forms a compound with one atom of silver. On examination we find that only one such compound of silver and acetic acid exists; and we find by experiment that 100 parts of silver acetate contain 64·68 parts by weight of silver; hence, the weight of the carbon, hydrogen, and oxygen, united with silver (Ag = 108), is $\frac{35·32 \times 108}{64·68} = 58·98$. In this silver acetate, however, one atom of hydrogen of the glacial acid was replaced by one of silver, so that the molecular weight of the glacial acetic acid is found to be $58·98 + 1 = 59·98$, or its formula is :

$$
\begin{aligned}
2\,C &= 24 \\
4\,H &= \ \ 4 \\
2\,O &= 32 \\
& \overline{\ 60\ }
\end{aligned}
$$

The slight difference observed between the found (59·98) and the calculated (60) molecular weight arises from unavoidable errors of experiment; the more analyses of the substance are made, the nearer will the mean result approach the calculated numbers.

In a similar manner the molecular weights of organic bases are determined by ascertaining the weight of the substance which unites with a known weight of hydrochloric acid to form a salt. In the case of certain organic acids and bases, two or more compounds containing different proportions of silver (or other metal) and hydrochloric (or other acid) are known, and it becomes a matter for consideration which of these is to be taken as containing one molecule of the organic compound to one atom of metal or acid; the choice in these cases is determined by the general properties of all the compounds, which never fail to point out the true character of the substance. The same method of decision also applies to many other bodies, such as sugar, turpentine, &c., which do not readily enter into combination with a metal or an acid.

There is one most important property by which the molecular weight of volatile organic bodies can be ascertained, viz., the *density* or *specific gravity of their vapors.* We have already seen that the vapor volume occupied by the molecule of almost all volatile compounds is twice that occupied by the atom of hydrogen. There are a very few exceptions to this general law, and these exceptions can frequently be explained by the fact that the substances decompose when heated, and that the vapor is not simply that of the original compound. The molecule of water, H_2O, weighing 18, occupies twice as large a volume as the atom of hydrogen weighing 1, or the density of water-gas is 9; so hydrochloric acid, HCl, weighing 36·5, occupies two volumes, and its density is 18·75, and ammonia, NH_3, weighing 17, has a density of 8·5.

This same simple relation also holds good in organic chemistry. *The molecule of every volatile organic compound occupies a volume twice as large as that occupied by an atom of hydrogen weighing* 1; *or the vapor density of an organic compound is half its molecular weight.*

The experimental determination of the vapor densities of organic compounds thus becomes an important matter as serving to control the correctness of the molecular weight ascertained by the foregoing methods. Thus, for instance, the density of the vapor of acetic acid is found by experiment to be 30·07 (H = 1); and this accordingly gives a molecular weight to acetic acid of 60·14, a number agreeing with that obtained from purely chemical considerations (*see ante*, p. 241).

Another example may serve to render evident the importance of this relation: the combustion of acetal shows that the simplest relation of its constituent atoms is represented by the formula $C_3 H_7 O$;* the determination of vapor density, however, gives the number 59·8 as the density of acetal gas, hence the molecular weight of acetal must be 59×2, and its formula not $C_3 H_7 O = 59$, but $C_6 H_{14} O_2 = 118$. It is, of course, possible, when the molecular weight of a compound has been otherwise ascertained, to calculate its vapor density; this calculated density will always differ slightly from that determined by experiment, owing to the unavoidable errors which occur; this, however, does not detract from the value of this method of controlling the molecular formula of a substance.

Determination of Vapor Density.

Two methods are employed for determining the vapor

* We see that as this formula contains an uneven number of hydrogen atoms the existence of this substance is impossible: we know that the true formula must be a multiple of this, which multiple we decide by the vapor density.

density of a compound. (1) By ascertaining the *weight* of a given volume of vapor. (2) By ascertaining the *volume* of a given weight of vapor. In the first of these processes, a thin glass globe is employed of about 200 to 300 cubic centimetres in capacity, having a finely drawn-out neck; the exact weight of the globe filled at a certain temperature and under an observed pressure having been found, a small portion of the substance whose density is to be determined is brought inside, and the globe then heated by plunging it into a water- or oil-bath (Fig. 59) raised to a temperature much above the boiling point of the substance. As soon as the

FIG. 59.

vapor has ceased to issue from the end of the neck, this end is hermetically sealed before a blowpipe, and the exact temperature as well as barometric pressure observed. The bulb thus filled with vapor is allowed to cool, and is next accurately weighed, and the point of the neck broken under mercury; the mercury rushes into the globe, owing to the vapor being condensed, and, if the experiment has been well conducted, completely fills it. From the volume of mercury which thus enters, the capacity of the globe is obtained.

We have now all the data necessary for the determination. In the first place we have to find the weight of a given volume of the vapor under certain circumstances of temperature and pressure, and we then have to compare this with the weight of an equal volume of hydrogen gas measured under the same circumstances. The following example of the vapor density of volatile hydro-carbon may serve to illustrate the method; weight of globe filled with dry air at 15·5° 23·449 grammes; weight of globe filled with vapor at 110° 23·720 grammes; capacity of the globe 178 cbc. As the barometric column stood near to 760 mm. and underwent no change from the beginning to the end of the experiment, no correction for pressure is necessary. In order to get the weight of the vacuous globe, the weight of air contained must be deducted from the weight of globe in air. Now 1 cbc. of air at 0°, and 760 mm. weighs 0·001293 gramme, and 178 cbc. of air at 15·5 would occupy $\frac{178 \times 273}{288 \cdot 5} = 168 \cdot 4$ at 0°, and the weight of this air is 0·218 gramme: hence the weight of the vacuous bulb is 23·231, and the weight of vapor 23·720—23·231 = 0·489 gramme. We must now find what 178 cbc. of hydrogen at 110° will weigh: 1,000 cbc. of hydrogen at 0° weigh 0·08936 gramme; 178 cbc. will contract to 126·9 cbc. at 0°. 126 cbc. of hydrogen at 0° weigh 0·01134 gramme, and this is therefore the weight of 178 cbc. of hydrogen at 110°. Hence $\frac{0 \cdot 489}{0 \cdot 01134} = 43 \cdot 13$ is the density of the vapor, as found by experiment. The formula of the substance is $C_6 H_{14}$, or its molecular weight is 86. In this example many minor corrections, such as the expansion of the glass globe, the error of the mercurial thermometer, &c., are not considered; the above method gives results which are sufficiently accurate when the object is to control the molecular weight of a compound.

The second method of vapor density determination consists in ascertaining the volume occupied by a given weight of substance when heated up to a temperature considerably above its boiling point. The mode of calculation is in principle the same as that of the former method. For the details of manipulation the reader must refer to the larger manuals.

Boiling Point and Fractional Distillation.

Another important physical property of organic compounds is the boiling point. Every volatile chemical compound has, under given circumstances of pressure, a fixed and constant boiling point; and this property is useful in ascertaining the purity of an organic liquid, as well as enabling us to separate the constituents of a mixture by means of fractional or continued distillation. The boiling points of the *homologous* series of hydrides, alcohols, chlorides, &c. (p. 231), rises with the increase of carbon, and frequently proportionally to this increase, although no general law connecting boiling point and chemical composition can be expressed. The arrangement used in the separation of liquids boiling at different points by means of fractional distillation is represented in Fig. 60. The large surface presented by the bulb tube (*a*) allows the vapor of the less volatile constituents to condense

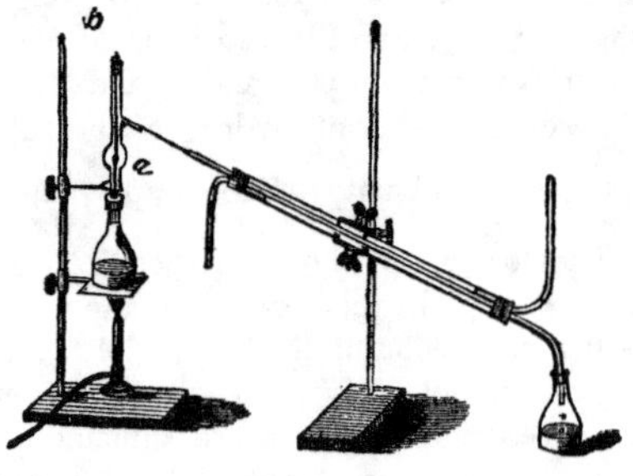

FIG. 60.

and flow back into the flask containing the mixture; the temperature of the vapor is indicated by the thermometer (*b*), and when the temperature rises beyond a given point the liquid already distilled over is removed and an empty flask substituted to collect the portion of liquid next coming over.

LESSON XXIX.

MONATOMIC ALCOHOL GROUP.

General Characteristics. The monatomic alcohols and their derivatives form a very large and important group of organic compounds. As an example of these alcohols we may take ethyl alcohol, $C_2 H_6 O$, known as spirits of wine; this substance, in common with all the other alcohols of this series, may be considered as water in which one atom of hydrogen is replaced by a radical having in this case the formula $C_2 H_5$; hence ethyl alcohol is $\left.\begin{matrix} C_2H_5 \\ H \end{matrix}\right\} O$. Ethyl alcohol is in constitution analogous to caustic potash, $\left.\begin{matrix} K \\ H \end{matrix}\right\} O$; and as, by adding hydrochloric acid to the latter, we get KCl (potassium chloride) and $\left.\begin{matrix} H \\ H \end{matrix}\right\} O$; in like manner the chlorides, iodides, and bromides of all the alcohol radicals can be obtained. The analogy of the ethyl with the potassium compounds is still further seen in the fact that an ethyl compound exists which stands to alcohol in the same relation as potassium monoxide to caustic potash; this compound is common- or ethyl-ether, $\left.\begin{matrix} C_2H_5 \\ C_2H_5 \end{matrix}\right\} O$. We also have analogous compounds to the potassium salts; thus we have, potassium nitrate . . . . $\left.\begin{matrix} K \\ NO_2 \end{matrix}\right\} O$;

Ethyl nitrate . . . . . . $\left.\begin{matrix} C_2H_5 \\ NO_2 \end{matrix}\right\} O$;

Hydrogen potassium sulphate $\left.\begin{matrix} K \\ H \end{matrix}\right\} SO_4$;

Hydrogen-ethyl sulphate . . $\left.\begin{matrix} C_2H_5 \\ H \end{matrix}\right\} SO_4$;

Potassium sulphate . . . . $\left.\begin{matrix} K \\ K \end{matrix}\right\} SO_4$;

Ethyl sulphate . . . . . $\left.\begin{matrix} C_2H_5 \\ C_2H_5 \end{matrix}\right\} SO_4$;

Potassium acetate . . . . $\left.\begin{matrix} C_2H_3O \\ K \end{matrix}\right\} O$;

Ethyl acetate . . . $\left.\begin{matrix} C_2\,H_3\,O \\ C_2\,H_5 \end{matrix}\right\}O.$

If ethyl alcohol be exposed to oxidizing agents, it first loses 2 atoms of hydrogen, and is converted into a new substance, $C_2\,H_4\,O$, called ethyl aldehyde; and if the oxidizing action continues longer, another product termed acetic acid is formed, which has the composition $C_2H_4O_2$. Both these substances may be regarded as containing an oxidized radical, or ethyl in which 2 atoms of hydrogen are replaced by 1 atom of oxygen; aldehyde thus becomes the hydride of this radical (called Acetyl), $\left.\begin{matrix} C_2H_3O \\ H \end{matrix}\right\}$, whilst acetic acid is water in which 1 atom of hydrogen is replaced by acetyl: thus, $\left.\begin{matrix} C_2H_3O \\ H \end{matrix}\right\}O.$ Aldehyde can be reduced again to alcohol directly by addition of 2 atoms of hydrogen, but acetic acid cannot be directly reduced to alcohol.

Every alcohol can thus be oxidized, and yields an aldehyde and an acid which stand in the same relation to one another as the above-mentioned bodies. All these acids are monobasic; that is, they contain only 1 atom of hydrogen replaceable by a metal. This hydrogen can also be replaced, not only by the ethyl and the other alcohol radicals, giving rise to bodies called the compound ethers, of which, $\left.\begin{matrix} C_2H_3O \\ C_2H_5 \end{matrix}\right\}O,$ acetic ether or ethyl-acetate, may be taken as an example, but also by acetyl itself or the other oxidized radicals; thus we obtain $\left.\begin{matrix} C_2H_3O \\ C_2H_3O \end{matrix}\right\}O,$ a substance which we shall term acetyl acetate, but which is frequently called acetic anhydride, or even anhydrous acetic acid.

Each alcohol also forms a series of compound ammonias; that is, ammonia, $\left.\begin{matrix} H \\ H \\ H \end{matrix}\right\}N$, in which one or more atoms of hydrogen are replaced by a radical: thus, for the ethyl series we have ethylamine $\left.\begin{array}{c}C_2H_5\\ H\\ H\end{array}\right\}N$; diethylamine, $\left.\begin{array}{c}C_2H_5\\ C_2H_5\\ H\end{array}\right\}N$; and triethylamine, $\left.\begin{array}{c}C_2\,H_5\\ C_2\,H_5\\ C_2\,H_5\end{array}\right\}N$. We can, indeed, go one step further in the addition of ethyl, and obtain a caustic substance resembling potash in its properties, and analogous to the ammonium hydrate, $\left.\begin{array}{c}NH_4\\ H\end{array}\right\}O$, but containing 4 of ethyl in place of the 4 of hydrogen ; thus, $\left.\begin{array}{c}N(C_2H_5)_4\\ H\end{array}\right\}O$; to this substance the name of tetra-ethyl-ammonium hydrate is given.

Compound ammonias are also known in which one or more atoms of the hydrogen of ammonia is replaced by the oxygenized radical of the acids, and these compounds are termed Amides ; thus we have with acetyl,

acetamide, $\left.\begin{array}{c}C_2H_3O\\ H\\ H\end{array}\right\}N$; diacetamide, $\left.\begin{array}{c}C_2H_3O\\ C_2H_3O\\ H\end{array}\right\}N$;

and ethyl-diacetamide, $\left.\begin{array}{c}C_2H_3O\\ C_2H_3O\\ C_2H_5\end{array}\right\}N$.

Compounds of the alcohol radicals analogous to arsenic and phosphorus trihydrides are also known ; thus, for instance, $\left.\begin{array}{c}C\,H_3\\ C\,H_3\\ C\,H_3\end{array}\right\}As$, trimethyl-arsine ; and, $\left.\begin{array}{c}C_2\,H_5\\ C_2\,H_5\\ C_2\,H_5\end{array}\right\}P$, triethyl-phosphine, are known. The alcohol radicals likewise combine with metals, such as zinc, tin, &c., to form bodies which in their turn combine with chlorine, &c., and have, therefore, been termed the organo-metallic radicals ; such substances are zinc ethyl, $\left.\begin{array}{c}C_2\,H_5\\ C_2\,H_5\end{array}\right\}Zn$, and cacodyl, $(C\,H_3)_2\,As$.

The following is a complete list of all the monatomic alcohols and acids now known, giving their formulæ, and boiling and melting points. In some cases the alcohol corresponding to a known acid has not yet been obtained ; a blank is then left in the alcohol series.

PRIMARY ALCOHOLS. General formula $C_n H_{2n+2}O$ } yield {			MONOBASIC ACIDS. General formula $C_n H_{2n} O_2$.			
Name.	Formula.	Boiling Point.	Name.	Formula.	Boiling Point.	Melting Point.
Methyl	$C H_4 O$	66°C	Formic . .	$C H_2 O_2$	100°	+ 1°
Deutyl or Ethyl .	$C_2 H_6 O$	78°·4	Acetic . .	$C_2 H_4 O_2$	118°	+ 17°
Trityl or Propyl .	$C_3 H_8 O$	96°	Propionic .	$C_3 H_6 O_2$	140°	below 20°
Tetryl or Butyl .	$C_4 H_{10} O$	109°	Butyric . .	$C_4 H_8 O_2$	162°	—
Pentyl or Amyl .	$C_5 H_{12} O$	132°	Valerianic .	$C_5 H_{10} O_2$	174°	—
Hexyl	$C_6 H_{14} O$	150°	Caproic . .	$C_6 H_{12} O_2$	199°	+ 5°
Heptyl	$C_7 H_{16} O$	164°	Œnanthylic	$C_7 H_{14} O_2$	219°	—
—	—	—	Capric . .	$C_8 H_{16} O_2$	236°	+ 14°
—	—	—	Pelargonic.	$C_9 H_{18} O_2$	260°	+ 18°
Decatyl	$C_{10} H_{22} O$	212	Rutic . .	$C_{10} H_{20} O_2$	—	27°·2
—	—	—	Lauric . .	$C_{12} H_{24} O_2$	—	43°·6
—	—	—	Myristic .	$C_{14} H_{28} O_2$	—	53°·8
Cetyl	$C_{16} H_{34} O$	melt. pnt. 20°	Palmitic .	$C_{16} H_{32} O_2$	—	62°·
—	—	—	Margaric .	$C_{17} H_{34} O_2$	—	—
—	—	—	Stearic . .	$C_{18} H_{36} O_2$	—	69°·2
—	—	—	Arachic .	$C_{20} H_{40} O_2$	—	75°
—	—	—	Behenic .	$C_{22} H_{44} O_2$	—	76°
—	—	—	Hyaenic .	$C_{25} H_{50} O_2$	—	77°
Ceryl	$C_{27} H_{56} O$	melt. pnt. 79°	Cerotic . .	$C_{27} H_{54} O_2$	—	78°
Melyl	$C_{30} H_{62} O$	85°	Melissic .	$C_{30} H_{60} O_2$	—	88°

Secondary Alcohols.

In the foregoing class of primary alcohols the group OH, hydroxyl, is attached to a carbon atom at the end of the chain. (See the figures on p. 272.) Other classes of alcohols are, however, known to exist: one of these is termed the class of secondary alcohols ;—in these the hydroxyl is attached to a carbon atom which is placed in the centre of the chain, or between two other carbon atoms. A glance at the figures at p. 272 will show that in the mono- and di-carbon series no secondary alcohols can occur. The first series in which such a compound exists is that containing the 3 carbon or propyl radical. The primary and secondary alcohols containing the same number of carbon atoms are *isomeric*, but differ in many of their properties and in the mode in which they undergo decomposition.

The secondary propyl alcohol is represented by the formula

$$\begin{cases} CH_3 \\ CH\,OH \\ CH_3 \end{cases} \quad \text{or} \quad C\begin{cases} CH_3 \\ CH_3 \\ H \\ OH \end{cases},$$

and is called Dimethyl carbinol: carbinol itself being $C\begin{cases} H \\ H \\ H \\ OH \end{cases}$, or methyl alcohol; and methyl carbinol, $C\begin{cases} CH_3 \\ H \\ H \\ OH \end{cases}$, being ethyl alcohol. On oxidation these substances lose 2 atoms of hydrogen, but yield no aldehyde, but a body termed a Ketone. Thus:

$$\underset{\text{Dimethyl carbinol.}}{\begin{cases} CH_3 \\ CH\,OH \\ CH_3 \end{cases}} - H_2 = \underset{\text{Dimethyl ketone.}}{\begin{cases} CH_3 \\ CO \\ CH_3 \end{cases}}.$$

The ketones take up hydrogen, forming the secondary alcohol, but on oxidation they do not yield the corresponding acid, forming acids containing a smaller number of carbon atoms.

The following is a list of the secondary alcohols at present known. They all contain the group CH_3, and on oxidation this is liberated in combination with one atom of carbon as acetic acid, whilst the remaining alcohol radical forms its corresponding acid. Thus from methyl-hexyl-carbinol we obtain acetic and caproic acids.

List of Secondary Alcohols.

		Boiling point.
Dimethyl carbinol	$C_3H_8O = C\left\{\begin{array}{l} CH_3 \\ CH_3 \\ O \\ OH \end{array}\right.$	84°
Methyl-ethyl-carbinol . . .	$C_4H_{10}O = C\left\{\begin{array}{l} CH_3 \\ C_2H_5 \\ H \\ OH \end{array}\right.$	97°
Methyl-propyl-carbinol . .	$C_5H_{12}O = C\left\{\begin{array}{l} CH_3 \\ C_3H_7 \\ H \\ OH \end{array}\right.$	108°
Methyl-butyl-carbinol . . .	$C_6H_{14}O = C\left\{\begin{array}{l} CH_3 \\ C_4H_7 \\ H \\ OH \end{array}\right.$	136°
Methyl-hexyl-carbinol . . .	$C_8H_{18}O = C\left\{\begin{array}{l} CH_3 \\ C_6H_{13} \\ H \\ OH \end{array}\right.$	181°

Tertiary Alcohols.

A third class of alcohols exists in which the hydroxyl (OH) is attached to a carbon atom which is placed between three other carbon atoms. These alcohols yield chlorides, &c., and on oxidation they at once split up into acids containing a lower number of carbon atoms. The first term of this class is that of the 4 carbon series, tertiary butyl alcohol, or trimethyl carbinol. The following are the members of the tertiary alcohol group as far as at present known:

		Boiling point.
Trimethyl carbinol . . .	$C_4H_{10}O = C\left\{\begin{array}{l} (CH_3)_3 \\ OH \end{array}\right.$	82°
Dimethyl-ethyl-carbinol .	$C_5H_{12}O = C\left\{\begin{array}{l} (CH_3)_2 \\ C_2H_5 \\ OH \end{array}\right.$	100°
Dimethyl-propyl-carbinol .	$C_6H_{14}O = C\left\{\begin{array}{l} (CH_3)_2 \\ C_3H_7 \\ OH \end{array}\right.$	120°

		Boiling point
Methyl-diethyl-carbinol .	$C_6H_{14}O = C\begin{cases} CH_3 \\ (C_2H_5)_2 \\ OH \end{cases}$	115°
Triethyl carbinol	$C_7H_{16}O = C\begin{cases} (C_2H_5)_3 \\ OH \end{cases}$	—
Diethyl-propyl-carbinol .	$C_8H_{18}O = C\begin{cases} (C_2H_5)_2 \\ C_3H_7 \\ OH \end{cases}$	—

MONOCARBON OR METHYL SERIES.

Radical Methyl, CH_3. *Methyl Alcohol*, $\left.\begin{matrix} CH_3 \\ H \end{matrix}\right\} O$, commonly called wood-spirit. It occurs in the dry distillation of wood, forming about one per cent. of the aqueous distillate; it is likewise met with in the oil of winter-green, derived from the *Gaultheria procumbens.* Methyl alcohol can likewise be synthetically built up from its constituent elements, but only by means of several complicated reactions, which will afterwards be mentioned.

Pure methyl alcohol is obtained from crude wood-spirit, in which it is contained mixed with a variety of other organic compounds, by forming a crystalline methyl oxalate; this, on treatment with water, is decomposed, and yields the alcohol in the pure state. Methyl alcohol is a colorless, mobile liquid, possessing a pure spirituous smell; the specific gravity of the liquid is 0·8142 at 0°, and its boiling point is 66°. It burns with a non-luminous flame, and is soluble in and miscible with water. Potassium dissolves in methyl alcohol with evolution of hydrogen and formation of potassium methylate, $\left.\begin{matrix} CH_3 \\ K \end{matrix}\right\} O$. Methyl alcohol, when acted on by oxidizing agents, yields methyl aldehyde and formic acid. By the action of bleaching powder on methyl alcohol, chloroform is obtained; acted upon by hydrochloric acid, the alcohol yields methyl chloride.

The action of strong sulphuric acid on methyl alcohol is remarkable, and is the type of a general reaction. These two substances must be mixed with care, as great heat is evolved when they come in contact. The first substances formed are hydrogen methyl-sulphate (by exchange of hydrogen for methyl) and water, $\left.\begin{matrix}H\\H\end{matrix}\right\}O$. When the former body comes in contact with another molecule of alcohol, we have another exchange of hydrogen and methyl occurring; but as this exchange can occur in two directions, we get either $\left.\begin{matrix}CH_3\\CH_3\end{matrix}\right\}O$ and $\left.\begin{matrix}H\\H\end{matrix}\right\}SO_4$, or $\left.\begin{matrix}H\\H\end{matrix}\right\}O$ and $\left.\begin{matrix}CH_3\\CH_3\end{matrix}\right\}SO_4$: in the first case, dimethyl ether and sulphuric acid; and in the second, water and methyl sulphate.

Methyl Chloride, $CH_3\ Cl$, is obtained as a colorless gas, condensing at -20°, by acting upon methyl alcohol with hydrochloric acid or phosphoric penta-chloride; it is also formed along with other substances by the action of chlorine upon marsh gas. The bromide and iodide are colorless liquids, prepared by acting on methyl alcohol with bromine and iodine in presence of phosphorus.

Methyl Hydride. Marsh gas, $CH_3\ H$. As we have seen, this gas occurs in nature as fire-damp and the gas of marshes. It can be obtained easily by heating sodium acetate with caustic alkali; the acetic acid splitting up into carbon dioxide and marsh gas, $C_2\ H_4\ O_2 = CO_2 + CH_4$.. Methyl hydride can also be obtained by passing the vapor of carbon disulphide, together with sulphuretted hydrogen gas, through a red-hot tube, and in this way it may be built up from its constituent elements. It may likewise be obtained by heating methyl iodide together with zinc and water. Methyl hydride is a colorless inflammable gas, which burns with a slightly luminous flame, and when mixed with air produces a dangerously explosive gas. Most oxidizing reagents do not act upon this hydride, but chlorine attacks it in the presence of sunlight with such violence, as to produce an explosion. By the slow action of chlorine, several *substitution products* are formed, amongst the chief of which are—methyl chloride, $CH_3\ Cl$, chloroform, $CH\ Cl_3$, and carbon tetra-chloride, $C\ Cl_4$.

Chloroform, $CH\ Cl_3$, is formed by the action of chlorine on marsh gas, but it is prepared by acting upon methyl or ethyl alcohols with bleaching powder. It is a mobile, heavy liquid, possessing a powerful and agreeable smell; its specific gravity is 1·525 at 0°, and it boils at 62°. Chloroform is much used in medicine, producing, when it is inhaled, a temporary but perfect insensibility to pain, and is therefore much valued in surgical operations. Many other organic volatile bodies act in a similar manner, but none so effectually and so harmlessly as chloroform. An iodine compound, analogous to the preceding, has been prepared; it is termed Iodoform, and is a yellow solid body.

Carbon Tetra Chloride, $C\,Cl_4$, is a colorless liquid, boiling at 77°, obtained as the last product of the action of chlorine on marsh gas. When this substance is brought into contact with an amalgam of sodium and water, an opposite substitution of hydrogen for chlorine occurs, marsh gas and all the intermediate products being formed.

Methyl Ether, $\left.\begin{matrix}CH_3\\CH_3\end{matrix}\right\}O$, a colorless and sweet-smelling gas at the ordinary temperature of the air; but condensing at -21° to a colorless liquid. It is prepared by heating the alcohol with sulphuric acid, as already described.

Methyl Cyanide, $\left.\begin{matrix}CH_3\\CN\end{matrix}\right\}$— is obtained by the distillation of ammonium acetate; thus,

$$C_2\,H_3\,(NH_4)\,O_2 - H_4\,O_2 = C_2\,H_3\,N;$$

or by heating potassium methyl-sulphate with potassium cyanide. It is a colorless liquid, possessing a disagreeable smell, and boiling at 77°. It is transformed into ethylamine in contact with zinc and dilute sulphuric acid, $C_2\,H_3\,N + H_4 - \left.\begin{matrix}C_2\,H_5\\H_2\end{matrix}\right\}N$; whilst acted upon by potash it forms acetic acid (*see* p. 274.)

Methyl (*Ethyl Hydride*), $C_2\,H_6$. This substance is obtained, 1st, by the action of methyl iodide upon zinc; 2ndly, by the decomposition of potassium acetate by a galvanic current; and 3dly, by the action of barium dioxide on acetic anhydride, thus:

$$2\left(\left.\begin{matrix}C_2\,H_3\,O\\C_2\,H_3\,O\end{matrix}\right\}O\right) + Ba\,O_2 = 2\,(C_2H_3\,O_2)\,Ba + C_2\,H_6 + 2\,C\,O_2$$

Acetic anhydride and barium dioxide } yield { Barium acetate and methyl and carbonic acid.

This hydrocarbon was formerly supposed to be the free radical of the methyl group, but no methyl compounds have in any way been derived from it, whilst by the action of chlorine upon it, *ethyl* chloride, $C_2 H_5 Cl$, is obtained as the first substitution product. Hence we must regard *methyl-* and *ethyl-hydride* as identical. The nitrate, nitrite, cyanate, and sulphate of the *methyl* series are also known.

LESSON XXX.

DI-CARBON OR ETHYL SERIES.

The starting point of this important series is common—or ethyl-alcohol, $C_2 H_6O$; this is the ethyl hydrate, and, like its numerous derivatives, contains the radical ethyl, $C_2 H_5$.

Ethyl Alcohol, $\left.\begin{matrix} C_2 H_5 \\ H \end{matrix}\right\} O$, is obtained in the vinous fermentation of sugar, a decomposition effected in aqueous sugar solutions in presence of yeast, in which alcohol and carbonic acid are chiefly formed ; the other products of fermentation are described under sugar (on p. 308). Alcohol can also be prepared from its elements by synthesis. This is done by obtaining acetylene (page 228), and combining this directly with hydrogen to form olefiant gas, $C_2 H_4$; this substance combines directly with strong sulphuric acid, forming hydrogen ethyl-sulphate, $\left.\begin{matrix} C_2 H_5 \\ H \end{matrix}\right\} SO_4$

and this, when boiled with water, forms sulphuric acid and alcohol by exchange of ethyl for hydrogen, thus :

$$\left.\begin{matrix} C_2 H_5 \\ H \end{matrix}\right\} SO_4 + \left.\begin{matrix} H \\ H \end{matrix}\right\} O = \left.\begin{matrix} H \\ H \end{matrix}\right\} SO_4 + \left.\begin{matrix} C_2 H_5 \\ H \end{matrix}\right\} O.$$

or, hydric ethyl-sulphate and water, yield sulphuric acid and alcohol.

Olefiant gas also combines with hydric iodide to form ethyl iodide, which forms alcohol when heated with caustic potash.

Alcohol and alcoholic liquids are prepared in large quantities by the fermentation of sugar derived from various sources. The fermented liquid is distilled, and the dilute aqueous spirit thus separated from non-volatile impurities; it is obtained in a more concentrated form by repeated rectifications, as it boils at a lower temperature than water. Alcohol cannot, however, be completely separated from water by simple distillation, the strongest spirit which can thus be prepared containing 10 per cent. of water. To withdraw all the water, the spirit must be distilled with some substance capable of combining with water, such as potassium carbonate or quicklime. The pure liquid thus obtained is termed *absolute alcohol:* it is a colorless, mobile liquid, possessing a pleasant, spirituous smell and burning taste; its specific gravity at 0° is 0·8095, and at 15°·5, 0·7939; and it boils at 78°·4, when the barometer stands at 760 mms. It has not been solidified, becoming only viscid at a temperature of -100°. Alcohol is very inflammable, burning with a slightly luminous blue flame. It absorbs moisture with great avidity, and mixes with water in all proportions, the mixture evolving heat and undergoing contraction.

Many salts, as well as gases, dissolve in alcohol; it likewise acts as a solvent for resins, organic bases, and essential oils, many of which do not dissolve in water. The determination of the strength of spirit, when free from sugar or other soluble matters, is ascertained by determining the specific gravity by means of delicate hydrometers, and reference to accurate tables, showing the percentage of water. In these estimations the temperature must be accurately observed, and corrections for deviations must be made, as alcohol expands considerably with increase of temperature, and the specific gravity is thereby altered. The "proof spirit" of the Excise contains 50·8 parts by weight of alcohol to 49·2 of water, and possesses a specific gravity of 0·920 at 15°·5. Owing to the high duty on pure spirit, the Government allows the sale of a mixture of ninety parts of strong alcohol with ten parts of wood-spirit for manufacturing and scientific purposes; this substance is called "methylated spirit," and is most useful to the scientific and manufacturing chemist. Spirits, wines, and beer also contain more or less alcohol, flavored with certain essential oils, sugar, or extracts. Brandy, whisky, and the other spirits, contain from 40 to 50 per cent. of alcohol; wines from 17 (Madeira and port) to 7 or 8 (light claret and hock) per cent., whilst strong ale and porter contain from 6 to 8 per cent.

Alcohol is decomposed when its vapor is passed through a red-hot tube; hydrogen, marsh gas, olefiant gas, naphthaline, benzole, and other products being formed. By oxidation alcohol is transformed first into aldehyde and then into acetic acid. This oxidation may be effected by the atmospheric oxygen in presence of finely divided platinum, or more slowly when certain fermentable bodies are present (*see* acetic acid, p. 274). The alkaline metals attack alcohol with rapidity, evolving hydrogen, and forming potassium or sodium ethylate, $\left.\begin{matrix} C_2H_5 \\ K \end{matrix}\right\} O$. Hydrochloric acid forms, with alcohol, ethyl chloride and water, and the corresponding bromine and iodine compounds act similarly. Strong sulphuric acid combines with alcohol to form hydric ethyl-sulphate, or sulphovinic acid, a substance which forms salts called the ethyl-sulphates; thus potassium-ethyl-sulphate is $\left.\begin{matrix} C_2H_5 \\ K \end{matrix}\right\} SO_4$

Ether, or *Ethyl-Ether*, $\left.\begin{matrix} C_2H_5 \\ C_2H_5 \end{matrix}\right\} O$. This important substance is formed in a variety of ways from ethyl compounds. The most simple reaction by which ether can be prepared is that of acting upon potassium ethylate with ethyl iodide, an exchange of ethyl and potassium taking place, thus:

$$C_2H_5I + \left.\begin{matrix}C_2H_5\\K\end{matrix}\right\}O = KI + \left.\begin{matrix}C_2H_5\\C_2H_5\end{matrix}\right\}O.$$

Ethyl iodide and potassium ethylate give potassium iodide and ether.

Another reaction by which ether is prepared on the large scale consists in heating a mixture of alcohol and sulphuric acid to 140°, when ether and water are given off. The decompositions which here take place are as follows: in the first place, alcohol and sulphuric acid form hydrogen ethyl-sulphate (sulphovinic acid) and water, by an exchange of hydrogen and ethyl, thus:

$$\underset{\text{Alcohol.}}{\left.\begin{matrix}C_2H_5\\H\end{matrix}\right\}O} + \underset{\text{Sulphuric Acid.}}{\left.\begin{matrix}H\\H\end{matrix}\right\}SO_4} = \underset{\text{Water.}}{\left.\begin{matrix}H\\H\end{matrix}\right\}O} + \underset{\text{Hydric Ethyl Sulphate.}}{\left.\begin{matrix}C_2H_5\\H\end{matrix}\right\}SO_4}$$

This hydrogen ethyl-sulphate next comes in contact with a second molecule of alcohol, another exchange of hydrogen for ethyl occurs, and ether and sulphuric acid are formed.

$$\underset{\text{Alcohol}}{\left.\begin{matrix}C_2H_5\\H\end{matrix}\right\}O} \text{ and } \underset{\text{Hydric Ethyl Sulphate}}{+ \left.\begin{matrix}C_2H_5\\H\end{matrix}\right\}SO_4} \text{ yield } \underset{\text{Ether}}{= \left.\begin{matrix}C_2H_5\\C_2H_5\end{matrix}\right\}O} \text{ and } \underset{\text{Sulphuric Acid.}}{+ \left.\begin{matrix}H\\H\end{matrix}\right\}SO_4.}$$

The water formed by the first decomposition, and the ether produced by the second, are given off as vapor, whilst the sulphuric acid remains behind, ready again to go through the same series of changes on meeting with two other molecules of alcohol. This process is called the *continuous etherification process*, as a current of alcohol may be passed continuously through the sulphuric acid heated to 140°, and a regular supply of ether and water thus obtained.

Ether is a colorless, very mobile liquid, possessing a strong and peculiar etherial smell. It is lighter than water, specific gravity 0·736, and is not miscible with this liquid. Ether boils at 34°·5, and its vapor is 37 times heavier than hydrogen, and can be poured from vessel to vessel like carbonic acid gas. It burns with a luminous flame, and explodes when mixed with air. From its low boiling point great care must be taken to avoid explosions when working with this substance, owing to the vapor becoming mixed with air. Ether is easily attacked by oxidizing agents, yielding the same products as alcohol, and it is also acted upon by chlorine, and a large number of substitution products formed.

Mixed Ethers containing two different radicals are obtained by acting, for instance, with ethyl-iodide upon potassium methylate, thus :

$$\underset{\text{Ethyl Iodide}}{C_2H_5I} \quad \overset{\text{and}}{+} \quad \underset{\text{Potassium Methylate}}{\left.\begin{matrix}CH_3\\K\end{matrix}\right\}O} \quad \overset{\text{yield}}{=} \quad \underset{\text{Potassium Iodide}}{KI} \quad \overset{\text{and}}{+} \quad \underset{\text{Methyl-Ethyl-Ether.}}{\left.\begin{matrix}CH_3\\C_2H_5\end{matrix}\right\}O},$$

or by acting on hydrogen methyl-sulphate, $\left.\begin{matrix}CH_3\\H\end{matrix}\right\}SO_4$, with ethyl alcohol. The following is a list of some of the more important simple and mixed ethers of this series :

Table of Simple and Mixed Ethers.

			Boiling Point.
Dimethyl Ether . .	C_2H_6O	$\left.\begin{matrix}CH_3\\CH_3\end{matrix}\right\}O$ . .	−21°
Methyl-ethyl Ether .	C_3H_8O	$\left.\begin{matrix}CH_3\\C_2H_5\end{matrix}\right\}O$ . .	+11°
Diethyl Ether . . .	$C_4H_{10}O$	$\left.\begin{matrix}C_2H_5\\C_2H_5\end{matrix}\right\}O$ . .	34°
Methyl-amyl Ether .	$C_6H_{14}O$	$\left.\begin{matrix}CH_3\\C_5H_{11}\end{matrix}\right\}O$ . .	92°

Ethyl-butyl Ether . .	$C_6\ H_{14}\ O$	$\left.\begin{matrix} C_2\ H_5 \\ C_4\ H_9 \end{matrix}\right\} O$ . .	80°
Ethyl-amyl Ether . .	$C_7\ H_{16}\ O$	$\left.\begin{matrix} C_2\ H_5 \\ C_5\ H_{11} \end{matrix}\right\} O$ . .	112°
Dibutyl Ether	$C_8\ H_{18}\ O$	$\left.\begin{matrix} C_4\ H_9 \\ C_4\ H_9 \end{matrix}\right\} O$ . .	104°
Ethyl-hexyl Ether .	$C_8\ H_{18}\ O$	$\left.\begin{matrix} C_2\ H_5 \\ C_6\ H_{13} \end{matrix}\right\} O$ . .	132°
Diamyl Ether . . .	$C_{10}\ H_{22}\ O$	$\left.\begin{matrix} C_5\ H_{11} \\ C_5\ H_{11} \end{matrix}\right\} O$ . .	176°

Ethyl Chloride, $C_2\ H_5\ Cl$, is obtained as a mobile liquid, having an etherial penetrating smell, by saturating alcohol with hydrochloric acid gas, or by acting with the phosphoric chlorides upon alcohol. On heating the mixture a volatile liquid is given off, which must be condensed in a freezing mixture. Ethyl chloride boils at 12°5. Ethyl iodide, $C_2\ H_5\ I$, and ethyl bromide, $C_2\ H_5\ Br$, are obtained by acting upon alcohol with iodine and bromine in presence of phosphorus. The *iodide* is much used for the preparation of other ethyl compounds, owing to the facility with which the iodine can be exchanged in double decompositions. It is a heavy, colorless liquid, boiling at 72°·2, and having a specific gravity of 1·946 at 16°.

Ethyl Cyanide, $C_2\ H_5\ C\ N$, is prepared by heating potassium ethyl sulphate with potassium cyanide. It may be considered as a nitrogen compound (nitril) of the next higher carbon series (propyl), as on heating with potash it yields propionic acid, thus:

Propionitril	and	Water	yield	Propionic Acid	and	Ammonia.
C_3H_5N	+	$2\ H_2O$	=	$C_3\ H_6\ O_2$	+	NH_3.

Propionitril, when acted upon by hydrogen, yields propylamine; $C_2\ H_5\ C\ N + 2\ H_2 = \left.\begin{matrix} C_3H_7 \\ H_2 \end{matrix}\right\} N.$ This reaction is important, as it is one which is common to all the series of alcoholic cyanides, and enables us to pass from a lower to a higher carbon series—in this case from the 2 to the 3 carbon series.

Ethyl Hydride, $C_2 H_5 H$. This substance, already described as methyl, is a colorless and tasteless gas, obtained by the action of zinc and water upon ethyl iodide when heated to 150° in a closed tube. It occurs in American petroleum. It is rapidly acted on by chlorine, and yields ethyl chloride, together with a series of further substitution products, the last of which is carbon trichloride, $C_2 Cl_6$.

Ethyl Nitrite, $C_2 H_5 N O_2$, is obtained as a sweet-smelling liquid by acting upon alcohol with nitric acid, and it is contained in the "sweet spirits of nitre" of the Pharmacopœia.

Ethyl Nitrate, $\left.\begin{matrix} C_2 H_5 \\ N O_2 \end{matrix}\right\} O$, is only formed by the action of nitric acid on alcohol when urea is present, as this body immediately destroys any nitrous acid which may be formed, which would prevent the production of nitrate.

Ethyl Sulphydrate, $\left.\begin{matrix} C_2 H_5 \\ H \end{matrix}\right\} S$. This compound, known as Mercaptan, is sulphur alcohol, *i. e.* alcohol in which the oxygen is replaced by sulphur. It is obtained by acting on potassium-sulph-hydrate, $\left.\begin{matrix} K \\ H \end{matrix}\right\} S$, with ethyl chloride, ethyl and potassium changing places. Mercaptan, like alcohol, can exchange its typical atom of hydrogen for metals: it forms with mercury an insoluble compound. This body boils at 36°, and possesses the nauseous garlic-like smell, characteristic of all the organic sulphur compounds.

Ethyl Sulphide, $\left.\begin{matrix} C_2 H_5 \\ C_2 H_5 \end{matrix}\right\} S$. This compound in the sulphur series is analogous to ether in the oxygen series; it is obtained by acting on potassium sulphide, $K_2 S$, with ethyl chloride. It is a colorless liquid, boiling at 91°, and possessing a strong, disagreeable odor.

Hydrogen Ethyl Sulphate, $\left.\begin{matrix} C_2 H_5 \\ H \end{matrix}\right\} SO_4$, or *Sulphovinic*

Acid, is formed when alcohol and strong sulphuric acid are mixed. It acts as an acid, and forms salts in which the typical hydrogen is replaced by a metal. The ethyl sulphates of the alkalies and alkaline earths are soluble salts, and crystallize well.

Ethyl Sulphate, $\left.\begin{matrix} C_2H_5 \\ C_2H_5 \end{matrix}\right\} SO_4$, is obtained by acting upon ether with sulphuric trioxide; it is a body which decomposes on distillation and on addition of water.

Ethyl Phosphates are known: they correspond to the tribasic alkaline phosphates in containing either 1, 2, or 3 molecules of ethyl, replacing hydrogen in tribasic phosphoric acid. Thus we have:

Dihydric Ethyl-Phosphate.	Hydric Diethyl-Phosphate.	Triethyl-Phosphate.
$\left.\begin{matrix} C_2H_5 \\ H_2 \end{matrix}\right\} PO_4$	$\left.\begin{matrix} C_2H_5 \\ C_2H_5 \end{matrix}\right\} PO_4$	$\left.\begin{matrix} C_2H_5 \\ C_2H_5 \\ C_2H_5 \end{matrix}\right\} PO_4$

Ethyl Carbonate, $\left.\begin{matrix} C_2H_5 \\ C_2H_5 \end{matrix}\right\} CO_3$, corresponding to sodium carbonate, $\left.\begin{matrix} Na \\ Na \end{matrix}\right\} CO_3$, can also be prepared.

Ethyl Cyanate, $\left.\begin{matrix} C\ N \\ C_2H_5 \end{matrix}\right\} O$, a colorless liquid, boiling at 60°, and possessing a powerful and irritating smell. It is formed by distilling potassium-ethyl-sulphate with potassium cyanate. In contact with caustic potash it forms ethylamine, thus:

$$\left.\begin{matrix} C_2N \\ C_2H_5 \end{matrix}\right\} O + 2\left(\left.\begin{matrix} K \\ H \end{matrix}\right\} O\right) = \left.\begin{matrix} C_2H_5 \\ H_2 \end{matrix}\right\} N + \left.\begin{matrix} K \\ K \end{matrix}\right\} CO_3$$

Ethyl Borates and Silicates likewise exist.

Ethyl, or *Butyl-Hydride*, C_4H_{10}. A substance having this composition, and formerly supposed to be the radical ethyl in the free state, is obtained by heating to 150° ethyl iodide with metallic zinc in a closed tube. From this body no ethyl compounds can in any way be prepared; but upon treatment with chlorine it yields the chloride of the *four* carbon radical, viz., C_4H_9Cl butyl chloride. Hence the so-called ethyl must be looked upon as butyl hydride, C_4H_9H.

TRICARBON SERIES. *Propyl alcohol* $\left.\begin{matrix} C_3H_7 \\ H \end{matrix}\right\} O$, has been found in the last products of distillation of French brandies; it boils at 96°, dissolves freely in water, but does not mix in all proportions. Propyl alcohol unites with sulphuric acid to form hydric propyl-sulphate, $\left.\begin{matrix} \dot{C}_3H_7 \\ \overset{\text{II}}{SO_2} \\ H \end{matrix}\right\} \begin{matrix} O \\ O \end{matrix}$

The propyl compounds have not been much studied; they closely resemble the foregoing ethyl series of bodies. Propyl alcohol, when oxidized, yields propionic acid (*see* p. 249).

TETRA-CARBON SERIES. *Butyl alcohol*, $\left.\begin{matrix} C_4H_9 \\ H \end{matrix}\right\} O$. Butyl alcohol is found in the oily liquid obtained at the end of the distillation of spirits from the beet-root molasses. It boils at 109°, and resembles common alcohol in its properties, as well as in those of its derivatives. The butyl, ether, chloride, iodide, nitrate, and other compounds, are known; they are prepared in the same manner as the ethyl compounds, and possess no specially note-worthy properties.

PENTA-CARBON SERIES. *Amyl alcohol*, $\left.\begin{matrix} C_5H_{11} \\ H \end{matrix}\right\} O$. This alcohol occurs commonly as the chief constituent of the Fousel oil obtained in the manufacture of potato brandy, from which it is obtained by washing with water and subsequent rectification. It is a colorless liquid, possessing a disagreeable, penetrating smell; it dissolves in alcohol and ether, but is not miscible with water. Amyl alcohol boils at 132°, and solidifies at −20°. Amyl alcohol, like the foregoing alcohols, forms, with sulphuric acid, hydric amyl-sulphate, which yields double salts, called the amyl-sulphates; it is also attacked by hydrochloric acid, amyl-chloride, $C_5H_{11}Cl$, being formed. Amyl alcohol, in presence of oxygen and finely divided platinum, undergoes oxidation to valeric acid, thus:

$$\underset{\text{Amyl Alcohol}}{\left.\begin{matrix} C_5\,H_{11} \\ H \end{matrix}\right\} O} + O_2 = \underset{\text{Valeric Acid.}}{\left.\begin{matrix} C_5\,H_9\,O \\ H \end{matrix}\right\} O} + H_2\,O.$$

Potassium and sodium can replace the typical hydrogen of this alcohol, forming potassium, or sodium amylate. The iodide and bromide are prepared in the same way as the corresponding ethyl compounds, with the substitution of amyl- for ethyl-alcohol.

Amyl Ether, $\left.\begin{matrix} C_5\,H_{11} \\ C_5\,H_{11} \end{matrix}\right\} O$, is a colorless liquid, boiling at 176°, obtained by the action of amyl iodide upon potassium or sodium amylate, thus:—

$$\underset{\text{Amyl Iodide}}{C_5\,H_{11}\,I} \underset{\text{and}}{+} \underset{\text{Sodium Amylate}}{\left.\begin{matrix} C_5\,H_{11} \\ Na \end{matrix}\right\} O} \underset{\text{give}}{=} \underset{\text{Amyl Ether}}{\left.\begin{matrix} C_5\,H_{11} \\ C_5\,H_{11} \end{matrix}\right\} O} \underset{\text{and}}{+} \underset{\text{Sodium Iodide.}}{Na\,I.}$$

Amyl Hydride, $C_5H_{11}\,H$. This substance is a volatile liquid, boiling at 30°, obtained by heating amyl iodide with zinc and water; it occurs, together with all the hydrides of this series of alcohol radicals, in American petroleum, and in the light oils from the distillation of coal.

The mixed methyl-amyl and ethyl-amyl ethers, have already been mentioned. Amongst the compound amyl ethers, the *acetate*, $\left.\begin{matrix} C_5\,H_{11} \\ C_2\,H_3O \end{matrix}\right\} O$, is prepared on a large scale, as it possesses the peculiar odor of Jargonelle pears, and it is used in flavoring cheap confectionery. This compound is obtained by distilling amyl alcohol with potassium acetate and sulphuric acid; it can also be prepared by heating the chloride with potassium acetate. If this amyl acetate be heated with potash, amyl-alcohol and potassium acetate are formed.

$$\underset{\text{Amyl Acetate}}{\left.\begin{matrix} C_5 H_{11} \\ C_2 H_3 O \end{matrix}\right\} O} + \underset{\text{and Potash}}{\left.\begin{matrix} K \\ H \end{matrix}\right\} O} = \underset{\text{give Amyl Alcohol}}{\left.\begin{matrix} C_5 H_{11} \\ H \end{matrix}\right\} O} + \underset{\text{and Potassium Acetate.}}{C_2 \left.\begin{matrix} H_3 O \\ K \end{matrix}\right\} O.}$$

Diamyl, $C_{10} H_{22}$. A substance having this composition is obtained by acting on amyl iodide with sodium. It is a colorless liquid, boiling at 158°, from which none of the amyl compounds can be obtained, but which yields decatyl chloride, $C_{10}H_{21} Cl$, on treatment with chlorine. We, therefore, consider this body as decatyl hydride, $C_{10}H_{21} H$, and as not belonging to the amyl group.

HIGHER ALCOHOLS. The alcohols containing 6 to 10 atoms of carbon resemble the foregoing series in their general properties. Hexyl and heptyl alcohols are found in certain fermented liquors; octyl alcohol is obtained by distilling castor oil with potash. The hydrides of these, as well as of all the higher and lower alcohol radicals, are found in American petroleum. This substance consists essentially of a mixture of these various hydrides from $C H_4$ (marsh gas), or even H_2 (hydrogen), up to hydrides which are solid, and contain a very large number of atoms of carbon, and to which the name of Paraffine has been given. The hydrides can be separated from each other by repeated rectifications, and obtained in the pure state. From these hydrides the corresponding chlorides can be prepared by the action of chlorine, and from the chlorides we can form the acetates and the alcohols themselves (*see* Amyl Acetate).

Cetyl Alcohol, $\left.\begin{matrix} C_{16}H_{33} \\ H \end{matrix}\right\} O$, is found combined with palmitic acid in spermaceti. It forms a white solid crystalline mass, but acts in its chemical properties like an alcohol: thus it forms a chloride, $C_{16} H_{33} Cl$, also a bromide and iodide; it likewise yields an ether, $\left.\begin{matrix} C_{16} H_{33} \\ C_{16} H_{33} \end{matrix}\right\} O$, obtained by the action of cetyl iodide upon potassium cetylate; and a compound with sulphuric acid, $\left.\begin{matrix} C_{16} H_{33} \\ SO_2 \\ H \end{matrix}\right\} O_2$. Cetyl alcohol undergoes oxidation when heated with caustic potash, yielding an acid in which one of oxygen replaces two of hydrogen of the alcohol, thus :

Cetyl Alcohol	and	Potash	yield	Potassium Palmitate	and Hydrogen.

$$\left.\begin{matrix} C_{16}\,H_{33} \\ H \end{matrix}\right\}O \;+\; \left.\begin{matrix} K \\ H \end{matrix}\right\}O \;=\; \left.\begin{matrix} C_{16}\,H_{31}\,O \\ K \end{matrix}\right\}C \;+\; H_4$$

This palmitic acid bears the same relation to cetyl alcohol as acetic acid does to common- or ethyl-alcohol.

Cerotyl Alcohol, $\left.\begin{matrix} C_{27}\,H_{55} \\ H \end{matrix}\right\}O$, is contained in Chinese wax ; it is a white solid solid, crystalline substance. When heated with potash it undergoes oxidation, and furnishes an acid, called cerotic acid, $\left.\begin{matrix} C_{27}\,H_{53}\,O \\ H \end{matrix}\right\}O$.

Melisyl Alcohol, $\left.\begin{matrix} C_{30}H_{61} \\ H \end{matrix}\right\}O$, a solid white substance contained in beeswax: when fused with potash it forms an acid termed melissic acid, $\left.\begin{matrix} C_{30}\,H_{59}\,O \\ H \end{matrix}\right\}O$.

LESSON XXXI.

Compounds of the Alcohol Radicals with the Nitrogen (Triad) group of Elements. N. P. As. Sb. Bi.

1. *Compound Alcoholic Ammonias.* The constitution of the primary monamines, as $\left.\begin{matrix} C_2\,H_5 \\ H_2 \end{matrix}\right\}N$, Ethylamine ; secondary monamines, as $\left.\begin{matrix} C_2\,H_5 \\ C_2\,H_5 \\ H \end{matrix}\right\}N$, Diethylamine ; and ter-

tiary monamines, as Triethylamine, $\left.\begin{matrix} C_2H_5 \\ C_2H_5 \\ C_2H_5 \end{matrix}\right\} N$, have already been mentioned (p. 248). These bodies are volatile; they all have a strong alkaline reaction and ammoniacal smell, and they combine with H Cl, &c., to form salts. These compound ammonias are formed in many ways, of which the most important are—

1. By the action of caustic alkalies on the cyanates of the alcohol radicals (*see* p. 260).

2. By the action of the iodides of these radicals on ammonia, we obtain the iodide of the compound ammonium, which, when treated by potash, yields the compound ammonia, thus,

Ethyl Iodide	and	Ammonia	give	Ethylamine and	Hydriodic Acid.

$$C_2H_5I + \left.\begin{matrix} H \\ H \\ H \end{matrix}\right\} N = \left.\begin{matrix} (C_2H_5) \\ H \\ H \end{matrix}\right\} N + HI.$$

Ethyl iodide acts similarly on ethylamine, giving rise to diethylamine and hydriotic acid, thus :

$$C_2H_5I + C_2H_5, H_2N = (C_2H_5)_2H, N + HI,$$

and also acts upon diethylamine in the same way, giving rise to triethylamine, thus :

$$C_2H_5I + (C_2H_5)_2HN = (C_2H_5)_3N + HI.$$

Ethyl iodide also combines with triethylamine to form tetra-ethylammonium iodide, $N(C_2H_5)_4I$. In practice all these compounds are formed together when ethyl iodide acts on ammonia. The compounds of mono-, di-, and tri-ethylamine with hydriodic acid are decomposed by caustic potash, and the volatile compound ammonias liberated; the case of the tetra-ethyl-ammonium iodide is different, as it is not decomposed by potash, but yields, when treated with silver hydroxide, a hydrated oxide, which is non-volatile without decomposition, and is analogous in constitution and similar in properties to caustic potash.

Tetra-ethylammonium Hydrate.	Potassium Hydrate.
$\left.\begin{matrix}(C_2H_5)_4N \\ H\end{matrix}\right\} O$	$\left.\begin{matrix}K \\ H\end{matrix}\right\} O.$

By acting on ethylamine with other iodides, such as methyl iodide, mixed amines can be prepared. The compound ammonias form double salts with platinic chloride ; the larger the number of organic radicals contained, the more soluble is the platinum salt. The following table gives the names, compositions, and boiling points of the most important of the compound ammonias. It will be seen that the boiling point increases with the increasing number of carbon atoms contained in the compound.

Primary Monamines.

		Boiling Point.
Methylamine	$\left.\begin{matrix}CH_3 \\ H \\ H\end{matrix}\right\} N$	below 0°
Ethylamine	$\left.\begin{matrix}C_2H_5 \\ H \\ H\end{matrix}\right\} N$	18°·7
Propylamine	$\left.\begin{matrix}C_3H_7 \\ H \\ H\end{matrix}\right\} N$	49°·7
Butylamine	$\left.\begin{matrix}C_4H_9 \\ H \\ H\end{matrix}\right\} N$	69°
Amylamine	$\left.\begin{matrix}C_5H_{11} \\ H \\ H\end{matrix}\right\} N$	94°
Caproylamine, or Hexylamine	$\left.\begin{matrix}C_6H_{13} \\ H \\ H\end{matrix}\right\} N$	126°

	Formula		Boiling Point.
Heptylamine	$\left.\begin{matrix} C_7\,H_{15} \\ H \\ H \end{matrix}\right\}$	N . .	146°
Octylamine	$\left.\begin{matrix} C_8\,H_{17} \\ H \\ H \end{matrix}\right\}$	N . .	170°

Secondary Monamines.

	Formula		Boiling Point.
Dimethylamine	$\left.\begin{matrix} C\,H_3 \\ C\,H_3 \\ H \end{matrix}\right\}$	N . .	8°·5
Methyl-ethylamine . . .	$\left.\begin{matrix} C\,H_3 \\ C_3\,H_5 \\ H \end{matrix}\right\}$	N . .	—
Diethylamine	$\left.\begin{matrix} C_2\,H_5 \\ C_2\,H_5 \\ H \end{matrix}\right\}$	N. .	57°·5
Diamylamine	$\left.\begin{matrix} C_5\,H_{11} \\ C_5\,H_{11} \\ H \end{matrix}\right\}$	N . .	170°

Tertiary Monamines.

	Formula		Boiling Point.
Trimethylamine	$\left.\begin{matrix} C\,H_3 \\ C\,H_3 \\ C\,H_3 \end{matrix}\right\}$	N . .	4°—5°
Triethylamine	$\left.\begin{matrix} C_2\,H_5 \\ C_2\,H_5 \\ C_2\,H_5 \end{matrix}\right\}$	N . .	91°
Diethyamylamine	$\left.\begin{matrix} C_2\,H_4 \\ C_2\,H_5 \\ C_5\,H_{11} \end{matrix}\right\}$	N . .	154°
Triamylamine	$\left.\begin{matrix} C_5\,H_{11} \\ C_5\,H_{11} \\ C_5\,H_{11} \end{matrix}\right\}$	N . .	257°

Boiling Point.

Methyl-ethyl-amylamine . $\left.\begin{matrix} CH_3 \\ C_2H_5 \\ C_5H_{11} \end{matrix}\right\} N$ . . . 135°

A comparison of these compounds shows that it is possible to have two or more bases of the same composition, but of different constitution; thus C_3H_9N stands for methyl-ethylamine and trimethylamine. In order to determine the constitution of a body having this composition, it is necessary to ascertain how many atoms of the replaceable hydrogen of the original ammonia it contains.

PHOSPHORUS BASES. Compounds corresponding to the preceding, but containing phosphorus instead of nitrogen, have been prepared; thus, triethyl-phosphine, $\left.\begin{matrix} C_2H_5 \\ C_2H_5 \\ C_2H_5 \end{matrix}\right\} P$, is obtained by acting upon zinc ethyl with phosphorus trichloride, the chlorine changing places with ethyl.

Triethylphosphine is a colorless liquid, boiling at 127°·5, possessing a powerful and disagreeable smell, It combines directly with oxygen, sulphur, and chlorine, in this respect differing from the foregoing nitrogen bases. With ethyl iodide it combines and forms iodide of tetra-ethyl phosphonium, $P(C_2H_5)_4I$, from which a strongly caustic hydrate, analogous to the corresponding nitrogen compound, can be obtained by the action of silver oxide (p. 265).

ARSENIC BASES. The compounds of arsenic with the alcohol radicals differ somewhat in constitution from the foregoing, inasmuch as we are acquainted in the methyl series with (1) tri-methyl-arsine $(CH_3)_3As$; (2) arsendimethyl $(CH_3)_2As$; (3) arsen-monomethyl $(CH_3)As$. The first of these is constructed on the type ammonia, and the two latter combine directly with one and two atoms of chlorine respectively, and then form compounds belonging to the general type, NH_3. We thus have the following compounds known:

As H H H Arsenic trihydride.

As CH_3 CH_3 CH_3 Trimethyl-arsine.

As CH_3 CH_3 Cl Arsendimethyl chloride.
As CH_3 Cl Cl Arsenmonomethyl dichloride.
As Cl Cl Cl Arsenic trichloride.

Trimethylarsine is a colorless liquid, boiling at 120°, formed by the action of methyl-iodide on an alloy of sodium and arsenic; it corresponds to trimethylamine and trimethyl phosphine.

Arsendimethyl, or *Cacodyl.* This substance is prepared by heating arsenic trioxide with potassium acetate. Cacodyl is a colorless liquid, boiling at 170°, which takes fire in contact with the air. It is extremely poisonous, and possesses a most disagreeable garlic-like smell, and must be prepared with great care. It combines with chlorine, oxygen, &c., and plays the part of an organo-metallic radical. One of the most important compounds is cacodylic acid, $\left.\begin{matrix}As\,(CH_3)_2\\H\end{matrix}\right\} O_2$; it is soluble in water, and is not poisonous. The formation of cacodyl and its oxide in the mode described, may be used as a delicate test for the presence of arsenic, from the strong and characteristic odor of this body.

ANTIMONY BASES. By acting on ethyl iodide with an alloy of antimony and potassium, a compound called tri-ethyl stibine, $\left.\begin{matrix}C_2H_5\\C_2H_5\\C_2H_5\end{matrix}\right\} Sb$, has been prepared; it is a colorless liquid, boiling at 158°·5, which takes fire and burns in contact with the air. It forms compounds with oxygen, sulphur, and chlorine.

Bismuth forms an analogous compound, Triethbismutine, $\left.\begin{matrix}C_2H_5\\C_2H_5\\C_2H_5\end{matrix}\right\} Bi$.

COMPOUNDS OF THE ALCOHOL RADICALS WITH BORON AND SILICON.

The ethyl compounds in this series are the only ones which are well investigated.

Bor ethyl, $\left.\begin{matrix}C_2H_5\\C_2H_5\\C_2H_6\end{matrix}\right\}B$, is a colorless liquid, boiling at 95°, possesses a very powerful acrid smell, and takes fire on exposure to the air, burning with a green flame. It is obtained by acting on ethyl borate, ($3(C_2H_5)\ BO_3$), with zinc ethyl.

Silicon ethyl, $\left.\begin{matrix}C_2H_5\\C_2H_5\\C_2H_5\\C_2H_5\end{matrix}\right\}Si$, is obtained by the action of zinc ethyl on silicon tetrachloride: it is a colorless liquid which boils at 150°, and is not attacked by nitric acid. It is acted on by chlorine, monochlorinated silicon ethyl, $Si\ C_8H_{19}Cl$, being the first product. This substance acts as the chloride of a monad radical: thus, when heated with acetate of potash, it yields an acetic ether, and this on treatment with potash forms a colorless liquid smelling like camphor, and acting as an alcohol, and having the formula $Si\ C_8H_{20}O$. Hence silicon ethyl may be regarded as nonyl hydride, C_9H_{20}, in which one atom of (tetrad) carbon has been replaced by one of (tetrad) silicon. A substance having the composition $Si\ H\ Cl_3$ has also been prepared. This body, it will be seen, is chloroform, $CHCl_3$, in which silicon replaces carbon.

		B. P.
Nonyl hydride	C_9H_{20}	137°
" chloride	$C_9H_{19}Cl$	196°
" acetate	$\left.\begin{matrix}C_9H_{19}O\\C_2H_3\end{matrix}\right\}O$	210°
" alcohol	$\left.\begin{matrix}C_9H_{19}\\H\end{matrix}\right\}O$	—
Silico-nonyl hydride	SiC_8H_{20}	150°
" chloride	$SiC_8H_{19}Cl$	115°
" acetate	$\left.\begin{matrix}SiC_8H_{19}O\\C\ H_3\end{matrix}\right\}O$	211°
" alcohol	$\left.\begin{matrix}SiC_8H_{19}\\H\end{matrix}\right\}O$	190°

Compounds of the Alcohol Radicals with Metals.

Zinc Ethyl, $\left.\begin{matrix}C_2H_5\\C_2H_5\end{matrix}\right\}Zn$. This important substance is obtained by the action of zinc upon ethyl iodide ; it is a colorless liquid, boiling at 118° ; it takes fire and burns with a greenish flame in contact with air or oxygen, and forms zinc ethylate, $\left.\begin{matrix}C_2H_5\\Zn\end{matrix}\right\}O$, when the oxidation goes on slowly. Zinc Ethyl is a valuable reagent, by means of which many other compounds can be obtained : thus, if we act with this substance on silicic chloride, we get zinc chloride and ***silicon ethyl;*** on mercuric chloride, we obtain ***mercuric ethyl;*** on lead chloride, we get *lead ethyl.* Zinc methyl and zinc amyl are also known. Compounds of tin, lead, and a few other metals with the alcohol radicals, can be prepared, possessing properties somewhat analogous to the foregoing substances.

Compounds of the *alkaline metals* with ethyl have been obtained by acting upon these metals with zinc ethyl. *Sodium Ethyl* combines directly with carbonic dioxide, and forms sodium propionate (p. 249), thus:

Sodium Ethyl and Carbonic Dioxide	yield	Sodium Propionate.
$Na(C_2H_5) + CO_2$	=	$C_3H_5NaO_2$.

LESSON XXXII.

COMPOUNDS DERIVED BY OXIDATION FROM THE ALCOHOLS.

Group of Fatty Acids and their derivatives. The mode in which the aldehydes and acids are connected with the corresponding alcohols has been already described (p. 247). These oxidized products contain a radical in which one atom of oxygen is substituted for two of hydrogen in the alcohol radical.

Ethyl alcohol, $\left.\begin{matrix}C_2H_5\\H\end{matrix}\right\}O$, gives acetic acid, $\left.\begin{matrix}C_2H_3O\\H\end{matrix}\right\}O$.

Amyl alcohol, $\left.\begin{matrix}C_5H_{11}\\H\end{matrix}\right\}O$, gives valeric acid, $\left.\begin{matrix}C_5H_9O\\H\end{matrix}\right\}O$.

These oxidized radicals form the starting point of a large number of compounds which in their properties resemble the alcoholic compounds, but differ by containing an atom of oxygen for 2 of hydrogen. Thus, by substituting the hydroxyl of an acid by an atom of chlorine, we get the chloride of the series; for instance, acetyl chloride, C_2H_3OCl: by replacing the hydrogen of the hydroxyl by metals or by alcohol radicals, we get:

Hydrogen Acetate.	Potassium Acetate.	Ethyl Acetate.	Acetyl Acetate.
$\left.\begin{matrix}C_2H_3O\\H\end{matrix}\right\}O$	$\left.\begin{matrix}C_2H_3O\\K\end{matrix}\right\}O$	$\left.\begin{matrix}C_2H_3O\\C_2H_5\end{matrix}\right\}O$	$\left.\begin{matrix}C_2H_3O\\C_2H_3O\end{matrix}\right\}O$.

If the oxygen of the hydroxyl be replaced by sulphur, we get:

Hydrogen Thiacetate.	Potassium Thiacetate.	Ethyl Thiacetate.
$\left.\begin{matrix}C_2H_3O\\H\end{matrix}\right\}S$	$\left.\begin{matrix}C_2H_3O\\K\end{matrix}\right\}S$	$\left.\begin{matrix}C_2H_3O\\C_2H_5\end{matrix}\right\}S$.

The monad acetyl can also replace hydrogen in ammonia, and we then get:

$$\text{Acetamide, } \left.\begin{matrix}C_2H_3O\\H\\H\end{matrix}\right\}N.$$

In many reactions the acids of this series decompose, with separation of one atom of carbon, as carbon dioxide. Thus if we decompose a solution of an acetate by a galvanic current, it splits up into carbon dioxide,

$$2\left.\begin{matrix}C_2H_3O\\H\end{matrix}\right\}O = CO_2 + \left.\begin{matrix}H\\H\end{matrix}\right\} + \left.\begin{matrix}CH_3\\CH_3\end{matrix}\right\},$$

hydrogen and methyl, which latter at once unites with another atom of methyl to form ethyl hydride or dimethyl.

The higher acids act similarly. Valeric acid yields dibutyl or octyl hydride:

$$2\left.\begin{matrix}C_5H_9O\\H\end{matrix}\right\}O = CO_2 + \left.\begin{matrix}H\\H\end{matrix}\right\} + \left.\begin{matrix}C_4H_9\\C_4H_9\end{matrix}\right\}.$$

So also when an acetate is heated with an alkali, carbon dioxide and methyl hydride are evolved:

$$\left.\begin{matrix} C_2H_3O \\ Na \end{matrix}\right\} + \left.\begin{matrix} Na \\ H \end{matrix}\right\} O = \left.\begin{matrix} CH_3 \\ H \end{matrix}\right\} + \left.\begin{matrix} Na \\ Na \end{matrix}\right\} CO_3.$$

Likewise sodium methyl and carbon dioxide combine directly to form sodium acetate:

$$\left.\begin{matrix} CH_3 \\ Na \end{matrix}\right\} + CO_2 = \left.\begin{matrix} C_2H_3O \\ Na \end{matrix}\right\} O.$$

Many other illustrations of this mode of decomposition might be given: the above will suffice to show that the formulæ already given for the acids do not explain or exhibit these reactions. In order to point out these relations we must write acetic acid, for instance, thus:

$$\left.\begin{matrix} CH_3CO \\ H \end{matrix}\right\} O;$$

and the general formula for the series is $\left.\begin{matrix} C_n H_{2n+1} CO \\ H \end{matrix}\right\} O.$ That is, the acid is a compound of an alcohol radical with the monad group, $\left.\begin{matrix} CO \\ H \end{matrix}\right\} O$, to which the name of *Carboxyl* has been given. This substance we thus regard contained in all the fatty acids; that it is formed by the oxidation of methyl we see from the following:

Ethyl Hydride.	Ethyl Alcohol.	Acetic Acid.
$\left\{\begin{matrix} CH_3 \\ CH_3 \end{matrix}\right.;$	$\left\{\begin{matrix} CH_3 \\ CH_2OH \end{matrix}\right.;$	$\left\{\begin{matrix} CH_3 \\ CO\,OH \end{matrix}\right.$

The hydrogen of the alcohol radical contained in the acid can be replaced by monad elements or radicals. Thus, when chlorine acts on acetic acid, the following chlorinated acids are formed:

Monochloracetic Acid.	Dichloracetic Acid.	Trichloracetic Acid.
$\left\{\begin{matrix} CH_2Cl \\ CO\,OH \end{matrix}\right.;$	$\left\{\begin{matrix} CH\,Cl_2 \\ CO\,OH \end{matrix}\right.;$	$\left\{\begin{matrix} C\,Cl_3 \\ CO\,OH. \end{matrix}\right.$

If the ether of a fatty acid be acted on with sodium, hydrogen is evolved and sodium takes its place: thus from ethyl acetate, $\left.\begin{matrix} CH_3 \\ CO.C_2H_5O \end{matrix}\right\}$, we get $\left.\begin{matrix} CH_2Na \\ CO.C_2H_5O \end{matrix}\right\}$. When this new body is acted upon by the iodide of an alcohol radical, sodium iodide is formed, and the radical replaces the metal. Ethyl iodide gives with the foregoing body ethyl-acetic-ether,

$$\left\{\begin{matrix} CH_2C_2H_5 \\ \left.\begin{matrix} CO \\ C_2H_5 \end{matrix}\right\} O \end{matrix}\right.,$$

which contains an acid identical with the butyric acid of butter, viz.:

$$\left\{\begin{matrix} C_3H_7 \\ \left.\begin{matrix} CO \\ H \end{matrix}\right\} O \end{matrix}\right. .$$

In this way all the higher members of the series of fatty acids may be prepared from acetic acid, as we only need to replace one atom of hydrogen in the radical by methyl. By further action of sodium upon ethyl acetate, 2 and 3 atoms of the hydrogen of the radical can be displaced by sodium; and these, when treated with methyl iodide, yield respectively:

$$(1) \left\{\begin{matrix} CH(CH_3)_2 \\ \left.\begin{matrix} CO \\ C_2H_5 \end{matrix}\right\} O \end{matrix}\right.; \text{ and } (2) \left\{\begin{matrix} C(CH_3)_3 \\ \left.\begin{matrix} CO \\ C_2H_5 \end{matrix}\right\} O \end{matrix}\right.;$$

the first being the ethyl ether of dimethyl acetic acid, or isobutyric acid, and the second that of trimethyl acetic acid, or tertiary valeric acid.

The following are the most important reactions, by means of which the monobasic acids can be obtained:

1. From the alcohol having the same number of carbon atoms by direct oxidation.

2. From the alcohol containing one atom less carbon, as follows:

(*a*) By decomposition of the alcoholic cyanide by potash, ethyl cyanide yielding propionic acid (see p. 258).

(*b*) By acting with the sodium compound on carbonic acid; sodium ethyl and carbonic acid yielding sodium propionate (see p. 270).

MONOCARBON SERIES. *Formic Acid,* $\left.\begin{matrix} CHO \\ H \end{matrix}\right\} O$. This acid occurs ready formed in the bodies of red ants, whence its name; it is likewise found in stinging-nettles. Formic acid is obtained by the oxidation of methyl alcohol, as well as of sugar, starch, and other organic bodies. It is formed synthetically by acting upon potash with carbonic oxide gas at 100°, thus:

Carbonic Oxide and Potash yield Potassium Formate.

$$CO + \left.\begin{matrix}H\\K\end{matrix}\right\}O = \left.\begin{matrix}CHO\\K\end{matrix}\right\}O$$

Formic acid, diluted with water, can be best prepared by decomposing oxalic acid, in presence of glycerine and water, into formic acid and carbonic dioxide, thus:

Oxalic Acid yields Formic Acid and Carbonic Dioxide.

$$C_2H_2O_4 = CH_2O_2 + CO_2.$$

In order to obtain formic acid in the pure glacial state, free from water, the lead formate is decomposed by a current of sulphuretted hydrogen gas, and lead sulphide and formic acid are produced. Formic acid is a colorless liquid, possessing a peculiar sharp smell and strong acid taste. It boils at 100°, and below 1° it solidifies to a white crystalline mass; its specific gravity at 0° is 1·235, and it is miscible in all proportions with water. Heated with sulphuric acid, it forms water and pure carbonic oxide gas, and oxidizing agents convert it easily into carbonic acid and water. A formate, heated with excess of baryta, yields oxalate, thus:

Formic Acid yields Oxalic Acid and Hydrogen.

$$2\,(CH_2\,O_2) = C_2\,H_2\,O_4 + H_2.$$

Formic acid is monobasic; and forms well-crystallizable salts called *Formates:* all the formates are soluble in water. It may be distinguished by its power of reducing metallic mercury and silver, as a gray powder, from the nitrates on boiling.

Formamide, $\left.\begin{matrix}CHO\\H\\H\end{matrix}\right\}N.$ Obtained by acting on ethyl formate with ammonia. It is a colorless liquid, boiling at 194°. The aldehyde of the monocarbon series is not known.

DICARBON SERIES.—*Acetyl Compounds.*

Aldehyde, $\left.\begin{matrix} C_2H_3O \\ H \end{matrix}\right\}$. Acet-aldehyde is obtained by oxidizing dilute alcohol by means of a mixture of manganese dioxide and sulphuric acid. It may also be prepared by distilling a mixture of an alkaline acetate and formate, thus :

$$\underset{\text{Potassium Acetate and Formate.}}{C_2H_3KO_2 + CHKO_2} = \underset{\text{Aldehyde.}}{C_2H_4O} + \underset{\text{Potassium Carbonate.}}{K_2CO_3}.$$

It is a colorless, suffocating-smelling liquid, boiling at 21° ; it has a specific gravity of 0·801 at 0°, and mixes in all proportions with water, alcohol, and ether. Aldehyde reduces metallic silver, depositing it as a bright mirror from solutions of the nitrate, and this reaction may be used to detect the presence of the substance. It unites directly with nascent hydrogen to form alcohol, $C_2H_4O + H_2 = C_2H_6O$; it likewise forms acetyl chloride when treated with chlorine, and acetic acid when acted upon by oxidizing agents. Aldehyde is capable of existing in three other peculiar states, or of undergoing *polymeric* modifications. If it is preserved in contact with excess of acid it remains unchanged, but if it be pure it soon deposits a solid substance having the same composition as aldehyde, and termed *Metaldehyde ;* this substance sublimes unchanged at 120°, but, when heated to 200° in a closed tube, it forms aldehyde again. *Paraldehyde* is another modification, and is a liquid boiling at 124° ; and a third modification termed *Acraldehyde,* boils at 110°. The molecular formula of paraldehyde appears to be, $C_6H_{12}O_3$, or $3(C_2H_4O)$; that of acraldehyde, $C_4H_8O_2$, or $2(C_2H_4O)$. Aldehyde is also isomeric with ethylene oxide (p. 283). Aldehyde forms a crystalline compound with ammonia, termed Aldehyde-ammonia, $\left.\begin{matrix} C_2H_3O \\ NH_4 \end{matrix}\right\}$, and it also unites with hydrogen sodium sulphite to form a solid compound. In many reactions aldehyde comports itself as the oxide of a dyad, $\overset{\text{II}}{C_2H_4}$

Acetal, $\left.\begin{matrix}\overset{II}{C_2 H_4} \\ (C_2 H_5)_2\end{matrix}\right\} O_2$. This substance is a derivative of aldehyde, in which this dyad radical aldehydene, $\overset{II}{C_2 H_4}$, occurs. It is obtained by heating aldehyde and alcohol together, and is formed together with aldehyde when alcohol is oxidized with sulphuric acid and manganese dioxide. A compound of a similar constitution, viz., *dimethyl acetal*, $\left.\begin{matrix}\overset{II}{C_2 H_4} \\ (C H_3)_2\end{matrix}\right\} O_2$, occurs in crude wood spirit. Acetal is isomeric with diethyl glycol (*see* p. 284).

Chloral, $\left.\begin{matrix}C_2 Cl_3 O \\ H\end{matrix}\right\}$ This substance may be considered as aldehyde, in which 3 of chlorine take the place of 3 of hydrogen; and although it has not been directly prepared from aldehyde, yet it resembles it in many properties, such as forming a crystalline compound with ammonia, which reduces silver salts. Chloral is obtained by the continued action of chlorine upon alcohol; it is a colorless powerfully-smelling liquid, boiling at 99°.

Acetic Acid, $C_2 H_4 O_2$. Dilute acetic acid has been known as vinegar from very early times; it occurs in the juices of certain plants and vegetables, but only in small quantities. The most important modes of preparing acetic acid are: (1) the method generally practically employed by the oxidation of alcohol; (2) the theoretically interesting processes, 1st,—the direct combination of carbonic acid and sodium methyl, thus:

$$C H_3 Na + CO_2 = \left.\begin{matrix}C_2 H_3 O \\ Na\end{matrix}\right\} O;$$

and 2d, by the action of potash on acetonitril, thus:

$$CH_3CN + 2 H_2O = \left.\begin{matrix}C_2 H_3 O \\ H\end{matrix}\right\} O + N H_3.$$

(3) Acetic acid is also prepared on a large scale by the dry distillation of wood; the crude acid thus obtained is commonly called Pyroligneous Acid.

Pure acetic acid is obtained by decomposing the acetates. The process by which alcoholic liquids (beer or wine) yield acetic acid by oxidation is termed the *Acetous Fermentation* (*see* p. 313): the liquids are exposed to the air at a temperature of about 25°, for a fortnight, when the alcohol is changed to vinegar. This change appears to be brought about by the presence of a peculiar vegetable growth (*mycoderma aceti*) which floats on the surface of the liquid, first absorbing the oxygen, and then giving it up to the alcohol.

Acetic acid in the pure state is obtained by heating sodium acetate with strong sulphuric acid: it is a colorless liquid, boiling at 118° and solidifying to an ice-like mass at 17°, and hence the name of Glacial Acetic Acid has been given to it. It possesses a peculiar sharp smell, and has a strong acid taste; it mixes in all proportions with water, but when distilled the mixture has no definite boiling point; the residue becomes stronger until glacial acid remains. Acetic acid may be recognized by its smell, and by the formation of ethyl acetate; also by the production of cacodyl when an acetate is heated with arsenious oxide. Acetic acid is monobasic, and forms a series of well-defined salts termed *Acetates*. The acetates of the alkalies are soluble crystallizable salts. Aluminium and ferric acetates are soluble compounds used in large quantities as mordants by dyers and calico-printers under the commercial names of Red Liquor and Iron Liquor. *Lead acetate*, or sugar of lead, and *copper acetate*, or verdigris, are the most important compounds of acetic acid and the heavy metals. The radicals methyl and ethyl, &c., can be substituted for the atom of typical hydrogen in acetic acid forming the compound ethers (*see ante*, page 247).

Acetyl Chloride, C_2H_3OCl, is obtained by the action of phosphorus trichloride upon acetic acid. It is a colorless liquid, fuming strongly in the air, and boiling at 55°. The corresponding bromide and iodide are known.

Acetyl Acetate, $\left.\begin{matrix} C_2 H_3 O \\ C_2 H_3 O \end{matrix}\right\} O$, or *Acetic Anhydride,* is a colorless liquid, boiling at 138°, formed by the action of acetyl chloride, or phosphoric oxychloride upon sodium acetate, thus:

$$4C_2H_3O_2Na + POCl_3 = 2\left(\left.\begin{matrix} C_2 H_3 O \\ C_2 H_3 O \end{matrix}\right\} O\right)\begin{matrix} + 3NaCl \\ + PO_3Na. \end{matrix}$$

It is decomposed by boiling with water into two molecules of acetic acid.

Chloracetic Acids. Chlorine acts upon acetic acid in replacing one, two, or three atoms of the hydrogen of the radical acetyl by chlorine: we thus obtain *monochloracetic acid,* $\left.\begin{matrix} CO\,CH_2Cl \\ H \end{matrix}\right\} O$; *dichloracetic acid,* $\left.\begin{matrix} COCHCl_2 \\ H \end{matrix}\right\} O$; and *trichloracetic acid,* $\left.\begin{matrix} COCCl_3 \\ H \end{matrix}\right\} O$. These three bodies are crystalline solids: the first fuses at 62°, and boils about 186°, the second boils at 195°; and the third boils about 200°. They form salts analogous to the acetate; and acetic acid may be regenerated from them by the action of nascent hydrogen.

Thiacetic Acid, $\left.\begin{matrix} C_2 H_3 O \\ H \end{matrix}\right\} S$. This substance stands to acetic acid in the same relation as mercaptan to alcohol (p. 259); it is prepared by the action of pentasulphide of phosphorus on acetic acid.

$$P_2 S_5 + 5\ C_2 H_4 O_2 = P_2 O_5 + 5\ C_2 H_4 O S.$$

It is a colorless liquid, possessing a peculiarly nauseous smell, and boiling at 93°. The anhydride, $\left.\begin{matrix} C_2 H_3 O \\ C_2 H_3 O \end{matrix}\right\} S$, is also known.

Acetyl Peroxide, $\left.\begin{matrix} C_2 H_3 O \\ C_2 H_3 O \end{matrix}\right\} O_2$, is a remarkable compound, obtained by the action of barium dioxide upon acetic acetate. It is a thick liquid, possessing energetic oxidizing properties, and on heating it decomposes with explosive violence.

Acetamide, $\left.\begin{matrix} C_2H_3O \\ H \\ H \end{matrix}\right\}N$, is the acetyl ammonia; it is obtained by the action of ammonia upon ethyl acetate by an exchange of acetyl for hydrogen, thus:

$$\left.\begin{matrix} C_2H_3O \\ C_2H_5 \end{matrix}\right\}O + \left.\begin{matrix} H \\ H \\ H \end{matrix}\right\}N = \left.\begin{matrix} C_2H_3O \\ H \\ H \end{matrix}\right\}N + \left.\begin{matrix} H \\ C_2H_5 \end{matrix}\right\}O.$$

It is also formed by the action of ammonia on acetyl chloride, and by the dry distillation of ammonium acetate. Acetamide is a colorless solid, fusing at 78°, and boiling at 222°.

Diacetamide, $\left.\begin{matrix} C_2H_3O \\ C_2H_3O \\ H \end{matrix}\right\}N$, and *Ethyl-Diacetamide*, $\left.\begin{matrix} C_2H_3O \\ C_2H_3O \\ C_2H_5 \end{matrix}\right\}N$, are also known. Corresponding compounds are likewise formed from the chloracetic acids.

Acetone, $\left.\begin{matrix} C_2H_3O \\ C_2H_3 \end{matrix}\right\}$. This compound, which may be regarded as methyl acetyl, is formed by replacing the chlorine in acetyl chloride by methyl, thus:

$$\left.\begin{matrix} CH_3 \\ CH_3 \end{matrix}\right\}Zn + 2\left(\left.\begin{matrix} C_2H_3O \\ Cl \end{matrix}\right\}\right) = 2\left(\left.\begin{matrix} C_2H_3O \\ CH_3 \end{matrix}\right\}\right) + Cl_2 Zn.$$

It is also obtained by the distillation of calcium acetate, or by passing the vapor of acetic acid through a red-hot tube. Acetone is a colorless liquid, boiling at 56°, forming, like aldehyde, a crystallizable compound with hydrogen sodium sulphite. By the action of sodium amalgam on a mixture of water and acetone, 2 atoms of hydrogen are taken up, and a substance formed which is *isomeric* with propylic alcohol, thus (p. 261):

$$C_3H_6O + H_2 = C_3H_8O.$$

Acetylene, C_2H_2. This remarkable substance is obtained by the direct combination of carbon and hydrogen at a very high temperature. For this purpose the carbon terminals of a powerful battery are brought together in an atmosphere of hydrogen; acetylene is formed and precipitated in combination with copper as a red powder, when the gas is passed through a solution of ammoniacal cuprous chloride. Acetylene is formed in all cases of incomplete combustion of carbon compounds, also when the vapor of alcohol is passed through a red-hot tube, and it is contained in small quantities in coal gas. It is a colorless gas, possessing a peculiar and unpleasant odor; it burns with a luminous smoky flame, and combines directly with chlorine and bromine.

HIGHER FATTY ACIDS. The names, composition, and boiling points of these acids have already been given (p. 249). In their general characteristics they closely resemble the first two of the series, formic and acetic acid. They occur in many natural fats, and they are all formed by the action of nitric acid upon mutton or beef stearine (p. 305).

These acids may be prepared synthetically by the following important reactions: (1) by the direct combination of carbonic acid with the sodium compound of the next lower alcohol radical (p. 274); (2) by the action of potash on the cyanide of the next lower alcohol radical (p. 258); and (3) by replacing 1 or 2 atoms of hydrogen in the radicals of the fatty acids by alcohol radicals. They are most of them oily liquids slightly soluble in water, easily soluble in alcohol, and each forms a well-defined series of salts. The higher members of the series, especially palmitic and stearic, occur in all fatty bodies; they are solid substances obtained by decomposing soaps made from palm-oil or beef-suet, which consist of sodium or potassium palmitate and stearate (*see* Fats, p. 305). These acids form anhydrides, compound ethers, chlorides, aldehydes, amides and acetones, corresponding in constitution and in general chemical characters with the same compounds in the acetyl series. For the description of the properties of these compounds a larger work on Organic Chemistry must be consulted.

It is, however, necessary to remember that a large number

of isomeric compounds exist amongst the acid as well as amongst the alcohol series. These abnormal acids are derived either from the corresponding abnormal alcohol or from some compound similarly constituted. The isomeric alcohols, acids, and hydrocarbons of the 4-carbon series are as follows :

Normal Butyl Hydride.	Normal Butyl Alcohol.	Normal Butyric Acid.	Secondary Butyl Alcohol.
CH_3	CH_3	CH_3	CH_3
|	|	|	|
CH_2	CH_2	CH_2	CH_2
|	|	|	|
CH_2	CH_2	CH_2.	$CH\,OH$
|	|	|	|
CH_3	CH_2OH	$CO\,OH$	CH_3

Isobutyl Hydride (Trimethyl formene).	Fermentation Butyl Alcohol.	Tertiary Butyl Alcohol.	Isobutyric Acid.
CH_3CH_3	CH_3CH_3	CH_3CH_3	CH_3CH_3
\\/	\\/	\\/	\\/
CH	CH	COH	CH
|	|	|	|
CH_3	CH_2OH	CH_3	$CO\,OH$

The *general reactions* of the group of monatomic alcohols and acids which offer the greatest theoretical interest are, certainly, those by which it is possible, in the first place, to prepare the most simple terms of the series synthetically from their elements, and, secondly, to pass directly by addition of carbon and hydrogen from these lower terms to the higher ones, and thus to mount up the series. Suppose that we begin with methyl alcohol obtained from inorganic sources: viz.: (1) Marsh gas prepared from sulphuretted hydrogen and carbonic disulphide, thus :

$$2\,SH_2 + CS_2 + Cu_8 = CH_4 + 4\,(Cu_2\,S).$$

(2) Methyl chloride, from this by the action of chlorine, thus :

$$CH_4 + Cl_2 = CH_3\,Cl + HCl.$$

(3) Methyl alcohol, from this by the action of potash, thus :

$$CH_3\,Cl + KHO = CH_4\,O + K\,Cl.$$

There are now several modes by which we can pass to the dicarbon series :—

(1.) From methyl alcohol we prepare methyl cyanide; this, on decomposition with potash, yields acetic acid (*see* p. 275), thus :

$$CNCH_3 + KHO + H_2O = C_2 H_3 KO_2 + NH_3.$$

It has recently been shown that we can directly reduce the acids to alcohols or aldehydes by the action of sodium upon the anhydrides ; and from aldehyde we can also obtain ethyl alcohol directly by the action of hydrogen (p. 273), thus :

$$C_2 H_4 O + H_2 = C_2 H_6 O.$$

(2.) From methyl alcohol we prepare methyl cyanide, and by acting upon this with hydrogen, we get ethylamine (p. 252), thus :

$$CNCH_3 + H_4 = \left.\begin{matrix} C_2 H_5 \\ H \\ H \end{matrix}\right\} N.$$

Ethylamine hydrochlorate acted upon by silver nitrite yields ethyl nitrite, which, on decomposition with potash, yields the alcohol, thus :

$$C_2 H_7 N + N_2 O_3 = C_2 H_5 NO_2 + H_2 O + N_2$$

and $$C_2 H_5 NO_2 + KHO = C_2 H_6 O + K NO_2.$$

(3.) From methyl alcohol, by the action of zinc on methyl iodide, we prepare the so-called radical methyl: this substance forms ethyl chloride when treated with chlorine ; from this we can pass through ethyl acetate to ethyl alcohol. The repetition of any of these three processes would enable us to pass to the tricarbon group, and so on.

LESSON XXXIII.

DIATOMIC ALCOHOLS AND THEIR DERIVATIVES.

As we have seen (p. 229), the hydrocarbons of the general formula, $C_n H_{2n}$, of which we may take ethylene, $C_2 H_4$, as an example, are non-saturated compounds, in which two of the combining powers of the carbon are not satisfied; hence these bodies combine directly with two atoms of chlorine, bromine, &c., to form a saturated compound. The lowest term of the series, CH_2, to which the name of methylene has been given, is not known in the free state, although its iodide, $C H_2 I_2$, has been isolated. The corresponding diatomic alcohol also has not been prepared, but the diacetate is known.

Ethylene, $C_2 H_4$. This substance, known as olefiant gas, has already been mentioned (p. 78); it is formed in the dry distillation of coal and various organic bodies; it is, however, best prepared by the action of hot sulphuric acid on alcohol; a mixture of 1 part of alcohol and 4 parts of sulphuric acid is heated in a flask with enough sand to form a pasty mass. The decomposition is a very simple one; alcohol loses 1 molecule of water, H_2O, and ethylene is formed. The chief physical properties of ethylene have already been mentioned (p. 79). It combines directly with 2 atoms of chlorine, also with hydrochloric and hydriodic acids; with chlorine it forms ethylene dichloride; with the hydracids it forms ethyl chloride, bromide, or iodide.

Ethylene dichloride, $C_2 \overset{II}{H_4} Cl_2$. Olefiant gas derives this name from its power of forming an oil when brought into contact with chlorine. On mixing these gases, drops are formed; and when collected, washed, and distilled, they yield the pure dichloride, This body boils at 82°·5, and is insoluble in water, but soluble in alcohol and ether. It is rapidly attacked by chlorine, and substitution products are formed, in which one, two, three, and lastly, four atoms of hydrogen are replaced by chlorine. Thus we have—

	Boiling Point.
$C_2 H_4 Cl_2$	82°·5
$C_2 H_3 Cl Cl_2$	115°
$C_2 H_2 Cl_2 Cl_2$	137°
$C_2 H Cl_3 Cl_2$	154°
$C_2 Cl_6$	182°

From ethyl chloride a series of chlorine substitution products are obtained, identical in composition, but differing in their properties from the foregoing; thus, the two sets of bodies boil at different temperatures; whilst those from ethylene are decomposed by alcoholic potash, those from ethyl chloride remain unchanged. These two series of bodies are said to be *isomeric.* The last term, $C_2 Cl_6$, is identical in both series.

Glycol or *Ethylene Alcohol*, $C_2H_4 \left\{ \begin{matrix} OH \\ OH \end{matrix} \right.$. This substance, which may be supposed to be 2 molecules of water, in which 2 atoms of the hydrogen are replaced by the dyad radical ethylene, is obtained by the action of ethylene dibromide upon silver acetate, silver bromide and glycol diacetate being formed, thus:

Ethylene Di-bromide and Silver Acetate yield Silver Bromide and Glycol Diacetate.

$$\overset{II}{C_2 H_4} Br_2 + 2\left(\left.\begin{matrix} C_2 H_3O \\ Ag \end{matrix}\right\} O \right) = 2 \, Ag \, Br + \left.\begin{matrix} C_2 H_3 O \\ C_2 H_3 O \\ \overset{II}{C_2 H_4} \end{matrix}\right\} O_2.$$

The pure glycol is obtained from the acetate by acting on it with baryta. Glycol is a colorless, inodorous, and sweetish-tasting, thick liquid; its specific gravity at 0° is 1·125, it boils at 197°·5, and it is soluble in all proportions in alcohol and water. When exposed to air in contact with water and platinum black, it absorbs oxygen rapidly, and is converted into glycolic acid, thus:

$$\left.\begin{matrix} CH_2OH \\ CH_2OH \end{matrix}\right\} + O_2 = H_2O + \left\{\begin{matrix} CH_2OH \\ CO\,OH \end{matrix}\right.$$

On treatment with hot nitric acid glycol oxidizes further to oxalic acid, thus:

$$\left.\begin{matrix} CH_2OH \\ CH_2OH \end{matrix}\right\} + 2O_2 = \begin{matrix} COOH \\ COOH \end{matrix} + 2H_2O.$$

From these reactions it would appear that glycolic and oxalic acids stand to glycol as acetic acid does to ethyl alcohol. Glycol likewise forms an aldehyde by loss of H_4, the substance having the composition $C_2 H_2 O_2$, and called Glyoxal. Glycol acts like alcohol in other respects; the typical hydrogen can be replaced by sodium, forming compounds analogous to sodium ethylate; it also forms a compound with sulphuric acid, called glycol-sulphuric acid, $\left.\begin{matrix} \overset{II}{C_2 H_4} \\ \overset{II}{SO_2} \\ H_2 \end{matrix}\right\} O_3$; and when heated with hydriodic acid, it forms ethylene iodide and water.

Glycol differs, however, from alcohol, inasmuch as it forms two acids, two chlorides, &c. Thus by the action of hydrochloric acid on glycol, the first product obtained is glycol chlorhydrine, that is, glycol in which 1 atom of Cl takes the place of the monad group, HO; whilst by the further action of chlorine a second replacement of the same kind occurs, and ethylene chloride is formed.

(1) Glycol.	(2) Glycol Chlorhydrine.	(3) Ethylene Chloride.
$\begin{cases} CH_2OH \\ CH_2OH \end{cases};$	$\begin{cases} CH_2Cl \\ CH_2OH \end{cases};$	$\begin{cases} CH_2Cl \\ CH_2Cl \end{cases}.$

The first of these bodies is built on the type of 2 molecules of water, the second on a mixed type of 1 of water and 1 of hydrochloric acid, and the third on the type of 2 of hydrochloric acid. There are also 2 acetates of glycol known, monoacetate and diacetate; two ethyl compounds, mono-ethyl glycol and di-ethyl glycol: this latter body is isomeric with acetal (p. 274).

Ethylene Oxide, $\overset{II}{C_2 H_4} O$. This substance is prepared by the action of potash on ethylene chlorhydrine, which loses a molecule of hydrochloric acid and forms glycol ether. Ethylene oxide is a volatile colorless liquid, boiling at 13·5, soluble in all proportions in water. It does not form, like its isomer aldehyde, a crystalline compound with ammonia, but it combines readily with hydrogen, chlorine, acids, &c. Alcohol, $C_2 H_6 O$, is formed by the direct union of ethylene oxide with H_2; and on oxidation glycolic acid is produced.

Ethylene oxide also unites directly with one molecule of water, forming glycol, and also with glycol to form polyethylene glycols.

Diethylene Glycol.

$$(1)\ C_2H_4O + C_2H_4\begin{cases} OH \\ OH \end{cases} = \begin{matrix} C_2H_4 \\ C_2H_4 \end{matrix}\begin{cases} OH \\ O \\ OH \end{cases};$$

Triethylene Glycol.

$$(2)\ C_2H_4O + C_2H_4\begin{cases} OH \\ O \\ OH \end{cases} = \begin{matrix} C_2H_4 \\ C_2H_4 \\ C_2H_4 \end{matrix}\begin{cases} OH \\ O \\ O \\ OH \end{cases}$$

It has been already mentioned that a dyad radical, ethylidene, is supposed to exist in aldehyde, which is isomeric with ethylene. The difference between these two series is that in ethylene two atoms of hydrogen are united to each atom of carbon, whereas in ethylidene the one carbon has one atom of hydrogen attached to it, whilst the second carbon is connected with the other three atoms of hydrogen. Thus:

Ethylidene Series.	Ethylene Series.
Aldehyde, $\left\{ \begin{matrix} CH_3 \\ CHO \end{matrix} \right.$;	Ethylene Oxide, $\left\{ \begin{matrix} CH_2 \\ CH_2 \end{matrix} \right\} O$;
Ethylidene Chloride, $\left\{ \begin{matrix} CH_3 \\ CHCl_2 \end{matrix} \right.$;	Ethylene Chloride, $\left\{ \begin{matrix} CH_2Cl \\ CH_2Cl \end{matrix} \right.$;
Acetal, $\left\{ \begin{matrix} CH_3 \\ CH \left\{ \begin{matrix} OC_2H_5 \\ OC_2H_5 \end{matrix} \right. \end{matrix} \right.$.	Diethyl Glycol, $\left\{ \begin{matrix} CH_2OC_2H_5 \\ CH_2OC_2H_5 \end{matrix} \right.$.

Many compounds of ethylene with the elements of the nitrogen group are known. The dyad ethylene replaces 2 atoms of hydrogen in 2 molecules of ammonia; and thus primary, secondary, and tertiary *diamines* and ammonium compounds are formed closely analogous to the compounds of ethyl. The ethylene diamines are volatile bases obtained by acting with ammonia on ethylene dibromide. Similar compounds in the phosphorus and arsenic series are also known.

Higher Diatomic Alcohols and Derivatives.

The higher carbon series yield olefines corresponding to ethylene.

The following is a complete list of the olefines and glycols, as far as they have been prepared:

Olefines.

		Boiling Point.
Ethylene	C_2H_4	
Propylene	C_3H_6	
Butylene	C_4H_8	+ 3°
Amylene	C_5H_{10}	+ 35°
Hexylene	C_6H_{12}	+ 69°

		Boiling Point.
Heptylene	$C_7 H_{14}$	+ 95°
Octylene	$C_8 H_{16}$	+ 125°
Decatylene (Diamylene)	$C_{10} H_{20}$	+ 160°
Cetene	$C_{16} H_{32}$	+ 275°
Cerotene	$C_{27} H_{54}$	
Melene	$C_{30} H_{60}$	

Glycols.

		Boiling Point.
Ethyl Glycol	$\left.\begin{matrix} C_2 H_4 \\ H_2 \end{matrix}\right\} O_2$	197°·5
Propyl Glycol	$\left.\begin{matrix} C_3 H_6 \\ H_2 \end{matrix}\right\} O_2$	188°
Butyl Glycol	$\left.\begin{matrix} C_4 H_8 \\ H_2 \end{matrix}\right\} O_2$	183°
Amyl Glycol	$\left.\begin{matrix} C_5 H_{10} \\ H_2 \end{matrix}\right\} O_2$	177°*
Hexyl Glycol	$\left.\begin{matrix} C_6 H_{12} \\ H_2 \end{matrix}\right\} O_2$	207°
Octyl Glycol	$\left.\begin{matrix} C_8 H_{16} \\ H_2 \end{matrix}\right\} O_2$	237°

Each of these combines with Cl_2 to form a di-chloride, and each forms a glycol, from which an aldehyde and *two* acids can be obtained by oxidation. The olefines above ethylene yield compounds with hydrochloric and hydriodic acids, which are not identical, but only *isomeric*, with the chlorides and iodides of the monatomic radicals.

* We have here to notice the remarkable exception which the first four of the glycols exhibit as regards their boiling points; in opposition to the general law, the higher carbon glycols boil at a lower temperature than those containing less carbon.

LESSON XXXIV.

Diatomic Acids, resulting from the Oxidation of the Glycols.

These bodies belong, like the glycols, to the type of $\left.\begin{matrix}H_2\\H_2\end{matrix}\right\}O_2$. There are two series of these acids, the first derived by the replacement of 2 atoms of hydrogen in the ethylene by 1 atom of oxygen, and the second by the replacement of all 4 atoms of hydrogen in ethylene by 2 atoms of oxygen. The first of these groups of acids may be termed the *Lactic Acid Series*, and the second the *Oxalic Acid Series*, from the substance best known in each. The relation of glycol to glycolic and oxalic acid, serving as a type of the general relations, is seen in the following :—

Glycol.	Glycolic Acid.	Oxalic Acid.
$\left\{\begin{matrix}CH_2OH\\CH_2OH\end{matrix}\right.$;	$\left\{\begin{matrix}CH_2OH\\CO\ OH\end{matrix}\right.$;	$\left\{\begin{matrix}COOH\\COOH\end{matrix}\right.$.

In like manner we have the following series of acids :—

Name of Acid.	Lactic Acid Series.
Carbonic Acid . . . (Hydrate)	$CH_2\,O_3$.
Glycolic	$C_2H_4O_3$.
Lactic	$C_3H_6O_3$.
Butylactic	$C_4H_8O_3$.

Valero-lactic . . . . $C_5H_{10}O_3$.

Leucic . . . . . . $C_6H_{12}O_3$.

Oxalic Series of Acids.

Type, $C_n H_{2n-2}O_4$.

Name of Acid.	Formula.
Oxalic.	$C_2H_2O_4$.
Malonic.	$C_3H_4O_4$.
Succinic.	$C_4H_6O_4$.
Pyrotartaric.	$C_5H_8O_4$.
Adipic.	$C_6H_{10}O_4$.
Pimelic.	$C_7H_{12}O_4$.
Suberic.	$C_8H_{14}O_4$.
Azelaic.	$C_9H_{16}O_4$.
Sebacic.	$C_{10}H_{18}O_4$.
Brassylic.	$C_{11}H_{20}O_4$.
Rocellic.	$C_{17}H_{32}O_4$.

Carbonic Acid, $CH_2 O_3$. This substance is only known

in its salts, the hydrate not having been prepared. These salts may be supposed to contain the *radical carbonyl*, $\overset{II}{CO}$. Carbonic acid contains 2 atoms of replaceable hydrogen; hence carbonates of the following composition are known:—

$$\left.\begin{matrix}\overset{II}{CO}\\K_2\end{matrix}\right\}O_2, \text{ and } \left.\begin{matrix}\overset{II}{CO}\\HK\end{matrix}\right\}O_2 \quad \left.\begin{matrix}\overset{II}{CO}\\C_2H_5\,K\end{matrix}\right\}O, \text{ and } \left.\begin{matrix}\overset{II}{CO}\\2(C_2H_5)\end{matrix}\right\}O.$$

Carbon Monoxide, C O (p. 76), is the isolated radical carbonyl; this substance combines directly with chlorine to form chlor-carbonyl (phosgene gas), C O Cl_2, and with $\left.\begin{matrix}K\\H\end{matrix}\right\}O$, to from $\left.\begin{matrix}COH\\K\end{matrix}\right\}O$, potassium formate.

Sulpho-Carbonic Acid, CH_2S_3. The potassium salt of this acid is obtained by the direct combination of carbonic disulphide (page 116) and potassium sulphide; and from this salt sulphocarbonic acid is prepared by the addition of hydrochloric acid. A large number of derivatives of this body are known: thus we have *di-methyl sulpho-carbonate*, $\left.\begin{matrix}CS\\2(CH_3)\end{matrix}\right\}S_2$; *hydric-methyl-sulpho-carbonate*, $\left.\begin{matrix}CS\\(CH_3)H\end{matrix}\right\}S_2$. For a description of the compounds intermediate between the sulpho- and the oxy-carbonates the larger manuals must be consulted.

The acids of the first series above carbonic are *monobasic*, that is, they have only one atom of hydrogen replaceable by a metal, while those of the second series are *dibasic*, or contain two of hydrogen thus replaceable; both series are diatomic.

Glycolic Acid, $C_2H_4O_3$. This substance is obtained by the oxidation of glycol; it may also be derived from the monatomic series of acids by the action of potash on mono-chloracetic acid, thus

Potassium Monochlor-Acetate.	and Potash	give	Potassium Glycolate.	and Potassium Chloride.
$C_2H_2ClKO_2$	$+ KHO.$	$=$	$C_2H_3KO_3$	$+ KCl.$

Glycolic acid forms a deliquescent crystalline mass, and forms salts called Glycolates, which contain only one atom of metal. An amide, called Glycolamide, is known, as well as a substance isomeric with it, termed Glycocolle.

Oxalic Acid, $C_2H_2O_4$. Oxalic acid is met with in the juice of many plants in the form of potassium or calcium salt. It is formed in a great variety of ways, chiefly by the oxidation of different organic bodies. The best way of preparing pure oxalic acid is by acting upon sugar with nitric acid; it has generally been manufactured in this way on a large scale, but at present it is prepared in very large quantities by the action of caustic potash on sawdust. Crude potassium oxalate is thus formed, and from this a pure oxalic acid is obtained by precipitating the insoluble calcium oxalate and decomposing this by sulphuric acid. Oxalic acid can also be prepared by the direct oxidation of glycolic acid.

Oxalic acid crystallizes in prisms which possess the composition $C_2H_2O_4 + 2H_2O$; these crystals lose their water of crystallization at 100°, or *in vacuo* over sulphuric acid. When heated to about 160° oxalic acid rapidly decomposes, forming carbon dioxide, carbon monoxide, and formic acid, whilst a small quantity of oxalic acid sublimes undecomposed. Heated with sulphuric acid, oxalic acid is decomposed into water and equal volumes of carbon monoxide and carbon dioxide gases (p. 76). Oxalic acid is a bibasic acid, and forms two classes of salts called Hydric Oxalates and Oxalates. The alkaline oxalates are all soluble in water; the oxalates of the other metals are generally insoluble. The potassium oxalates are:

$C_2K_2O_4 + H_2O$. Potassium oxalate (neutral oxalate).

$C_2HKO_4 + H_2O$. Hydric potassium oxalate (binoxalate).

$C_2\,H\,K\,O_4\,C_2\,H_2\,O_4 + 2\,H_2\,O$. Acid potassium oxalate (quadroxalate).

Calcium oxalate is a very insoluble salt, and is the form in which this metal is obtained for quantitative estimation. Methyl and ethyl oxalates are obtained by distilling the respective alcohols with oxalic acid; the first boils at 161°, and has the formula $\left.\begin{matrix}\overset{\prime\prime}{C_2\,O_2}\\(C\,H_3)_2\end{matrix}\right\}O_2$; the second boils at 186°, and has the composition $\left.\begin{matrix}\overset{\prime\prime}{C_2\,O_2}\\(C_2\,H_5)_2\end{matrix}\right\}O_2$.

Oxalic Amides. By heating neutral ammonium oxalate, a white powder called *Oxamide* is left; the composition of this substance is $\left.\begin{matrix}\overset{\prime\prime}{C_2\,O_2}\\H_2\\H_2\end{matrix}\right\}N_2$, and it may be considered as being 2 molecules of ammonia, in which 2 atoms of hydrogen are replaced by $\overset{\prime\prime}{C_2\,O_2}$. By heating hydric ammonium oxalate, a substance called *Oxamic Acid* is obtained, having the formula

$$\left.\begin{matrix}H\\H\end{matrix}\right\}N,$$

$$\left.\begin{matrix}\overset{\prime\prime}{C_2\,O_2}\\H\end{matrix}\right\}O,$$

belonging to a mixed type, one of water and one of ammonia, in each of which one atom of hydrogen is replaced by the dyad $C_2\,O_2$.

Lactic Acid, $\left.\begin{matrix}C_3\,H_4\,O\\H_2\end{matrix}\right\}O_2$. This acid is contained in sour milk, and is formed from sugar by a peculiar change called the Lactic Fermentation; an acid of the same composition is contained in the flesh of animals: this is, however, not identical with that obtained by the fermentation of sugar,—hence the former is termed para-lactic-acid. It can also be formed artificially—

(1.) By the direct oxidation of propyl glycol.

(2.) By the decomposition of mono-chlor propionic acid by alkalies.

(3.) By allowing aldehyde, hydrocyanic acid, and hydrochloric acid to remain in contact for several days.

$$C_2 H_4 O + H C N + H Cl \ 2(H_2 O) = N H_4 Cl + C_3 H_6 O_3.$$

Lactic acid is a syrupy liquid of specific gravity 1·215, which cannot be distilled without decomposition, but, when heated, forms *lactide* (the anhydride), $C_3 H_4 O_2$, and *dilactic acid*, $C_6 H_{10} O_5$. When it is heated with hydriodic acid, lactic acid forms propionic acid. The lactates form a well-defined class of salts, containing as a rule one atom of metal—the other atom of hydrogen being replaceable only by an organic radical: thus we have ethyl-lactic acid, $\left.\begin{matrix} C_3 H_4 O \\ C_2 H_5 \\ H \end{matrix}\right\} O_2$, forming also a definite series of salts. All the lactates are soluble in water and alcohol: the zinc lactate is the most characteristic of the salts; it crystallizes in shining needles.

Lactyl Chloride, $C_3 H_4 O Cl_2$, is formed by the action of phosphorus pentachloride on calcium lactate.

Lactamide, $C_3 H_7 O_2 N$; lactic monamide is obtained by the action of ammonia on lactide. It is isomeric with *alanine*, a substance formed by the union of aldehyde, hydrocyanic acid, and water.

The higher acids of the lactic series do not possess sufficient general interest to entitle them to consideration in an elementary work. We therefore pass to the higher acids of the oxalic series.

Malonic Acid, $C_3 H_4 O_4$, is obtained by the oxidation of malic acid (p. 293); it is an unimportant substance.

Succinic Acid, $\left.\begin{matrix} C_4 H_4 O_2 \\ H_2 \end{matrix}\right\} O_2$. This acid is obtained by the distillation of amber; it occurs in certain resins, and in small quantities in various animal juices, and it is produced by the fermentation of sugar (p. 308). It can be artificially prepared (1.) by the action of hydriodic acid upon malic and tartaric acids (p. 293).

(2.) By action on ethylene di-cyanide by potash, thus :

$$C_2 H_4 (CN)_2 + 4 H_2 O = C_4 H_6 O_4 + 2 NH_3.$$

(3.) By the action of potash on cyanopropionic acid, thus :

$$C_3 H_5 (CN) O_2 + 2 H_2 O = C_4 H_6 O_4 + N H_3.$$

(4.) By the action of nitric acid on butyric acid, thus :

$$C_4 H_8 O_2 + 3 O = H_2 O + C_4 H_6 O_4.$$

Succinic acid forms large colorless crystals, which fuse at 180°, and begin to boil at 235°, the vapor decomposing into succinic anhydride and water. It forms a chloride as well as anhydride when heated with phosphorus pentachloride. Bromine substitution products are also known, viz., mono-brom-succinic acid, $\left.\begin{matrix} C_4 H_3 Br O_2 \\ H_2 \end{matrix}\right\} O_2$, and di-brom-succinic acid, $\left.\begin{matrix} C_4 H_2 Br_2 O_2 \\ H_2 \end{matrix}\right\} O_2$; these acids, when treated with water and silver oxide, are respectively converted into malic and tartaric acids. Succinic acid forms two classes of salts, and is bibasic ; the salts of the alkaline metals are soluble, and these form an insoluble brown precipitate with ferric salts.

Succinic Anhydride, $C_4 H_4 O_3$, is also known.

The ammonia derivatives of this acid are :

Succinamide, $\left.\begin{matrix} C_4 H_4 O_2 \\ H_2 \\ H_2 \end{matrix}\right\} N_2$, and *Succinimide*, $\left.\begin{matrix} C_4 \overset{II}{H_4} O_2 \\ H \end{matrix}\right\} N$.

For the special properties of the higher carbon acids of this series, the reader must consult a larger work on the subject.

Connected with succinic acid very intimately are two acids of much importance, viz., malic and tartaric acids.

Malic Acid, $C_4 H_6 O_5$. This acid occurs in the juice of most fruits, from which it can be obtained; it can also be prepared by the substitution of H O for Br in monobrom-succinic acid (p. 292). Malic is a triatomic acid, but only two of the three typical atoms of hydrogen can be replaced by metal; hence it is bibasic. The malates are soluble in water; malic acid itself crystallizes in needles. When malic acid is heated to about 180° it loses $H_2 O$, and is converted into a new acid, $C_4 H_4 O_4$, which exists in two isomeric states, forming *fumaric* and *maleic* acids. These substances both unite directly with hydrogen and yield succinic acid, $C_4 H_6 O_4$.

Tartaric Acid, $C_4 H_6 O_6$. Tartaric acid exists in the juice of many fruits (grape, tamarind, &c.); it is deposited as potassium salt during the fermentation of wine, and this salt is known as tartar. Several interesting isomeric conditions of tartaric acid exist; thus the ordinary acid possesses the power of turning the plane of polarized light round to the right, and therefore is termed *Dextro-tartaric acid*, whilst another form obtained from certain specimens of tartar does not affect the ray of polarized light in any way, and is said to be inactive. This inactive tartaric acid, termed *Racemic Acid*, can be divided into the common, or dextro-tartaric, and a new acid possessing the opposite power of deviating the plane of polarization to the *left*, and hence called *Levro-tartaric acid.* There also appears to be a fourth modification of this acid which is distinguished by being inactive like racemic, but not capable of being split up into the two active varieties. The inactive variety of tartaric acid can be prepared artificially by the action of lime-water on bibromo-succinic acid, $C_4 H_4 Br_2 O_4$, each of the atoms of bromine being replaced by H O, and yielding tartaric acid, $C_4 H_6 O_6$. Tartaric acid is also formed by the action of nitric acid on sugar of milk.

Tartaric acid (dextro) crystallizes in large oblique rhombic prisms belonging to the monoclinic system, which dissolve easily in water. When heated to 180° it fuses and undergoes decomposition, evolving a peculiar odor of caramel. In presence of oxidizing agents tartaric acid is converted into carbonic, formic, and oxalic acids; and when fused with caustic potash it forms acetic and oxalic acids. When tartaric acid is heated with hydriodic acid for several hours it is first reduced to malic, and afterwards to succinic acid, by losing first one and then two atoms of oxygen. Tartaric acid is a bibasic acid, containing two atoms of typical hydrogen which can be replaced by metals; hence there are two classes of alkaline tartrates; thus we have—

Hydrogen Potassium Tartrate (Cream of Tartar), $\left.\begin{matrix} K \\ H \end{matrix}\right\} C_4 H_6 O.$

Potassium Tartrate . . . $\left.\begin{matrix} K \\ K \end{matrix}\right\} C_4 H_4 O_6.$

Tartaric acid forms with antimony a remarkable compound termed *Tartar Emetic;* this compound may be considered as potassium tartrate, in which one atom of potassium is replaced by a monatomic radical, Sb O. We then have tartar emetic, $C_4 H_4 (Sb O) KO_6$. This body is obtained by boiling a solution of cream of tartar with antimonious oxide; the oxide dissolves, and, on cooling, tartar-emetic is deposited in crystals. This salt is much used in medicine, and acts as a violent poison when taken in quantity. Tartaric acid and citric acid are largely used by the calico printer to act as a discharge or solvent for the mordant, thus giving white spots on a colored ground.

Citric Acid, $C_6 H_8 O_7$. This acid is tribasic, and it is found in the juice of the lemon, and occurs in many other fruits, together with malic acid. Citric acid obtained from these sources crystallizes in large colorless crystals, which dissolve very easily in water. Three series of citrates exist, in which one, two, or three atoms of hydrogen are replaced by metal. The citrates of the alkaline metals are soluble, those of the alkaline-earth metals, of lead, and silver are insoluble in water.

LESSON XXXV.

CYANOGEN COMPOUNDS.

We have already seen that a series of compounds exists containing a monatomic radical, CN, called Cyanogen, standing in close relation to carboxyl. The following are some of the most important of the cyanogen compounds :

Hydrocyanic Acid	$\left.\begin{matrix}H\\CN\end{matrix}\right\}$	Cyanic Acid	$\left.\begin{matrix}CN\\H\end{matrix}\right\}O.$
Cyanogen Gas	$\left.\begin{matrix}CN\\CN\end{matrix}\right\}$	Sulpho-Cyanic Acid	$\left.\begin{matrix}CN\\H\end{matrix}\right\}S.$
Cyanogen Chloride	$\left.\begin{matrix}CN\\Cl\end{matrix}\right\}$	Cyanamide	$\left.\begin{matrix}CN\\H\\H\end{matrix}\right\}N.$

The cyanogen compounds may also be considered as derivatives of ammonia; thus cyanic acid acts in many cases as if it were $\left.\begin{matrix}\overset{II}{CO}\\H\end{matrix}\right\}$ N, and this group becomes connected with the oxalic acid series of bodies. The cyanogen compounds are remarkable for forming series of *polymeric* modifications: thus we have CN Cl, and $C_3N_3Cl_3$; $\left.\begin{matrix}CN\\H\end{matrix}\right\}$ O, and $\left.\begin{matrix}C_3N_3\\H_3\end{matrix}\right\}$ O_3.

Cyanogen, $\left.\begin{matrix}CN\\CN\end{matrix}\right\}$. This substance is obtained by heating the mercury, gold, or silver cyanides; it is found in small quantities in the gases of the iron blast furnace. Its properties have already been mentioned (p. 84.) It is formed by the action of heat on oxamide, and ammonium oxalate, and is thus connected with the oxalic group, as cyanogen is oxamide, minus 2 molecules of water. Cyanogen forms with potash a mixture of potassium cyanide and cyanate.

Hydrocyanic Acid, HCN. Hydrocyanic acid has quite

recently been obtained by the direct union (without condensation) of nitrogen and acetylene, when a series of electric sparks is passed through a mixture of these gases. The mode of preparation and chief properties of this substance have already been mentioned. This acid easily undergoes decomposition, and cannot therefore be kept for a length of time either in the pure state or in aqueous solution. It yields ammonium formate thus: $HCN + 2\,H_2O = \left.\begin{matrix} CHO \\ NH_4 \end{matrix}\right\} O$, as aceto-nitril yields acetic acid (p. 252). With chlorine and bromine it yields cyanogen chloride and bromide. The best method of detecting hydrocyanic acid is founded on the formation of Prussian blue. To the liquid containing the acid a few drops of a ferrous and a ferric salt are added; then excess of caustic soda; and lastly, an excess of hydrochloric acid, the formation of a deep blue liquid, from which a deep blue precipitate separates either at once or after a little time, indicates the presence of hydrocyanic acid. The presence of this substance may also be recognized by evaporating some of the solution on a watch glass with ammonium sulphide to dryness on a water-bath; on adding a drop of ferric chloride, a deep red coloration of ferric sulpho-cyanide is produced, if hydrocyanic acid be present.

The simple metallic Cyanides are formed by the direct action of hydrocyanic acid upon a metallic oxide; in addition to these a large number of *double Cyanides* are known.

Potassium Cyanide, KCN, is formed when potassium is burnt in cyanogen or in hydrocyanic acid gas, or when potash is added to aqueous hydrocyanic acid. It is prepared on a large scale by fusing potassium ferro cyanide (p. 297) with potassium carbonate. The iron is thus separated, and potassium takes its place. Potassium cyanide is a white salt, very soluble in water and hot alcohol; it fuses easily without decomposition, and acts as a violent poison. Potassium cyanide is largely used in photography for dissolving the unaltered silver salts; also in still greater quantity in the art of electrotyping in gold and silver, serving as a solvent for these metals.

The sodium and ammonium cyanides are also soluble, very poisonous salts.

Mercuric Cyanide, $Hg\ C_2\ N_2$, is a soluble, easily crystallizable salt, formed by dissolving mercuric oxide in aqueous hydrocyanic acid. When heated it decomposes into gaseous cyanogen (C_2N_2), mercury, and a brown substance, isomeric with cyanogen gas, and called *Paracyanogen.*

The other simple cyanides are insoluble in water; amongst the most important is the white silver cyanide, and the brownish-red copper cyanide. In writing the formulæ of these compounds, it is useful to express cyanogen by the symbol Cy. Amongst the numerous compound cyanides, those of potassium and iron are the most important; in these the iron is contained in combination in a different mode to that in the ordinary iron salts, inasmuch as it is not precipitated from the cyanide solution by such re-agents as ammonia, or ammonium sulphide. The same remark applies to cobalt, and in a less degree to several other metals.

Potassium Ferrocyanide, K_4FeCy_6. This salt, commonly called yellow prussiate of potash, is made on a large scale by heating nitrogenous organic matter with potashes and iron filings. On dissolving the mass in water, and evaporating the solution, large yellow quadratic crystals of potassium ferrocyanide, containing 3 atoms of water of crystallization, are deposited; it is not poisonous, acting as a mild purgative. When heated strongly, it yields potassium cyanide and iron carbide, and when treated with dilute sulphuric acid, hydrocyanic acid is formed. By the action of strong and hot sulphuric acid, the salt is decomposed, and carbonic oxide gas evolved. Solutions of this salt produce with ferric salts, a deep blue precipitate of Prussian blue, a ferric-ferrocyanide having the composition $Fe_4(FeCy_6)_3$. With ferrous salts the yellow prussiate produces a white precipitate.

Hydrogen Ferrocyanide, or *Ferrocyanic Acid*, $H_4\ Fe\ Cy_6$. This acid is formed by adding hydrochloric acid to a strong solution of the foregoing salt. It acts as a strong acid, and is tetrabasic, forming a series of salts in which the four typical atoms of hydrogen of the acid are replaced by an equivalent of metal.

Potassium Ferricyanide, K_3 Fe Cy_6. This salt, called red prussiate of potash, is obtained by passing chlorine gas through a solution of the yellow prussiate, which loses one atom of potassium; the action must be allowed to continue until a drop of solution produces no blue precipitate with a ferric salt. On evaporation, the salt separates out in deep red prismatic crystals. This substance produces in solutions of ferrous salts a deep blue precipitate of Turnbull's blue, $Fe_3(FeCy_6)_2$.

Nitro-Ferrocyanides are a peculiar class of salts, obtained by the action of nitric acid on potassium ferrocyanide. The sodium salt crystallizes in red prisms, and produces with the slightest trace of an alkaline sulphide a deep purple color.

Cyanogen Chlorides. Cyanogen forms with chlorine a chloride which exists in 2 polymeric modifications; they are both obtained by the action of chlorine upon hydro-cyanic acid.

		Boiling Point.	Melting Point.
Gaseous Cyanogen Chloride .	Cy Cl	— 12°	— 15°
Solid	Cy_3 Cl_3	190°	140°

Cyanic Acid, $\left.\begin{matrix}Cy\\H\end{matrix}\right\}O$. The salts of this acid, termed cyanates, are readily formed by the direct oxidation of cyanides, and by the action of cyanogen gas upon potash. Cyanic acid tself cannot be prepared in the free state from its salts, as on liberation it at once changes into polymeric modifications called *cyanuric acid* and *cyamelide.* It can, however, be obtained by heating cyanuric acid in a retort, and collecting the volatile cyanic acid in a freezing mixture; it forms a colorless mobile liquid, but it immediately changes into solid cyamelide when taken out of the freezing mixture. Cyanic acid in aqueous solution combines at once with water to form ammonium carbonate, $\left.\begin{matrix}CN\\H\end{matrix}\right\}O + H_4\,O_2 = NH_4\,H\,CO_3$, and with ammonia to form urea, $\left.\begin{matrix}CN\\H\end{matrix}\right\}O + NH_3 = CO\,N_2\,H_4$.

Cyanic acid is a monobasic acid.

Ammonium Cyanate, $\left.\begin{matrix}CN\\NH_4\end{matrix}\right\}O$, is formed by bringing together dry ammonia and cyanic acid; but this salt undergoes gradually at ordinary temperatures, and at once at 100°, a remarkable molecular change, becoming *urea*,

$$\left.\begin{matrix}CN\\NH_4\end{matrix}\right\}O = \left.\begin{matrix}\overset{II}{CO}\\H_2\\H_2\end{matrix}\right\}N_2.$$

Urea, or *Carbamide*, $\left.\begin{matrix}\overset{II}{CO}\\H_2\\H_2\end{matrix}\right\}N_2$. This important substance is found in large quantity in the urine of mammalia, and in small amount in various animal juices. It is obtained artificially—(1.) From ammonium cyanate,

$$CN\,NH_4\,O = \left.\begin{matrix}\overset{II}{CO}\\H_2\\H_2\end{matrix}\right\}N_2.$$

(2.) By the action of ammonia on ethyl carbonate, thus:

$$\underset{\text{Ethyl Carbonate.}}{\left.\begin{matrix}\overset{II}{CO}\\2\,(C_2\,H_5)\end{matrix}\right\}O_2} + \left.\begin{matrix}H_2\\H_2\\H_2\end{matrix}\right\}N_2 = \underset{\text{Urea.}}{\left.\begin{matrix}\overset{II}{CO}\\H_2\\H_2\end{matrix}\right\}N_2} + 2\,\underset{\text{Alcohol.}}{\left.\begin{matrix}C_2\,H_5\\H\end{matrix}\right\}O}.$$

(3.) By the action of mercuric oxide on oxamide:

$$\underset{\text{Oxamide.}}{\left.\begin{matrix}\overset{II}{C_2\,O_2}\\H_2\\H_2\end{matrix}\right\}N_2} + Hg\,O = \underset{\text{Carbamide.}}{\left.\begin{matrix}\overset{II}{CO}\\H_2\\H_2\end{matrix}\right\}N_2} + CO_2 + Hg.$$

The first of these methods is that by which urea is best prepared. For this purpose yellow prussiate of potash is mixed with manganese dioxide, and the mixture heated on an iron plate; potassium cyanate is thus formed, and this salt is dissolved in water and mixed with ammonium sulphate. On evaporating to dryness the urea can be extracted with alcohol. Urea thus prepared crystallizes in long striated needles, which dissolve in their own weight of cold water, and to the same extent in hot alcohol. When heated to 120°, urea fuses and begins to decompose, forming substances termed *ammeline* and *biuret;* whilst at a higher temperature cyanuric acid is produced. When heated with water in closed tubes to 100°, urea forms carbonic acid and ammonia, showing that it is an amide of carbonic acid. Nitrous acid decomposes urea instantly into carbonic acid, nitrogen, and water. Urea is the product of the oxidation of the nitrogenous constituents of the body, and the quantity of urea excreted is a measure of the activity of the changes going on. Urea forms compounds with acids and with bases. Urea nitrate and oxalate are the most important salts. With mercuric oxide, urea forms an important insoluble compound, which is employed as a means of estimating the quantity of urea in a solution.

Compound Ureas. These compounds are formed by acting on cyanic acid with a compound ammonia. They may be considered as urea, in which one or more atoms of hydrogen are replaced by methyl, ethyl, &c. Compound ureas, containing the oxidized radicals, acetyl, butyryl, &c., are also known.

Cyanuric Acid, $\left.\begin{matrix}Cy_3\\H_3\end{matrix}\right\}O_3$. This polymer of cyanic acid is a solid crystalline substance formed on heating urea, or by acting with water on the solid cyanogen chloride. It is a tribasic acid, and is formed on the type of 3 molecules of water.

Sulphocyanic Acid, $\left.\begin{matrix}Cy\\H\end{matrix}\right\}S$. The potassium salt of this acid is easily prepared by heating potassium ferrocyanide with sulphur; on dissolving and crystallizing, potassium sulphocyanide, $\left.\begin{matrix}Cy\\K\end{matrix}\right\}S$, is deposited. The acid may be obtained by acting on mercuric sulphocyanide with sulphuretted hydrogen.

Cyanamide, $\left.\begin{matrix}CN\\H\\H\end{matrix}\right\}N$; *Dicyanamide*, $\left.\begin{matrix}CN\\CN\\H\end{matrix}\right\}N$; and *Tricyanamide*, $\left.\begin{matrix}CN\\CN\\CN\end{matrix}\right\}N$, are obtained by the action of ammonia on cyanogen chloride. Several other amidic compounds of cyanogen exist, for description of which the larger manuals must be consulted.

Uric Acid, $C_5\,H_4\,N_4\,O_3$. This substance, found in the urine of birds, serpents, &c., is connected with the foregoing compounds. From uric acid a large number of derivatives have been obtained, amongst which may be named *alloxan*, $C_4\,H_2\,N_2\,O_4$; *dialuric acid*, $C_4\,H_4\,N_2\,O_4$; *alloxatine*, $C_8\,H_4\,N_4\,O_7$; and *parabanic acid*, $C_3\,H_2\,N_2\,O_3$: these derivatives of uric acid can be generally regarded as amides, containing the radical, $C_2\,O_2$, of oxalic acid.

LESSON XXXVI.

TRIVALENT ALCOHOLS AND THEIR DERIVATIVES.

THE hydrocarbon group having the general formula $C\,H_{2n-1}$, act in accordance with the views already expressed (p. 231) as triatomic radicals, to which the generic name of *Glycerines* has been given from the special name of one of the series, viz. $C_3H_5\left\{\begin{matrix}OH\\OH\\OH\end{matrix}\right.$. This substance is arranged on the type of 3 molecules of water, and contains 3 atoms of typical hydrogen, each replaceable by a metal. From this formula it is clear that the possible number of derivatives of the triatomic alcohols is much larger than that of either of the preceding classes. The relation existing between the composition of the mono-, di-, and triatomic alcohols of the same carbon series is a very simple one, as is seen by the following comparison of the three carbon series :

Propyl-hydride, $C_3 H_8$.
Monovalent propyl-alcohol, $C_3 H_7 OH$.

Divalent propyl-glycol, $C_3 \left.\begin{matrix} \overset{\text{II}}{H_3} \\ H_2 \end{matrix}\right\} O_2$, or $C_3 H_8 O_2$.

Trivalent propyl-glycerine, $C_3 \left.\begin{matrix} \overset{\text{III}}{H_5} \\ H_3 \end{matrix}\right\} O_3$, or $C_3 H_8 O_3$.

The glycerines of the mono- and dicarbon series have not been prepared; that of the tri-carbon series is best known and may be taken as the type ; amyl glycerine has also been prepared.

Glycerine, $C_3 H_5 \left\{\begin{matrix} OH \\ OH \\ OH \end{matrix}\right.$. This substance is contained in most oils and fats, both vegetable and animal, which consist of triatomic ethers of the higher terms of the fatty acid series ; thus beef suet, or stearine, is glycerine tri-stearate, or glycerine in which the 3 atoms of typical hydrogen have been replaced by 3 molecules of the radical, $C_{18} H_{35} O$, of stearic acid (p. 249). Glycerine is also found in small quantities in the fermentation of sugar. Glycerine is formed from fats by the process of *saponification*, or treatment of the oil with caustic alkali, which decomposes the compound forming an alkaline stearate (soap), and liberating the glycerine which remains in solution when the soap is separated by throwing in common salt. In order to obtain pure glycerine the fat may be decomposed by lead oxide ; the glycerine remains in solution, and the lead-soap or plaister is precipitated. Another and better method is to decompose the fats with high pressure steam, free stearic acid and glycerine being produced.

Glycerin is a colorless thick syrupy liquid, of specific gravity 1·28; it possesses a very sweet taste (whence its name), and is soluble in water and alcohol. It can be distilled in presence of aqueous vapor and *in vacuo*, but it undergoes decomposition when heated in the air. When mixed with dilute nitric acid, glycerin undergoes oxidation and forms *glycerinic acid*, $C_3\,H_6\,O_4$, by exchange of H_2 for O; this acid, therefore, stands to glycerin as acetic acid does to ethyl alcohol. Glycerin is reduced by hydriodic acid to isopropyl iodide, thus:

$$\underset{\text{Glycerin.}}{C_3\,H_8\,O_3} + 5\,H\,I = \underset{\text{Iso-propyl Iodide.}}{C_3\,H_7\,I} + 2\,I_2 + 3\,H_2\,O.$$

Propylglycol is also reduced to the same substance, and thus we can pass from the di- and triatomic series of alcohols to the monatomic series of iso-alcohols.

If the nitric acid employed to act on glycerin be concentrated, a new compound called *Trinitrin*, or *Trinitro-glycerin*, is formed; this is glycerin in which the 3 atoms of typical hydrogen are replaced by $N\,O_2$; thus, $\left.\begin{matrix} C_3H_5 \\ 3(NO_2) \end{matrix}\right\} O_3$. This substance explodes violently when heated, and is used somewhat for blasting and other purposes. Heated with hydrochloric acid, glycerin forms compounds termed *Chlorhydrins*, of which there are three formed by the replacement of 1, 2, or 3 molecules of hydroxyl (OH) by chlorine, thus:

$$\underset{\text{Glycerin.}}{C_3H_5\left\{\begin{matrix} OH \\ OH \\ OH \end{matrix}\right.};\quad \underset{\text{Chlorhydrin.}}{C_3H_5\left\{\begin{matrix} Cl \\ OH \\ OH \end{matrix}\right.};\quad \underset{\text{Dichlorhydrin.}}{C_3H_5\left\{\begin{matrix} Cl \\ Cl \\ OH \end{matrix}\right.};\quad \underset{\text{Trichlorhydrin.}}{C_3H_5\left\{\begin{matrix} Cl \\ Cl \\ Cl \end{matrix}\right.}.$$

Glycerin Ethers of the Fatty Acids.—The *acetins* are prepared by the action of strong acetic acid upon glycerin; they are three in number:

Mono-acetin.	Di-acetin.	Tri-acetin.
$\left.\begin{array}{c} C_3H_5 \\ H_2 \\ C_2H_3O \end{array}\right\} O_3;$	$\left.\begin{array}{c} C_3H_5 \\ H \\ C_2H_3O \\ C_2H_3O \end{array}\right\} O_3;$	$\left.\begin{array}{c} C_3H_5 \\ C_2H_3O \\ C_2H_3O \\ C_2H_3O \end{array}\right\} O_3.$

These substances, which resemble the fats in constitution, are obtained by acting upon glycerine with glacial acetic acid. They are thick oily liquids, only sparingly soluble in water, and boiling at a high temperature.

The Stearic-, Palmitic-, and Oleic Ethers of Glycerine, or *Stearines, Palmitines, and Oleines,* are of great importance, as forming the natural fats. The stearines may be prepared artificially by heating glycerine with stearic acid.

Monostearine.	Distearine.	Tristearine.
$\left.\begin{array}{c} C_3H_5 \\ (C_{18}H_{35})OH_2 \end{array}\right\} O_3.$	$\left.\begin{array}{c} C_3H_5 \\ 2(C_{18}H_{35}O)\,H \end{array}\right\} O_3.$	$\left.\begin{array}{c} C_3H_5 \\ 3(C_{18}H_{35}O) \end{array}\right\} O_3.$

Tri-stearine can be obtained by melting beef or mutton suet, and separating the fibrous matter by filtration and crystallizing the stearine from solution in hot ether. It forms bright white shining plates, insoluble in alcohol and water, but readily soluble in ether. The melting point of stearine appears to undergo changes, and hence it is probable that this substance is capable of existing in several distinct modifications.

Similar glycerine-ethers have been prepared with many of the other members of the series of fatty acids. By the action of the mono-, di-, and tri-chlorhydrines upon sodium ethylate, the three ethyl-glycerine ethers, or *ethylines,* have been prepared; these are—

Ethyline.	Di-ethyline.	Tri-ethyline.
$\left.\begin{array}{c} C_3H_5 \\ (C_2H_5)\,H_2 \end{array}\right\} O_3.$	$\left.\begin{array}{c} C_3H_5 \\ 2(C_2H_5)H \end{array}\right\} O_3.$	$\left.\begin{array}{c} C_3H_5 \\ 3(C_2H_5) \end{array}\right\} O_3.$

Poly-glycerines are known, corresponding to the polyatomic glycols (p. 283). Thus we have—

Di-glycerine, $\left.\begin{matrix} C_3 H_5 \\ C_3 H_5 \\ H_4 \end{matrix}\right\} O_5.$ Tri-glycerine $\left.\begin{matrix} C_3H_5 \\ C_3H_5 \\ C_3H_5 \\ H_5 \end{matrix}\right\} O_7.$

Natural Fats and Oils. The natural oils and fats are all compounds of glycerine chiefly with palmitic, oleic, or stearic acids, and they are contained in the bodies both of plants and animals. The fats cannot be distilled without decomposition, and when heated, give rise to a powerfully smelling substance, called acroleine (p. 306). The oils are separated into the drying and non-drying; the former become dry and resinous on exposure to air from oxidation, whilst the others remain unaltered. The drying oils are generally glycerides of acids not belonging, but nearly related, to the fatty acid series; such, for instance, is the acid of linseed oil, called linoleic acid, $C_{16} H_{28} O_2$: oleic acid, $C_{18} H_{34} O_2$, is found in almost all oils and fats, the compound of this acid with glycerine constituting the liquid portions of the fats.

When the oils or fats are acted upon by nitric acid they are decomposed, and the series of fatty acids formed. Fatty bodies when boiled with alkali undergo the remarkable change termed *Saponification* (p. 302); the fat is decomposed into a fatty acid which combines with the alkali, and glycerine which is liberated passes into solution. Fats may also be saponified or separated into acid and glycerine by distillation with steam alone.

Allyl Compounds. Intimately connected with glycerine are the compounds of a non-saturated monovalent radical, $C_3 H_5$, called allyl. By the action of phosphorus iodide upon glycerine, a monovalent iodide, $C_3 H_5 I$, is obtained, from which a number of bodies have been derived, whilst the acrid substance acroleine formed in the destructive distillation of glycerine is the aldehyde of this series.

Allyl Alcohol, $\left.\begin{matrix} C_3H_5 \\ H \end{matrix}\right\}$ O, obtained by acting with ammonia on allyl oxalate.

$$\left.\begin{matrix} C_2O_2 \\ 2(C_3H_5) \end{matrix}\right\} O_2 + 2NH_3 = \left.\begin{matrix} C_2O_2 \\ H_2 \\ H_2 \end{matrix}\right\} N_2 + 2\left(\left.\begin{matrix} C_3H_5 \\ H \end{matrix}\right\} O.\right)$$

Allyl oxalate and ammonia yield oxamide and allyl alcohol.

It is a colorless liquid, boiling at 103°, possessing a pungent smell. It is oxidized in presence of air and platinum to acroleine and acrylic acid, which stand to this alcohol in the same relation as aldehyde and acetic acid stand to ethyl alcohol; thus *acroleine* is $\left.\begin{matrix} C_3H_3O \\ H \end{matrix}\right\}$ and *acrylic acid* $\left.\begin{matrix} C_3H_3O \\ H \end{matrix}\right\}$ O. Sodium dissolves in allyl alcohol, forming sodium allylate, one atom of typical hydrogen in the alcohol being replaced by sodium; when this substance acts upon allyl iodide, an exchange of allyl and sodium takes place, *allyl ether,* $\left.\begin{matrix} C_3H_5 \\ C_3H_5 \end{matrix}\right\}$ O, being formed. The *allyl sulphide* $\left.\begin{matrix} C_3H_5 \\ C_3H_5 \end{matrix}\right\}$ S, is remarkable as occurring in nature as the essential oil of garlic, and the sulphide artificially prepared is identical in properties with the natural essence. In like manner, *allyl sulphocyanide,* $\left.\begin{matrix} C_3H_5 \\ CN \end{matrix}\right\}$ S, is found as the essential oil of black mustard seed; it can also be artificially prepared by treating allyl iodide with silver sulphocyanide. It boils at 148°. Allyl sulphide boils at 140°.

Acrolein, C_3H_4O, is the aldehyde of allyl alcohol, and is formed when the alcohol is oxidized, 2 atoms of hydrogen being removed. Acrolein is also produced by the abstraction of 2 molecules of water from glycerin:

$$C_3H_8O_3 - 2H_2O = C_3H_4O.$$

Acrolein is a colorless liquid, boiling at 52·4°, and possessing a most pungent odor, which attacks the mucous membrane of the nose and eyes. It rapidly oxidizes to acrylic acid, $\left.\begin{matrix}C_3H_3O\\H\end{matrix}\right\}O$, a substance possessing close analogy with acetic acid, which combines with hydrogen to form propionic acid.

Acrylic acid is the first term of a series of monobasic acids, whose corresponding alcohols, with the exception of allyl alcohol, have not yet been prepared. They differ from the series of fatty acids in containing 2 atoms of hydrogen less.

Acrylic Acid	$C_3H_4O_2$
Crotonic Acid	$C_4H_6O_2$
Angelic Acid	$C_5H_8O_2$
Pyroterebic Acid	$C_6H_{10}O_2$
Hypogæic Acid	$C_{16}H_{30}O_2$
Oleic Acid	$C_{18}H_{34}O_2$
Erucic Acid	$C_{22}H_{42}O_2$

Crotonic acid occurs in croton oil, and angelic acid in the archangel-root, whilst angelic aldehyde, C_5H_8O, is contained in the essential oil of chamomile. Oleic acid exists, as has been said, in many oils, especially in almond oil, olive oil, and lard: this acid when acted upon by nitrous acid forms a new solid acid isomeric with oleic acid, and called elaïdic acid. Erucic acid is contained in rapeseed oil.

Hydrocarbons of the Acetylene Series.

A series of non-saturated hydrocarbons isomeric with acetylene (p. 78) exist; they combine directly with 2 and 4 atoms of chlorine or bromine, and in the latter case form saturated compounds. These hydrocarbons are closely related to those of the ethylene series (p. 337). Thus, by acting upon the iodides or bromides of the ethylene series with alcoholic potash, we get the hydrocarbon of the acetylene series, thus:

$$\underset{\text{Ethylene dibromide.}}{C_2H_4Br_2} + 2\,KHO = \underset{\text{Acetylene.}}{C_2H_2} + 2\,KBr + 2\,H_2O.$$

The following is a list of the acetylene series of hydrocarbons:

		B. P.
Acetylene.	C_2H_2	—
Allylene	C_3H_4	—
Crotonylene.	C_4H_6	18°
Valerylene	C_5H_8	45°
Hexoylene	C_6H_{10}	80°
Œnanthylidene . . .	C_7H_{12}	107°
Caprylidene	C_8H_{14}	133°
Rutylene	$C_{10}H_{18}$	150°
Benylene.	$C_{15}H_{28}$	225°

Acetylene is formed whenever a substance containing carbon and hydrogen undergoes incomplete combustion (p. 93). The compounds of acetylene with certain metals are very remarkable: if this gas be passed into an ammoniacal solution of cuprous chloride, a red precipitate of cupro-acetyl oxide, $\left.\begin{matrix}C_2Cu_2H\\C_2Cu_2H\end{matrix}\right\}O$, is formed; whilst if an ammoniacal solution of a silver salt be used, a similar compound, $\left.\begin{matrix}C_2Ag_2H\\C_2Ag_2H\end{matrix}\right\}O$, is precipitated as a white powder. Both these substances explode when heated or struck with a hammer; and both when heated with hydrochloric acid evolve acetylene gas.

If acetylene be led over fused potassium, the metal takes the place of the hydrogen, forming the compounds C_2HK and C_2K_2: these bodies decompose violently in contact with water, forming potash and acetylene.

Allylene, $C_3 H_4$, is formed by the action of potash upon propylene dichloride. The other members of this series are powerfully smelling liquids which combine with 2 and 4 atoms of bromine.

Tetravalent Alcohols and their Derivatives.

The only tetravalent alcohol as yet known is *erythrite*, a solid white substance found in certain lichens and fungi; its composition is $C_4 H_{10} O_4$. Treated with hydriodic acid, erythrite forms secondary *butyl iodide*. Thus:

$$C_4 H_{10} O_4 + 7 H I = C_4 H_9 I + 4 H_2 O + 3 I_2.$$

Hexavalent Alcohols and their Derivatives.

The best-defined member of this series is *mannite*, $C_6 H_{14} O_6$, which is the alcohol of a hexavalent radical, $C_6 H_8$. Mannite is a solid, sugar-like substance contained in manna, the exudation from several species of ash. Mannite can be artificially prepared from certain varieties of sugar, which take up H_2 when treated with water and sodium amalgam, $C_6 H_{12} O_6 + H_2 = C_6 H_{14} O_6$. By oxidation of mannite, the reverse change occurs, and a fermentable sugar, $C_6 H_{12} O_6$, is produced: this change can also be effected by a peculiar ferment. The chief reasons for giving a hexavalent character to mannite are, (1) that when this substance is acted on by nitric acid, a compound, called *nitro-mannite*, is formed, which is mannite containing the 6 typical atoms of hydrogen, replaced by NO_2: thus its composition is $\left.\begin{matrix}\overset{\text{VI}}{C_6 H_8} \\ 6(NO_2\end{matrix}\right\} O_6$, and is the nitrate of a hexatomic radical, which corresponds to ethyl nitrate in the monatomic series: (2) that mannite is attacked by hydriodic acid in a similar manner to glycerine (p. 302), and erythrite (p. 306), the monatomic iodide of the same number of carbon atoms being formed; thus we have

$$\underset{\text{Mannite}}{C_6H_{14}O_6} + 11\,HI = \underset{\text{Iso-hexyl Iodide.}}{C_6H_{13}I} + 5\,I_2 + 6\,H_2O.$$

In like manner the 6 of hydrogen can be replaced by 6 atoms of the radical of stearic acid; we have then a compound mannite hexastearate, $\left.\begin{matrix}\overset{\text{VI}}{C_6H_8}\\ 6(C_{18}H_{35}O)\end{matrix}\right\}O_6.$

The substances having the composition of the iodides of the monatomic radicals, and obtained by deoxidation from glycerine (p. 302), erythrite (p. 306), and mannite, are *isomeric*, and *not identical* with these iodides; they may be considered to be compounds of the olefine with hydriodic acid. By the action of silver oxide and water, the bodies yield a hydrate having the composition of the monatomic alcohols, but differing from these in their boiling points, and in their chemical relations; thus these so-called *iso-alcohols* readily yield the olefines from which they are derived, and, on oxidation, do not produce the corresponding acid, but form an acetone by loss of hydrogen; indeed these iso-alcohols can be obtained by the action of hydrogen upon the acetones, thus—

$$\underset{\text{Acetone.}}{C_3H_6O} + H_2 = \underset{\text{Iso-propyl alcohol.}}{C_3H_8O_2}.$$

LESSON XXXVII.

SACCHARINE BODIES.

THESE substances are frequently termed carbo-hydrates, inasmuch as they contain hydrogen and oxygen in the proportion to form water, united with carbon. They form an important class of substances as occurring widely diffused in the bodies of plants. They may be divided into three classes: (1) Sucroses, or the sugars proper; (2) Glucoses, or the grape sugars; (3) Amyloses, or starch and woody fibre.

Each of these three classes contain several distinct substances.

1. *Sucroses.*	2. *Glucoses.*	3. *Amyloses.*
$C_{12}H_{22}O_{11}$.	$C_6H_{12}O_6$.	$C_6H_{10}O_5$.
Sucrose + (or cane sugar).	Dextrose + (or grape sugar).	Starch + Glycogene +
Lactose + (or milk sugar).	Levulose — (or fruit sugar).	Dextrine + Inuline —
Melitose, Melezitose, Mycose. } +	Galactose. —	Gums. Cellulose. Tunicine.

The most important distinguishing physical property of these bodies is their action on polarized light. Like tartaric acid (p. 293), and many other substances, these saccharine bodies possess the power of turning the plane of polarization, some to the right-hand and some to the left: thus dextrose, or grape sugar, turns it to the right, levulose, or fruit sugar, to the left. The right-handed substances are marked in the preceding list with a +, the left-handed with a —.

SUCROSES. *Sucrose*, or *Cane Sugar*, $C_{12}H_{22}O_{11}$.—This important substance occurs in the juice of certain plants, especially the sugar-cane, beetroot, mallow, and sugar-maple; also, in smaller quantity, in honey and various kinds of fruit, together with a mixture of dextrose and levulose. Sugar is prepared from the sugar-cane, which contains about 18 per cent. of sugar, by crushing out the juice by passing the cane between rollers; the juice is at once heated to about 60°, and a small quantity of milk of lime added for the purpose of precipitating the albuminous matter derived from the cane, the presence of which renders the juice liable to quick fermentation. The juice is then raised to the boiling point, the scum which rises to the surface removed, and the clear liquid remaining is boiled down in copper pans until it attains a certain consistence, when it is filtered through linen bags, and again evaporated to a syrup, which, on cooling, deposits crystals of moist or brown sugar. The mother liquor is again evaporated, and again allowed to cool and deposit another crop of crystals; the dark-colored, uncrystallizable sugar is termed molasses, or treacle. The raw sugar is dissolved, and again boiled with lime and filtered. The filtered liquor is then decolorized by flowing through a thick bed of animal charcoal, and the colorless filtrate evaporated down to the point of crystallization, under diminished pressure, in vacuum-pans. The object of this is to enable the syrup to boil at a lower temperature than it would do under the ordinary pressure, and thus to prevent formation of uncrystallizable sugar, and to avoid the charring and coloring of the syrup which then takes place. The concentrated syrup is then either allowed to crystallize in moulds, giving loaf sugar, or the small crystals are freed from adhering mother liquor by quick drying in a hydro-extractor, or rapidly revolving sieve. Much saving is attained by the use of the vacuum-pans, and if its employment were universal in the colonies, where the sugar is first prepared, the formation of much treacle, or uncrystallizable sugar, would be avoided, and the profit to the planter proportionately increased. A method of treating the cane-juice has lately been proposed, which bids fair to revolutionize the manufacture of raw sugar. It depends upon the fact, that by a peculiar plan of rapid evaporation, the whole of the water can be got rid of, without charring the sugar, which is thus obtained as a solid mass, and all formation of treacle avoided.

Sugar crystallizes in monoclinic prisms, which phosphoresce when broken: its specific gravity is 1·606. It is soluble in one-third of its weight of cold, and in rather more of hot, water, and is nearly insoluble in alcohol and ether. Its specific power of rotation is 73°8′ to the right. Sugar melts at 160° to a colorless liquid, which solidifies on cooling to a colorless transparent mass (barley-sugar), and, on standing, becomes crystalline and opaque. When more strongly heated, water is given off, and a dark colored mass, called caramel, is left behind. When acted on by nitric acid, either saccharic or oxalic acid is formed, according to the strength of the acid and the heat employed. Strong sulphuric acid converts sugar into a black mass, with evolution of sulphuric dioxide. A mixture of these two acids in the cold acts on sugar to form a nitro-substitution product, $C_{12}H_{18}(NO_2)_4O_{11}$, an amorphous mass, liable to explode on percussion. Solutions of sucrose easily reduce the noble metals from their solutions on warming, whilst cupric salts are only slowly decomposed in alkaline solution of sucrose. Cane-sugar is not directly fermentable, but, in presence of yeast, it takes up a molecule of water and forms a mixture of dextrose and levulose, both capable of undergoing fermentation :—

$$\underset{\text{Sucrose.}}{C_{12}H_{22}O_{11}} + H_2O = \underset{\text{Dextrose.}}{C_6H_{12}O_6} + \underset{\text{Levulose.}}{C_6H_{12}O_6}.$$

By the action of dilute sulphuric acid, the same change is produced; and also by long continued boiling of a solution of sugar. Sucrose combines with certain metallic oxides to form definite compounds ; thus we have $C_{12}H_{22}O_{11}\,CaO$, whilst other metals replace some of the hydrogen of the sugar ; thus lead forms a compound having the formula $C_{12}H_{18}Pb_2O_{11}$.

Lactose, or milk sugar, occurs only in the milk of mammalia, from which it is obtained in the crystalline state by evaporation. The crystals, which are rhombic, contain an atom of water of crystallization, given off at 140°. Lactose dissolves in 6 parts of cold and 2 parts of boiling water; it does not possess nearly so sweet a taste as sucrose, and feels gritty in the mouth, and its specific power of rotation is +59°3′. Lactose does not ferment itself, but when much yeast is added, fermentation occurs, after some time, mannite being formed. In presence of cheese, &c., the lactic fermentation sets in. Dilute acids convert lactose into a peculiar glucose, called *galactose*, which is directly fermentable, and yields mucic acid when treated with nitric acid. Lactose reduces an alkaline copper solution in the cold, precipitating cuprous oxide; but the quantity of this substance formed is not so great as when the same weight of glucose is employed. Lactose, when oxidized, yields mucic, saccharic, tartaric, and oxalic acids.

GLUCOSES, $C_6H_{12}O_6$. *Dextrose*, or right-handed glucose, grape- or starch-sugar, is found in many kinds of fruit, in manna and honey mixed with levulose, or left-handed glucose. It forms a normal constituent of blood, white of egg, and exists in small quantity in healthy urine, whilst it is excreted in large quantities in that liquid in the disease termed diabetes.

Dextrose is formed in many ways.

(1.) By boiling starch or dextrine with diluted acids.

(2.) By the action of malt upon starch (*see* Dextrine).

(3.) By the action of dilute acids upon sucrose (when it is formed together with levulose.

(4.) By the action of acids upon many glucosides.

Dextrose is prepared by boiling starch with dilute sulphuric acid, adding chalk to neutralize the acid, and evaporating the liquid to a syrup when the sugar crystallizes. It may also be easily prepared by washing honey with dilute alcohol; the levulose being more soluble is thus removed. Dextrose turns the plane of polarization to the right: its permanent rotation-power is $+56°$. It is soluble in its own weight of water, and dissolves easily in dilute alcohol, and is not nearly so sweet as sucrose; the crystals contain one molecule of water, which they lose at 60°. Dextrose immediately precipitates red cuprous oxide from alkaline cupric solutions; and the quantity of dextrose present in a solution can be ascertained by employing a standard solution of alkaline copper salt. From silver salts the metal is deposited by dextrose, in the form of a mirror. Nitric acid oxidizes dextrose to saccharic or oxalic acid.

Levulose, or left-handed glucose. This forms an incrystallizable colorless syrup ; it is more soluble in water and alcohol than dextrose, and is therefore sweeter. Its action on polarized light changes remarkably with the temperature : thus at a temperature of 14°C. its rotatory power is 106°, whereas at 90°C. it is reduced to 53°. Levulose reduces cupric salts like dextrose ; it is obtained by neutralizing with lime the mixture of glucoses obtained by the action of sulphuric acid on sucrose. The levulose lime-compound is a solid, whilst dextrose forms a liquid substance. By decomposing this lime-compound with oxalic acid pure levulose is obtained.

The acids having the composition $C_6H_{10}O_8$ (*mucic* and *saccharic*) obtained by the action of dilute nitric acid on the different sugars, must be regarded as products of oxidation of mannite, the hexatomic alcohol : levulose yields mannite when acted on by nascent hydrogen, and hence stands to this substance as aldehyde to alcohol.

$$\text{Mannite, } \left.\begin{matrix} C_6H_8 \\ H_6 \end{matrix}\right\} O_6.$$

$$\left.\begin{matrix} C_6H_6 \\ H_6 \end{matrix}\right\} O_6. \qquad \left.\begin{matrix} C_6H_6O \\ H_6 \end{matrix}\right\} O_6. \qquad \left.\begin{matrix} C_6H_4O_2 \\ H_6 \end{matrix}\right\} O_6.$$

Levulose. Mannitic Acid. Mucic and Saccharic Acids.

$$\text{Alcohol, } \left.\begin{matrix} C_2H_5 \\ H \end{matrix}\right\} O.$$

$$\left.\begin{matrix} C_2H_3 \\ H \end{matrix}\right\} O. \qquad \left.\begin{matrix} C_2H_3O \\ H \end{matrix}\right\} O.$$

Aldehyde. Acetic Acid.

FERMENTATION. This name has been given to a peculiar and interesting class of decompositions, which have long been known, but differ altogether from the ordinary chemical actions. Many organic bodies are capable of undergoing fermentation in presence of certain complicated substances termed *ferments*, giving rise to several products differing according to the nature of the fermented body and the ferment. Careful investigation has shown that the process of fermentation entirely depends upon the presence and growth of certain *living organisms* forming the ferment. Different kinds of ferments give rise to different products. Thus we have one ferment (yeast) which effects the alcoholic fermentation, another which sets up the lactic fermentation, a third producing the acetous fermentation, &c. Most of these ferments are vegetable growths of a low kind, but one at least, viz., that causing the butyric fermentation, is an animal; and this, strange to say, cannot live in contact with free oxygen, but flourishes in an atmosphere of hydrogen. In order that the ferment should grow, it must be supplied with proper food, especially with ammoniacal salts and alkaline phosphates; these are contained in the albuminous matter generally present in the liquid about to be fermented. In order that the fermentation should go on well, the temperature should be from 20° to 40°; at much higher, as at much lower temperatures, the vitality of the ferment is destroyed.

In many cases, spontaneous fermentation sets in without the apparent addition of any ferment: thus wine, beer, milk, urine, &c., when allowed simply to stand exposed to the air, become sour or otherwise decompose. These changes are, however, not effected without the presence of vegetable or animal life, and are true fermentations; the *sporules*, or seeds of these living bodies, always float about in the air, and on dropping into the liquid begin to propagate themselves, and in the act of growing evolve the products of the fermentation. If the above liquids be left only in contact with air which has been passed through a red-hot platinum tube, and thus the living sporules destroyed; or if the air be simply filtered by passing through cotton wool, and the sporules prevented from coming into the liquid, it is found that these fermentable liquids may be preserved for any length of time without undergoing the slightest change.

The following are the five principal forms of fermentation:

1. The alcoholic fermentation, producing chiefly alcohol and carbonic acid.

2. The acetous fermentation, producing acetic acid.

3. The lactic fermentation, yielding chiefly lactic acid.
4. The butyric fermentation, yielding chiefly butyric acid.
5. The mucous fermentation, giving rise to gum and mannite.

Alcoholic Fermentation. The glucoses are able, when dissolved in presence of yeast, to undergo fermentation, evolving mainly alcohol and carbonic acid,—

$$C_6 H_{12} O_6 = 2\ C_2 H_6 O + 2\ CO_2.$$

About 6 per cent. of the glucose undergoes a different change, part being used as nourishment for the yeast, and another part forming glycerine and succinic acid. From 100 parts of glucose about 3·5 parts of glycerine are produced, and 0·6 to 0·7 of succinic acid, whilst 1·2 to 1·5 parts of cellulose and fatty matter are formed by the growth of the yeast. The alcoholic fermentation occurs best at a temperature of between 25° and 30°.

LESSON XXXVIII.

Amylaceous Bodies and Gums.

Dextrine, $C_6 H_{10} O_5$. This substance, called British Gum, is prepared by heating starch to about 150°; if a small quantity of nitric or hydrochloric acid is added to the starch, the transformation takes place much more rapidly. Dextrine is also formed together with dextrose by the action of malt extract upon starch. It deviates the plane of polarization strongly to the right, its rotatory power being + 138°7′. Dextrine is very soluble in water and insoluble in alcohol, and on boiling with dilute acids dextrine is converted into dextrose.

Gum Arabic. The natural exudation from several species

of acacias; it consists chiefly of the potassium and calcium salts of arabic acid, $C_{12} H_{20} O_{10}$.

Inulin. A substance contained in the roots of many plants; it is intermediate between gums and starch; it yields lævulose when boiled with dilute acids.

Glycogen, or *Animal Starch,* is an insoluble powder formed in the liver and placenta; it is easily converted into glucose.

Starch, $C_6 H_{10} O_5$ (or some multiple of these numbers). This important substance exists most widely diffused throughout the vegetable world. It consists of a white powder composed of *granules* which have a distinctly *organized* structure, and are of various sizes. The following are the diameters of the granules of some of the most important varieties of starch:

Potato	0·185 mm.	Wheat . .	0·050 mm.	Millet .	0·010 mm.
Sago .	0·070 "	Indian corn	0·030 "	Beetroot	0·004 "

Starch granules are insoluble in cold water, alcohol, and ether, but when they are heated with water to between 70° and 72° they swell up and split open, forming a thick mass called Starch Paste. If this paste be boiled with a larger quantity of water, the particles of starch become so finely divided that they pass through a filter, and if boiled for a length of time, the solution becomes clear, and the starch is rendered soluble; and from this solution alcohol precipitates a white amorphous powder of soluble starch. When heated to 160° starch is converted into dextrine. Starch, in its insoluble and soluble modifications, forms with free iodine a deep blue compound, the color of which is destroyed a little below 100°, but appears again on cooling. This color is characteristic of starch, and is not produced with dextrine or the other isomers of starch. When the soluble azotized matter contained in malt, called *diastase,* acts upon starch, it forms dextrine and dextrose; and by a longer action the dextrine is also converted into dextrose.

$$3(C_6 H_{10} O_5) + H_2 O = 2(C_6 H_{10} O_5) + C_6 H_{12} O_6,$$

Starch. Dextrine. Dextrose.

The action of dilute sulphuric acid upon starch is similar to that of diastase. Strong sulphuric acid in the cold dissolves starch, forming a compound acid; nitric acid also dissolves starch, and on adding water to the solution a white substance called *xyloidine* is precipitated; this is a substitution product, being starch in which one atom of hydrogen is replaced by $N O_2$, thus: $C_{12} H_{19} (NO_2) O_{10}$.

Cellulose, $C_6 H_{10} O_5$. This is the colorless material of the woody fibre of young plants; it may be obtained in the pure state from cotton or linen fibre by boiling out the impurities with alkali, alcohol, ether, &c. Cellulose is a white substance insoluble in water, alcohol, or ether, but dissolving in an ammoniacal solution of cupric oxide. By the action of strong sulphuric acid, cellulose is converted either into an insoluble substance which colors blue with iodine, or into a soluble body like dextrine; if this acid solution be diluted with water and boiled, dextrose is formed by fixation of one molecule of water. A useful substance is obtained under the name of parchment paper, by dipping sheets of paper into strong sulphuric acid.

Gun Cotton. The action of strong nitric acid upon cellulose is interesting; if cotton wool be thrown in small portions at a time into a mixture of equal volumes of strong sulphuric and nitric acids, it does not undergo any apparent change, but on drying it is found to be very inflammable. It is a substitution product, being cellulose in which two or more atoms of hydrogen are replaced by $(N O_2)$ thus: $C_6 H_7 (N O_2)_3 O_5$. By the action of ferrous chloride nitric oxide is evolved and cellulose again formed. The use of gun-cotton as a substitute for gunpowder has been proposed, as it offers many advantages :—

(1) The explosive force of gun-cotton is, weight for weight, greater than that of gunpowder. (2) The products of combustion of gun-cotton, being chiefly carbonic acid and nitro-

gen, are not so apt to foul the gun. (3) When moistened it becomes inexplosive, and only requires drying to render it again explosive.

The reasons which render the general adoption of this substance doubtful are (1) its liability to explode on percussion; (2) the possibility of its spontaneous decomposition when kept for a length of time.

Gun-cotton, or certain forms of this substance, dissolves readily in a mixture of ether and alcohol, and yields a solution which is termed *Collodion*, and is largely used for the purpose of forming a thin coating on glass to receive silver salts, upon which the photographic image is formed.

GROUP OF GLUCOSIDES.

The numerous substances constituting this class occur in the bodies of many plants, and yield a glucose on decomposition, together with other bodies; they may be considered as kinds of compound ethers of glucose. The most important are amygdaline, salicine, and tannine.

Amygdaline, $C_{20}H_{27}NO_{11} + 3H_2O$. Found in bitter almonds, and obtained by dissolving out by alcohol, and precipitating the amygdaline with ether; it forms small white crystals which are soluble in water. The most remarkable decomposition which amygdaline undergoes is that which is brought about in the bruised almond by the presence of an albuminous substance called Synaptase, by which bitter almond oil, hydrocyanic acid, and glucose, are produced:—

$$\underset{\text{Amygdaline.}}{C_{20}H_{27}NO_{11}} + 2H_2O = \underset{\text{Hydride of Benzoyl.}}{C_7H_6O} + \underset{\text{Hydrocyanic Acid.}}{HCN} + \underset{\text{Glucose.}}{2C_6H_{12}O_6}.$$

Salicine, $C_{13}H_{18}O_7$, contained in the pith of the willow and poplar, and also found in the castoreum contained in a gland of the beaver. Salicine crystallizes in bright white needles; it is soluble in water and alcohol, but insoluble in ether, and its solution possesses a strongly bitter taste. In presence of certain ferments it is decomposed as follows:

$$\underset{\text{Salicine.}}{C_{13}H_{18}O_7} + H_2O = \underset{\text{Saligenine.}}{C_7H_8O_2} + \underset{\text{Glucose.}}{C_6H_{12}O_6}.$$

Tannin, or *Tannic Acid*, $C_{27}H_{22}O_{17}$. This substance is contained widely diffused in certain parts of plants; it is distinguished by forming an insoluble compound with gelatin, and by producing a black color (ink) with ferric compounds. Tannic acid occurs in largest quantities in gall-nuts (an excrescence formed on the oak by an insect); it is extracted by aqueous ether from the powdered gall-nut. Tannin thus prepared is an uncrystallizable mass, soluble in water and alcohol, but insoluble in pure ether. Tannin forms glucose and gallic acid when it is exposed to the air, or when treated by dilute acids.

$$\underset{\text{Tannin.}}{C_{27}H_{22}O_{17}} + 4H_2O = 3\underset{\text{Gallic Acid.}}{C_7H_6O_5} + \underset{\text{Glucose.}}{C_6H_{12}O_6}.$$

Tannin heated to 215° yields pyrogallic acid.

Myronic Acid, $C_{10}H_{19}NS_2O_{10}$.—The potassium salt of this acid exists in the seeds of black mustard; it decomposes into the oil of mustard (allyl-sulphocyanate), glucose, and hydrogen-potassium-sulphate, in contact with an albuminous ferment found in the seeds, thus:

$$\underset{\text{Potassium Myronate.}}{C_{10}H_{18}KNS_2O_{10}} = \left.\begin{matrix}H\\K\end{matrix}\right\}SO_4 + \underset{\text{Oil of Mustard.}}{\left.\begin{matrix}CN.\\C_3H_5\end{matrix}\right\}S} + \underset{\text{Glucose.}}{C_6H_{12}O_6}.$$

LESSON XXXIX.

THE GROUP OF AROMATIC COMPOUNDS.

It has already been stated that in these bodies the carbon atoms are more closely combined together than is the case in the foregoing group, or, in other words, that the aromatic hydro-carbons contain relatively less hydrogen than those which we have hitherto studied. Another peculiarity of these substances is, that they contain at least 6 atoms of carbon, and that the more complicated compounds split up into those containing 6 carbon atoms. It appears in fact that all the aromatic bodies contain a group of 6 carbon atoms, in which 18 of the combining powers of the carbon are taken up by union of carbon with carbon (p. 232), whilst 6 remain open to saturation.

The simplest combination amongst the aromatic series is that of benzol, $C_6 H_6$, and from this body a large number of other substances can be derived by substitution of one or more atoms of hydrogen by more or less complicated molecules. Thus, for instance, we are acquainted with four hydrocarbons homologous with benzol but differing by $C H_2$; these substances are in fact benzol, in which 1, 2, or 3 atoms of hydrogen have been replaced by methyl, $C H_3$. We thus have:

			Boiling Point
(1)	Benzol	$C_6 H_6$	82°
(2)	Toluol, or Methyl-Benzol	C_7H_8, or $C_6H_5(CH_3)$	111°
(3)	Xylol, or Di-methyl Benzol	C_8H_{10}, or $C_6H_4 \begin{cases} CH_3 \\ CH_3 \end{cases}$	139°
(4)	Cumol, or Tri-methyl Benzol	C_9H_{12}, or $C_6H_3 \begin{cases} CH_3 \\ CH_3 \\ CH_3 \end{cases}$	168°

Each of these methylated benzols yields an important series of derivatives, corresponding to those obtained from benzol. Thus in each, one or more atoms of hydrogen can be replaced (1) by chlorine, giving chlorine substitution products; (2) by the monad NO_2 yielding nitro-substitution products; (3) by the monad $N H_2$ yielding amido-compounds; or (4) by the monad HO yielding a peculiar set of alcohol-like bodies, termed Phenols, as well as a series of true alcohols, isomeric with the phenols. The following table gives the names and formulæ of some of the derivatives of benzol, and methyl-benzol, or toluol:—

Benzol	$C_6 H_6$
Monochlor-benzol .	$C_6 H_5 Cl$
Nitro-benzol . . .	$C_6 H_5 (NO_2)$

Amido-benzol, or Aniline .	$C_6 H_5 (NH_2)$
Phenol, or Carbolic Acid .	$C_6 H_5 (HO)$
Toluol	$C_7 H_8$, or $C_6 H_5 (CH_3)$
Monochlor-toluol . .	$C_7 H_7 Cl$, or $C_6 H_4 Cl (CH_3)$
Nitro-toluol . . .	$C_7 H_7 NO_2$, or $C_6 H_4 (NO_2) (CH_3)$
Amido-toluol, or Toluidine .	$C_7 H_9 N$, or $C_6 H_4 (NH_2) (CH_3)$
Cressol	$C_7 H_8 O$, or $C_6 H_4 (HO) (CH_3)$

A series of bodies, isomeric with these toluol compounds, exists, in which the substitution takes place with the hydrogen of the methyl. This is termed the *Benzyl series.*

Toluol. $C_7 H_8$, or $C_6 H_5 (CH_3)$ — Benzyl Chloride. $C_7 H_7 Cl$, or $C_6 H_5 (CH_2 Cl)$

Benzylamine.

$$\left.\begin{matrix} C_7 H_7 \\ H \\ H \end{matrix}\right\} N, \quad \text{or} \quad \left.\begin{matrix} C_6 H_5 (CH_2) \\ H \\ H \end{matrix}\right\} N.$$

Benzyl Alcohol.

$$\left.\begin{matrix} C_7 H_7 \\ H \end{matrix}\right\} O, \quad \text{or} \quad \left.\begin{matrix} C_6 H_5 (CH_2) \\ H \end{matrix}\right\} O.$$

By oxidation benzyl alcohol yields an aldehyde, $C_7H_6 O$, *oil of bitter almonds*, and an acid, $C_7 H_6 O_2$, *benzoic acid*, both derived from this alcohol, as ethyl aldehyde and acetic acid are derived from ethyl alcohol.

The di- and tri-methyl benzols also furnish similar double series of isomeric derivatives. In like manner all the higher alcohol radicals, ethyl, propyl, butyl, &c., can be substituted for one or more atoms of hydrogen in benzol, and thus an almost unlimited number of isomeric bodies may be prepared; thus ethyl benzol, $C_6 H_5 (C_2H_5)$, is *isomeric*, but not *identical* with di-methyl benzol or xylol, $C_6 H_4 \left\{\begin{matrix} CH_3 \\ CH_3 \end{matrix}\right.$.

Benzol, $C_6 H_6$. Found in the light oils obtained by the destructive distillation of coal; it is a colorless liquid, refracting light powerfully, boiling at 82°, and freezing at —4°·5. It is also obtained by distilling benzoic acid with slacked lime. Benzol is attacked by chlorine, and several chlorides formed; when treated with nitric acid, an interesting substance called *nitro-benzol*, $C_6 H_5 (NO_2)$, is produced, a substitution product in which one of hydrogen of benzol is replaced by $N O_2$; and we also know a solid substance called *di-nitro-benzol*, $C_6 H_4 2 (NO_2)$. In contact with reducing agents, nitro-benzol undergoes the following important reduction to *aniline*, in which the monad group (NO_2) is replaced by the monad $(N H_2)$, thus:

$$\underset{\text{Nitro-benzol.}}{C_6 H_6 (NO_2)} + 3 H_2 = \underset{\text{Aniline.}}{C_6 H_5 (NH_2)} + 2 H_2 O.$$

Phenol, or *Carbolic Acid*, $C_6 H_5 (HO)$. This is a white solid crystalline body, fusing at 35°, and boiling at 188°, found in the heavy coal-oils. It dissolves in the alkalies, forming a phenate, but it does not possess an acid reaction. The most important property of this body is its powerful antiseptic qualities, and it is much used as a disinfectant, both alone and when combined with lime.

Monochlor Benzol, $C_6 H_5 Cl$, is formed by the direct action of chlorine on benzol, or when phosphorus penta-chloride acts upon phenol.

Picric Acid, $C_6 H_3 (NO_2)_3 O$. When phenol is acted upon by nitric acid, 1, 2, or 3 atoms of hydrogen may be substituted by NO_2. Tri-nitro carbolic acid, or picric acid, is a bright yellow crystalline body, very soluble in water; it is obtained by the action of nitric acid upon many other substances besides carbolic acid and its derivatives. Picric acid is employed in the arts as a yellow dye for silk and woollen goods.

Aniline, or *Amido-benzol*, $C_6 H_5 (NH_2)$. This important body is benzol, in which one atom of hydrogen is replaced by the monad group (NH_2), and it is, therefore, properly called

Amido-benzol. The mode of preparing aniline from benzol has just been described, the reduction of nitro-benzol being generally effected by a mixture of iron filings and acetic acid. It may also be obtained by the action of potash on isatine (p. 328).

$$\underset{\text{Isatine.}}{C_8 H_5 NO_2} + 4\,(KHO) = \underset{\text{Aniline.}}{C_6 H_7 N} + \underset{\text{Potassium Carbonate.}}{2\,(K_2 CO_3)} + H_2.$$

It is also found amongst the products of the destructive distillation of coal.

Aniline is a colorless liquid, possessing a peculiar smell; its specific gravity at 0° is 1·036, and it boils at 185°. It is nearly insoluble in water, but dissolves in alcohol and ether; it unites with acids to form definite salts, but it does not turn red litmus paper blue. Crude aniline is manufactured now on a very large scale for the preparation of the so-called aniline colors, so generally used in calico-printing, and woollen and silk dyeing. The smallest trace of aniline may be easily detected by adding to an aqueous aniline solution an aqueous solution of an alkaline hypochlorite, when a splendid red coloration is formed: this is prepared in quantity by adding to aniline sulphate a dilute solution of potassium bichromate. This substance forms one of the important aniline colors, and is called *mauve;* it contains a base of complicated constitution, termed mauveine. The color mauve can be prepared by many other methods; the best of these is by heating aniline with a double chloride of sodium and copper. The other coloring matters derived from aniline are noticed on the next page. Aniline gives rise to a very large number of derivatives; thus we have a series of compounds in which one or more atoms of hydrogen are replaced by ethyl and other radicals. We are also acquainted with di-amido benzol, thus:

$$\underset{\text{Aniline or Amido-benzol.}}{C_6 H_5 (NH_2)} \qquad \underset{\text{Di-amido-benzol.}}{C_6 H_4\, 2\,(NH_2)}$$

Aniline has been called Phenylamine, $\left.\begin{matrix} C_6 H_5 \\ H \\ H \end{matrix}\right\} N$; and in certain respects it resembles the compound ammonias: thus, for instance, it is capable of forming a hydroxide analogous to $\left.\begin{matrix} NH_4 \\ H \end{matrix}\right\} O$, a body which is a non-volatile strongly alkaline base, $\left.\begin{matrix} N\,C_6H_5(C_2H_5)_3 \\ H \end{matrix}\right\} O$, called Triethyl-phenyl-ammonium-hydroxide. If one atom of hydrogen is replaced by an oxidized radical, as acetyl, we get $\left.\begin{matrix} C_6H_5NH_2H \\ C_2H_3O \end{matrix}\right\} O$, aniline acetate; and this on heating loses a molecule of water, yielding an amide called acetanilide, $C_6H_5N\left\{\begin{matrix} H \\ C_2H_3O \end{matrix}\right.$, just as ammonium acetate yields acetamide (p. 277).

Azo Compounds of Aniline.—Pure aniline when treated with nitrous acid is decomposed into phenol, nitrogen, and water:

$$C_6H_5(NH_2) + NO_2H = C_6H_5(OH) + H_2O + N_2.$$

If, however, an aqueous solution of aniline nitrate be acted on by nitrous acid, one atom of nitrogen replaces two atoms of hydrogen, and a substance termed *Diazo-benzol nitrate*, $C_6H_5N_2NO_3$ is formed. This compound crystallizes in colorless needles, and either by heat or on percussion it decomposes with explosive violence. Diazo-benzol combines with acids, forming compounds analogous to the salts of ammonia; but it also combines with hydroxides, and with amido derivatives. Thus it combines directly with aniline to form *Diazo-amido-benzol*, $C_{12}H_{11}N_3$, or $C_6H_5N_2C_6H_5(NH_2)$, a body crystallizing in golden yellow scales, which also decomposes with explosion on percussion. In the Azobenzol compounds N_2 replace H_2: the group N_2 acts, therefore, as a dyad, two of the three bonds of each atom of nitrogen being connected with two bonds of the other atom; thus, $- N = N -$.

Aniline Yellow is an isomeric and non-explosive modification of Diazo-amido-benzol.

Oxyphenol, or *Pyrocatechin*, $C_6H_4\left\{\begin{matrix} OH \\ OH \end{matrix}\right.$.—This substance stands in the same relation to phenol that ethylene alcohol does to ethyl alcohol; and is therefore a diatomic phenol.

It is obtained by the action of potash on iodo-phenol, $C_6H_4I(OH)$, and is produced by the dry distillation of catechu, many resins, and wood.

BENZYLIC, OR TOLUIC SERIES. *Toluol*, or *Methyl-benzol*, $C_7H_8 = C_6H_5(CH_3)$. This hydrocarbon occurs likewise in the coal-oils, and it boils at 111°; it is also formed by the distillation of toluic acid with excess of lime. It can be prepared synthetically from benzol by replacing one atom of hydrogen by methyl (p. 320). By the action of oxidizing agents it is converted into benzoic acid, thus: $C_7H_8 + O_3 = C_7H_6O_2 + H_2O$.

Nitro-toluol, $C_7H_7(NO_2)$, is obtained by the action of nitric acid upon toluol, and by reduction a basic substance is obtained analogous to aniline, and called *amido-toluol*, or *Toluidine*, C_7H_9N or $C_6H_5(NH_2)(CH_3)$. This is a solid substance, which exists always in commercial aniline, and in fact is a necessary ingredient for the purpose of the manufacture of the red and violet aniline colors. Toluidine fuses at 40°, and boils at 198°. Toluidine is isomeric with benzylamine.

Cressol, $C_7H_7(HO)$, a crystallizable solid, homologous with phenol, contained in the crude carbolic acid; it boils at 203°.

Rosaniline, $C_{20}H_{19}N_3$. The compounds of this substance form the splendid red-aniline color, known as *magenta*. The color may be obtained in various ways, from crude aniline; the best process consists in heating this crude substance and arsenic acid together to a temperature of from 120° to 140°; the quantities taken being 12 parts of dry arsenic acid of commerce (containing 13·5 per cent. of water) to 10 parts of aniline. The color cannot, however, be prepared at all from *pure aniline;* the presence of toluidine is necessary for its formation. It is a singular fact, that the pure base, rosaniline, is a colorless substance, and that it is only in its salts that its magnificent coloring powers become visible. The crystals exhibit by reflected light the metallic green color of beetles' wings, but are of a deep red color when viewed by transmitted light; they are soluble in alcohol, yielding splendid red solutions. By the action of nascent hydrogen on rosaniline, a new base is formed which forms colorless salts ; to this the name of *Leucaniline* has been given. It contains 2 atoms more hydrogen than rosaniline, these two bodies standing in the same relation as blue and white indigo (p. 328).

An Aniline blue is obtained by the replacement of 3 atoms of hydrogen in rosaniline by phenyl, $C_6 H_5$, whilst a violet is obtained by substituting 3 of methyl, ethyl, or any of the alcohol radicals.

BENZYL GROUP.

Benzyl Alcohol, $\left.\begin{matrix} C_7H_7 \\ H \end{matrix}\right\} O$, obtained by the action of alcoholic potash, or nascent hydrogen, on oil of bitter almonds (which is the aldehyde of the series). It is an oily colorless liquid, boiling at 207°. Oxidizing agents converted first into the aldehyde, $C_7 H_6 O$, and lastly into the acid of the series, $C_7 H_6 O_2$, benzoic acid.

Benzoic Aldehyde. Oil of bitter almonds, $\left.\begin{matrix} C_7 H_5 O \\ H \end{matrix}\right\}$. This oil does not exist already formed in bitter almonds, but is the result of a decomposition of the amygdaline contained in the almond (p. 318).

It can likewise be obtained by distilling a benzoate and a formate—in this respect resembling the aldehyde of the alcohol group ; it also forms a crystalline compound with sodium-hydric-sulphite. Bitter almond oil is a colorless strongly-smelling liquid, boiling at 180° ; the commercial substance (used in cookery) is very poisonous, as it invariably contains an admixture of hydrocyanic acid. On exposure to air or oxygen, or when acted upon by oxidizing agents, it is converted into benzoic acid. Benzoic aldehyde may be regarded as toluol or methyl benzol, in which two atoms of hydrogen of the methyl are replaced by one atom of oxygen, $C_6 H_5 (C H O)$; whilst *Benzoyl Chloride,* $C_7 H_5 O Cl$, is the last substance in which the one remaining atom of hydrogen in the methyl is replaced by chlorine. The vapor of bitter almond oil decomposes into benzol and carbon monoxide when passed through a red-hot tube; and benzoyl chloride is formed by the direct action of carbonyl chloride on benzol, the monad group (C O Cl) changing places with one atom of hydrogen, thus:

$$\underset{\text{Chlor-carbonyl.}}{CO\,Cl_2} + \underset{\text{Benzol.}}{C_6\,H_6} = \underset{\text{Benzoyl Chloride.}}{C_6\,H_5\,(C\,O\,Cl)} + H\,Cl.$$

Benzoyl chloride can also be formed by the action of phosphorus pentachloride upon benzoic acid; it is a colorless liquid, boiling at 199°.

Benzoic Acid, $C_7\,H_6\,O_2$. Found in many resins, especially in gum benzoin; it also occurs in the urine of cows, and in the putrefied urine of man and other animals; it can be obtained by the oxidation of benzyl alcohol and bitter almond oil. Benzoic acid may be easily prepared by heating gum benzoin, when the acid sublimes in pearly white plates; it fuses at 121° and boils at 250°. Benzoic acid forms a series of salts, most of which are soluble; the ferric benzoate falls as an insoluble red precipitate when sodium benzoate is added to ferric chloride.

Benzoic Peroxide, $\left.\begin{matrix}C_7\,H_5\,O\\C_7\,H_5\,O\end{matrix}\right\}O_2$. A well-crystallized substance, obtained by the action of barium peroxide on benzoyl chloride; it explodes when heated, and resembles acetyl peroxide (p. 277).

Benzoic Benzoate, or *Benzoic Anhydride,* $\left.\begin{matrix}C_7\,H_5\,O\\C_7\,H_5\,O\end{matrix}\right\}O.$ obtained by acting upon potassium benzoate with benzoyl chloride, thus:

$$\left.\begin{matrix}C_7\,H_5\,O\\K\end{matrix}\right\}O + C_7\,H_5\,O\,Cl = \left.\begin{matrix}C_7\,H_5\,O\\C_7\,H_5\,O\end{matrix}\right\}O + K\,Cl.$$

It is a solid substance, fusing at 24°, and boiling at about 310°; it is soluble in alcohol and ether, and several mixed anhydrides are also known — thus we have acetyl benzoate, $\left.\begin{matrix}C_2\,H_3'O\\C_7\,H_5\,O\end{matrix}\right\}O.$

Benzylamine, $\left.\begin{matrix} C_7\,H_7 \\ H \\ H \end{matrix}\right\} N$; a colorless liquid, isomeric with toluidine, boiling at 182°, obtained by the action of ammonia upon benzyl chloride. It is a true amine, and gives rise to corresponding secondary and tertiary amines.

Hippuric Acid, $C_9\,H_9\,N\,O_3$, is contained in the urine of horses and herbivorous animals. It can also be artificially prepared from zinc glycocine and benzoyl chloride; it is in fact glycocine, in which one atom of hydrogen is replaced by the radical $C_7\,H_5\,O$ of benzoic acid, thus :

$$\underset{\text{Glycocine.}}{C_2\,H_3\,(NH_2)\,O_2.} \qquad \underset{\text{Glycocine-Benzoic Acid, or Hippuric Acid.}}{C_2\,H_2\,(C_7\,H_5\,O)\,(NH_2\,O_2.}$$

Benzoic acid is converted by passing through the animal body into hippuric acid.

SALICYLIC GROUP. This group contains the dyad radicals, $\overset{\text{II}}{C_7}\,H_6$, or $\overset{\text{II}}{C_7}\,H_4\,O$. Salicylic acid stands to benzoic as glycolic to acetic, thus:

$$\underset{\text{Acetic.}}{\left.\begin{matrix} C_2\,H_3\,O \\ H \end{matrix}\right\} O.} \qquad \underset{\text{Glycolic.}}{\begin{matrix} \left.\begin{matrix} H \\ C_2\,\overset{\text{II}}{H_2}\,O \end{matrix}\right\} O. \\ \left.\begin{matrix} \\ H \end{matrix}\right\} O. \end{matrix}}$$

$$\underset{\text{Benzoic.}}{\left.\begin{matrix} C_7\,H_5\,O \\ H \end{matrix}\right\} O.} \qquad \begin{matrix} H \\ \left.\begin{matrix} C_7\,\overset{\text{II}}{H_4}\,O \end{matrix}\right\} O. \\ \left.\begin{matrix} H \end{matrix}\right\} O. \end{matrix}$$

And as glycolic acid is obtained from monochloracetic acid by the action of potash, so salicylic acid can be prepared from monochlor-benzoic acid. A glycol of this series called Saligenine also exists, from which the acid is obtained by oxidation. Salicylic acid is formed as a methyl compound in the essential oil of *Gaultheria procumbens.* For a description of the remaining aromatic compounds a larger work must be consulted.

INDIGO. This substance is the blue coloring matter derived from several species of *Indigofera*. The leaves are macerated in water, when they undergo oxidation, forming a yellow substance which, on exposure to air, deposits indigo in the form of a dark blue powder. This, when evaporated to dryness and cut into small cakes, constitutes the indigo of commerce. The pure coloring matter termed Indigotine is obtained from commercial indigo in crystals by sublimation; its composition is $C_8 H_5 NO$. Indigo is insoluble in water and in cold alcohol and ether; strong, or fuming sulphuric acid, dissolves indigo, forming a deep blue solution. Indigo occurs sometimes in healthy urine in small quantities. When indigo is exposed in contact with alkalies to reducing agents it passes into a soluble and colorless substance by absorption of hydrogen. The substance thus produced is called *white indigo;* its formula is $C_8 H_6 N O$. This property is largely employed in indigo dyeing. An indigo-vat being prepared, containing 1 part of indigo, 2 parts of ferrous sulphate, and 3 parts of slaked lime to about 200 parts of water, these are allowed to stand for some time in a closed vessel. The cloth is then dipped into the liquid, and on exposure to air becomes permanently dyed by the deposition of insoluble blue indigo in the fibre of the tissue.

Isatine, $C_8 H_5 NO_2$. By the careful oxidation of indigo this substance is formed; it crystallizes in large deep yellow crystals. By the action of potash it is converted into aniline.

$$C_8 H_5 N O_2 + 4 K H O = C_6 H_7 N + 2 (K_2 C O_3) + H_2.$$

Isatine and Caustic Potash. Aniline.

LESSON XL.

TURPENTINES AND CAMPHOR GROUP. This series of bodies appears to contain a common group of ten carbon atoms, and affording a large number of isomeric derivatives. It is particularly difficult to distingush between many of these bodies which appear identical in their chemical relations, but differ in their physical properties, and hence are said to be *physical isomers.* The following are the hydrocarbons from which these substances are derived:

$C_{10}H_{20}$	$C_{10}H_{18}$	$C_{10}H_{16}$	$C_{10}H_{14}$.
Di-amylene	Camphene or Menthene.	Terebene and its Isomers.	Cymol.

These hydrocarbons yield oxidized products termed Camphors; we thus have—

$C_{10}H_{20}O$	$C_{10}H_{18}O$	$C_{10}H_{16}O$	$C_{10}H_{14}O$.
Menthen-camphor.	Borneo-camphor.	Laurel-camphor.	Thymol and Carvol.

The camphors stand in the same relation to the above hydrocarbons as benzyl alcohol stands to toluol. By a further process of oxidation acids are formed; thus we have—

$C_{10}H_{16}$ terebene, $C_{10}H_{16}O_2$, camphinic acid.
$C_{10}H_{16}O$, laurel camphor, $C_{10}H_{16}O_4$, camphoric acid.

Turpentines and Isomers, $C_{10}H_{16}$. Oil of turpentine of commerce generally consists of a mixture of several isomeric modifications of this hydrocarbon. It is obtained from several species of pine; that from *Pinus nigra*, *abies*, and *sylvestris* constitutes common turpentine—that from the larch is known as venice turpentine. On distillation with water a volatile aromatic liquid comes over, and rosin or colophony remains in the retort.

The best-known natural varieties are *terebenthene* from the *Pinus maritina*, boiling at 161°, and possessing a power of left-handed polarization of —42°·3; *austra-terebenthene*, from the *Pinus Australis*, boiling also at 161°, but possessing a right-handed polarizing power of + 21°·5. These turpentines, when heated, or when acted on by sulphuric acids and other re-agents, form isomers differing in their action on the ray of polarized light, some being right- and some left-handed, whilst others are inactive. Terebenthene combines with hydrochloric acid, and forms isomeric compounds; it also combines with water to form a solid hydrate. On oxidation, the turpentines pass into resins.

Many *essential oils* are isomers of turpentine; of these may be mentioned essential oil of lemons, of bergamotte, neroli, lavender, pepper, camomile, carraway, cloves, &c. These often contain other oxidized oils in addition to the terebenes. Of these bodies laurel, or common camphor, $C_{10}H_{16}O$, is the most important; it is yielded chiefly by the *Laurus camphora* of China and Japan, although it can be obtained from other plants. *Camphor* is a white, crystalline, semi-transparent mass; it fuses at 175°, and boils at 204°; it is soluble in alcohol, and its solution deviates the plane of polarization to the right + 47°·4. Camphor dissolves in alcoholic potash unaltered, but, on heating, is first converted into Borneo camphor, $C_{10}H_{18}O$,and afterwards into camphinic, $C_{10}H_{16}O_2$, and campholic acid, $C_{10}H_{18}O_2$. On boiling with nitric acid, it is oxidized to camphoric acid, $C_{10}H_{16}O_4$. Like the turpentines, camphor also exists in several physical isomeric modifications, which chiefly differ in their action on polarized light. The camphoric acid obtained from these different camphors also exhibit differences in their properties.

Resins and Balsams.—Resin, or colophony, is obtained in the distillation of crude turpentine; the other resins, such as lac, mastic, copal, &c., have a similar composition. They are oxidation products of the terebenes.

Caoutchouc, or *India Rubber, and Gutta Percha.*—The first of these is a compound of hydrogen and carbon, and is an invaluable substance to the chemist; it is the hardened juice of several tropical trees, and in the pure state is white.

Caoutchouc combines with sulphur in various proportions, forming the vulcanised caoutchouc of commerce, which contains from 2 to 3 per cent. of sulphur. If heated more strongly with sulphur, a black, horny mass called ebonite or vulcanite, is formed. Gutta percha is also the hardened juice of a species of a *sapotacea*, growing in Borneo, Singapore, &c.

The pure substance is white, and insoluble in alcohol, but soluble in ether.

Naphthalin, $C_{10}H_8$, is a solid hydrocarbon, formed in large quantities in the process of the distillation of coal, and found in the heavy oils of coal tar. It is also formed when the vapors of many organic bodies, such as alcohol and acetic acid, are passed through a red-hot tube. Naphthalin crystallizes in pearly white plates, which dissolve in hot alcohol and cold ether; it fuses at 79°·2 and boils at 218°, but may be sublimed at a lower temperature. When heated, and a light applied, it burns with a luminous and smoky flame. Naphthalin forms with strong sulphuric acid two compound acids, corresponding to sulphovinic acid. Nitric acid attacks naphthalin, forming nitro-substitution products, and also an acid termed *phthalic acid*, $C_8H_6O_4$; this substance is connected with the benzol series, inasmuch as on heating with excess of lime, or baryta, it is converted into benzol:—

$$\underset{\text{Phthalic Acid.}}{C_8H_6O_4} = C_2O_4 + \underset{\text{Benzol.}}{C_6H_6}.$$

The nitro-substitution products of naphthalin are four in number, being bodies in which 1, 2, 3, or 4 atoms of hydrogen are replaced by the group NO_2. From nitro-naphthalin, $C_{10}H_7(NO_2)$, an amido-naphthalin is formed by substitution of NO_2 by NH_2; this substance, $C_{10}H_7(NH_2)$, is generally termed *naphthylamine.* Di- and tri-amido-naphthalin are also known. The chlorine substitution products of naphthalin are very numerous bodies, in which 1, 2, 3, 4, 6, and 8 atoms of hydrogen in the original hydrocarbon are replaced by chlorine. A second series of chlorine compounds is known, which ar substitution products of naphthalin di-chloride, $C_{10}H_8Cl_2$.

Nearly connected with the naphthalin group is the important coloring matter derived from madder, and termed *alizarin*, $C_{10}H_6O_3$. This substance appears not to be contained ready-formed in madder, the root of *Rubia tinctoria*, but to be produced, together with glucose, from a body termed *rubian* by the action of acids, alkalies, and ferments. Alizarin is deposited in long, red, needle-shaped crystals. It is but very slightly soluble in cold, but more soluble in hot water, and easily dissolves in alcohol. Alizarin produces insoluble red-colored compounds with alumina and stannic oxide, which are termed lakes, and a purple or black compound with ferric oxide. Hence, in *calico printing* solutions of these oxides are used as *mordants*, and are printed in pattern on the cotton cloth which, after undergoing certain preparatory processes, is then boiled in the "dye-beck," containing the ground madder-root mixed with water. The alizarin of the madder forms with the mordanted cloth an insoluble compound, which is colored pink, purple, black, or chocolate, according as the mordant has been pure alumina, or pure iron, or a mixture of the two. Animal fabrics, such as silk or wool, do not require the application of mordants; they are able alone to fix and render insoluble the coloring matter.

A substance isomeric with alizarin has been artificially prepared from the naphthalin compounds ; it crystallizes in yellow needles, insoluble in water, but soluble in ether. Mordanted cotton cloth cannot be dyed with this compound, but silk and wool are colored yellow.

VEGETO-ALKALOIDS.

Under this name a series of bodies containing carbon, hydrogen, oxygen, and nitrogen is grouped, which act as bases, and are found in certain plants. These bodies, for the most part, have not been artificially prepared, and although it is believed that they belong to the class of compound ammonias, yet their constitution is at present unknown. Some few of the alkaloids are liquid and volatile, and contain only carbon, hydrogen, and nitrogen: these have a more simple constitution. The alkaloids exert a powerful influence on the ray of polarized light, some deviating the plane to the right and some to the left; they also combine with acids to form salts, in this respect resembling ammonia, thus:—

Type, $NH_3 + HCl = NH_4Cl$ or NH_3HCl.
$C_{17}H_{19}NO_3 + HCl = C_{17}H_{20}NO_3Cl$ or $C_{17}H_{19}NO_3HCl$.
Morphine. Morphine Hydrochlorate.

They also form double crystallizable salts, with platinic chloride, in this respect again resembling ammonia. The alkaloids act most violently on the animal economy; some, such as strychnine, nicotine, &c., form the most violent poisons with which we are acquainted, whilst others, such as quinine and morphine, act as most valuable medicines.

Alkaloids containing Carbon, Hydrogen, and Nitrogen.

Piperidine, $C_5H_{11}N$. This alkaloid is obtained by distilling piperin, $C_{17}H_{19}NO_3$, the base contained in black pepper. Piperidine contains one atom of hydrogen, which can be replaced by an alcoholic group, hence its formula is $\left.\begin{matrix} C_5H_{10} \\ H \end{matrix}\right\} N$. It is a colorless liquid, boiling at 106°, and possessing a strong ammoniacal odor of pepper.

Conine, $C_8H_{15}N$, contained in the hemlock, *Conium maculatum;* it is a colorless liquid, boiling at 212°, and has a strong alkaline reaction, forming salts with acids. Conine acts as a narcotic poison. Under certain circumstances, conine yields butyric acid, by oxidation.

Nicotine, $C_{10}H_{14}N_2$, is the chief alkaloid, contained in tobacco, which contains varying quantities, from 2 to 8 per cent. of this substance. Nicotine boils about 240°, undergoing partial decomposition, but it may be distilled in an atmosphere of hydrogen without loss.

Nicotine is soluble in water, alcohol, and ether, and it acts as one of the most violent poisons with which we are acquainted; a small quantity acting on the motor nerves, and producing convulsions and afterwards paralysis. Nicotine does not contain any hydrogen replaceable by an alcohol radical, and when treated with iodide of ethyl, a salt corresponding to ammonium iodide is produced:

$$\left.\begin{matrix} \overset{III}{C_5 H_7} \\ \overset{III}{C_5 H_7} \end{matrix}\right\} N_2 \qquad \left.\begin{matrix} \overset{III}{C_5 H_7} \\ \overset{III}{C_5 H_7} \\ 2\,(C_2 H_5) \end{matrix}\right\} N_2 I_2.$$

Nicotine. Ethyl-nicotine Iodide.

Alkaloids containing Carbon, Hydrogen, Oxygen, and Nitrogen.

Alkaloids of Opium. Opium is the dried juice of the head of the poppy (*Papaver somniferum*); it is prepared largely in Asia Minor, Turkey, Egypt, and India; the Smyrna opium is most esteemed, and contains from 10 to 15 per cent. of morphine. There are no less than six different alkaloids contained in opium; of the semorphia and narcotine are found in largest quantity:

Morphine,	$C_{17} H_{19} NO_3$.	Papaverine,	$C_{20} H_{21} NO_4$.
Codeine .	$C_{18} H_{21} NO_3$.	Narcotine .	$C_{22} H_{23} NO_7$.
Thebaine .	$C_{19} H_{21} NO_3$.	Narceine .	$C_{23} H_{29} NO_9$.

In addition to these substances, opium contains a neutral crystallizable substance called Meconine, $C_{10} H_{10} O_4$, and an acid called Meconic Acid, with which the alkaloids are chiefly combined, as well as many other substances in small quantities, besides vegetable matter, &c. These alkaloids, although possessing a very closely analogous composition, have not yet been converted one into the other. Opium acts as a most valuable medicine, in small doses acting as a sedative, although heightening the pulse and the action of the heart. Taken in larger doses it acts as a narcotic poison, a stupor and prostration soon ensuing, resulting in loss of all voluntary power of motion, complete coma, and death. It appears that thebaine is the most powerful of the alkaloids, then papaverine, narcotine, codeine, and morphine.

Morphine, $C_{17} H_{19} NO_3 + H_2O$. In order to prepare morphia, the opium is extracted with water, and the meconic acid precipitated by calcium chloride; on evaporating the filtrate, crystals of morphine-hydrochlorate separate out. Morphine dissolves in 1,000 parts of cold and 400 of boiling water; hot alcohol dissolves it easily, whilst it is insoluble in ether. It forms crystalline salts soluble in water, and it appears to contain no replaceable hydrogen; as an ammonium-iodide is obtained when it is acted upon with ethyl iodide. Small quantities of morphine can easily be detected by the formation of a deep blue coloration, when this substance comes in contact with ferric chloride.

Codeine, $C_{18} H_{21} NO_3 + H_2O$, is left in the mother liquors from which the morphine has crystallized. Codeine is much more soluble in water than morphine, and is contained in opium in much smaller quantities; it has a strong alkaline reaction and neutralizes acids.

Thebaine, $C_{19}H_{21} NO_3$, is contained in very small quantities in opium; its poisonous properties are more violent than any other of these alkaloids; it produces tetanus.

Papaverine, $C_{20} H_{21} NO_4$. Distinguished from the other opium bases by giving a deep blue color with strong sulphuric acid.

Narcotine, $C_{22} H_{23} NO_7$, remains insoluble when opium is treated with water, and it is obtained by dissolving it out from the "marc" or insoluble portion of the opium with hydrochloric acid. It dissolves in 128 parts of boiling alcohol and 19 of boiling ether. Narcotine when heated with potash furnishes ammonia and methylamine, as well di- and tri-methylamine; and when treated with hydriodic acid, it furnishes 3 molecules of methyl iodide for every molecule of narcotine.

Alkaloids of the Strychnos. Two alkaloids possessing most powerful poisonous properties, and called Strychnine and Brucine, are found in the seeds of the *Strychnos nux vomica*, and in the *Strychnos Ignatius* or the St. Ignatius's bean. *Strychnine*, $C_{21} H_{22} N_2O_2$, is a base forming crystallizable salts, of which 1½ per cent. is contained in the St. Ignatius's bean. It acts as a violent poison, producing tetanic convulsions; it however is sometimes given in very small doses in medicine; its salts are all extremely bitter and tart. Strychnine can be detected when present in minute quantities, by yielding, with sulphuric acid and potassium bichromate, an intense purple color, which passes rapidly into a red, and then into a yellow color.

Brucine, $C_{23} H_{26} N_2 O_4 + 4 H_2O$, is found alone in false angustura bark, and together with strychnine in *nux vomica;* it is more soluble in water and alcohol than strychnine. Brucine and its salts are less poisonous and less bitter than the strychnine compounds; it can be distinguished from strychnine by the bright red color produced when this substance is moistened with nitric acid; indeed this reaction may also be employed as a most delicate test for the presence of nitric acid.

Alkaloids of the Chinchonas. The bark of this species of trees, originally grown in Peru, but now transplanted to Java and India, contains 2 alkaloids, quinine and cinchonine, and each of these yields 2 isomeric modifications, quinidine and quinicine; cinchonidine and cinchonicine.

The Alkaloids are combined in the bark with a peculiar acid termed quinic acid. Quinine is a most valuable medicine, acting as a febrifuge; cinchonine does not possess the same valuable properties.

Quinine, $C_{20} H_{24} N_2O_2$. This alkaloid may be precipitated from the solution of its sulphate as a white crystalline powder. It dissolves in 350 parts of cold water, and in 2 parts of alcohol. Its solution has a strong bitter taste, and deviates the plane of polarization to the *left.* Quinine may be detected by adding chlorine-water, and afterwards an excess of ammonia, to solutions of the sulphate, when a green color is produced; another characteristic reaction consists in the deep red color produced when finely powdered potassium ferrocyanide is thrown into the solution of quinine in chlorine water. Quinine appears to possess no replaceable hydrogen, as when treated with ethyl iodide a salt of an ammonium compound is formed. Quinine sulphate is the salt used in medicine; it is not very soluble in water, but dissolves easily when a drop or two of sulphuric acid is added. Its solution possesses very strongly the property of *fluorescence.*

Quinidine and *Quinicine.* The first of these isomers of quinine is found in the bark, and it resembles quinine in its febrifuge qualities, but it deviates the plane of polarization strongly to the *right.* Quinicine is obtained by acting upon quinine by heat. It is a bitter substance, possessing a semi-solid resinous consistency, and deviating the plane of polarization feebly to the right.

Cinchonine, $C_{20} H_{24} N_2 O$. This body is separated from the quinine, which accompanies it, by its less solubility in alcohol; thus cinchonine requires 30 parts of boiling alcohol for solution, and therefore crystallizes out whilst the quinine remains in solution. Cinchonine is not nearly so powerful a febrifuge as quinine; it is, however, used as a medicine in some countries. Although it only differs from quinine by containing one atom less oxygen, it has not yet been transformed into the latter. It does not produce a green color with chlorine-water and ammonia like quinine; it acts as a strong base, and forms salts which are more soluble in water and alcohol than those of quinine.

Cinchonidine—Cinchonicine. The first of these isomers is found, together with quinidine, in the brown resinous mass left after the extraction of the two chief alkaloids. It produces a left-handed rotation on a polarized ray, whilst cinchonine produces a right-handed polarization. Cinchonicine is obtained by heating a cinchonine sulphate to 120° or 130°; it deviates the polarized ray feebly to the right: hence we have

Quinine	exerting a	powerful	left-handed	rotation.
Quinidine	"	powerful	right-handed	"
Quinicine	"	feeble	right-handed	"
Cinchonine	"	powerful	right-handed	"
Cinchonidine	"	powerful	left-handed	"
Cinchonicine	"	feeble	right-handed	"

Theobromine, $C_7 H_8 N_4 O_2$, the crystallizable alkaloid contained in cocoa (*Theobroma Cacao*). If in this substance one atom of hydrogen be replaced by methyl, *Cafeine* is formed.

Cafeine, or *Theine,* $C_8 H_{10} N_4 O_2 + H_2O$. The active principle of tea and coffee; also found in the leaves of *Ilex Paraguayensis*, which the South Americans much use in place of tea; also in guarana, a kind of chocolate made from the fruit of *Paulinia Sorbilis*. The quantity of the alkaloid contained in tea is about 2 per cent.; in coffee, 0·8 to 1 per cent.; in guarana 5 per cent.; and in the Paraguay tea about 1·2 per cent.

A description of the numerous alkaloids of less general interest will be found in the larger manuals.

LESSON XLI.

Albuminous Substances.

Under this head we class a number of peculiar compounds forming a characteristic and essential portion of the bodies of animals, and occurring also in certain parts, especially the seeds, of vegetables. These compounds possess a very complicated constitution, and our knowledge of their true chemical relations is most incomplete. They do not crystallize, and exist in an amorphous jelly-like form; hence it is very difficult to obtain them in the pure state, so that there is some doubt even about their chemical composition. They all contain sulphur, and most of them phosphorus, in addition to carbon, hydrogen, oxygen, and nitrogen, and in their different forms possess nearly the same composition.

Albumin is seen in one of its purest forms in the white of egg; it is also contained in the serous or liquid portion of the blood. It may be obtained by adding acetic acid to white of egg and diluting with water, when a white flocculent precipitate of albumin is formed. When dried, it forms a yellow, transparent, gum-like mass; this, on addition of cold water, remains as a white insoluble powder, which, like the precipitated albumin, dissolves in water containing a small trace of free alkali. One of the most characteristic properties of albumin is its power of coagulation; if soluble white of egg be heated to about 65° C., it becomes solid and opaque; in this state it is insoluble in water, but dissolves in dilute alkali.

Fibrin. This substance exists in solution in the blood, but immediately becomes solid when the blood leaves the living body; it can be obtained by washing the clot, or thick part of blood, until the red color has disappeared, or it may be obtained by agitating fresh blood with twigs. It then is obtained in the form of colorless filaments, which are tasteless and insoluble in water; on drying, it forms a horny mass like albumin. The fibrin of flesh appears to differ from that of blood; and differences have been observed between the fibrin from arterial and that from venous blood.

Casein is the nitrogenous substance contained in milk and cheese; it closely resembles albumin in its properties, being coagulated by acids. Casein is insoluble in pure water, but dissolves in a very dilute solution of an alkali. In milk the casein is not coagulated by boiling, but an acid, or a portion of the inner coating of the calf's stomach, called *rennet*, at once separates out the casein and butter as curds, and leaves the milk-sugar, and salts in solution as whey.

Vegetables contain similar substances, which are scarcely to be distinguished from the bodies derived from an animal source. *Glutin*, or the sticky, elastic substance contained with starch in wheaten flour, is vegetable fibrin, whilst vegetable albumin and casein occur in the juices and seeds of plants. The following table shows the percentage composition of the albuminous bodies; it is impossible to give any formulæ for these complicated substances:

	Albumin.	Fibrin.	Casein.
Carbon . . .	53·5	52·7	53·8
Hydrogen . .	7·0	6·9	7·2
Nitrogen . . .	15·5	15·4	15·6
Oxygen . . .	22·0	23·5	22·5
Sulphur . . .	1·6	1·2	0·9
Phosphorus . .	0·4	0·3	0·0
	100·0	100·0	100·0

Gelatin is a nitrogenous substance obtained from the animal body; it is prepared by boiling the tissues, and is then known as glue, isinglass, or gelatin; its composition is the same as that of the tissue from which it is prepared.

Animal Chemistry is a most important branch of chemical science, and one which unfortunately is but very slightly advanced: our knowledge of the composition and chemical constitution of the substances contained in the animal body is very incomplete, and concerning many of the chemical changes which occur in the different parts of the animal we are almost entirely ignorant.

The *Bones* of animals consist principally of tribasic calcium phosphate, together with a kind of gelatin; the earthy phosphate dissolves in hydrochloric acid, leaving the bone as an elastic gelatinous mass; when burnt, the friable and earthy matter alone remains. Bone contains—

Animal matter	33
Calcium phosphate	57
Calcium carbonate	8
Calcium fluoride	1
Magnesium phosphate	1
	100

The *Blood* of animals is the channel by means of which their bodies not only receive all the requisite supply of materials for their growth and for the repair of waste, but by means of which they are able to get rid of the worn-out matters which need immediate removal. In vertebrate animals the blood has a red color and a temperature above the medium in which the animal lives; in mammalia, and especially in birds, this artificial warmth is plainly noticed. The temperature of the blood is singularly constant in different animals under the most varying conditions of climate; it is 36°9 (98° F.) in man and 42°·8 (or 109° F.) in birds. The chief peculiarity of blood is the existence in it of very small round or oblong discs, differing in size and shape in different animals (diameter 0·0075 mm. in man, and four times as large in frogs). These are called the blood-globules, or corpuscles; they are of a red color, and float in a colorless liquid; when the fibrin coagulates it carries down with it mechanically the red globules.

Healthy human blood possesses the following average composition, and its *specific gravity* is 1·055:

Coagulum, or clot.	Fibrine	0·30	13·0.
	Corpuscles	12·70	
Serum.	Water	79·00	87·0.
	Albumin	7·00	
	Fatty matters	0·06	
	Salts	0·94	

The color of the blood discs is due to a substance called *hæmatin;* this contains about 7 per cent. of iron, but the iron can be withdrawn by sulphuric acid without any apparent alteration of the red coloring matter. Blood is charged with dissolved gases, especially oxygen, nitrogen, and carbonic acid; and the oxidation of the tissues is effected by the presence of the former gas: the arterial blood (freshly oxidized from the lungs) contains in 100 volumes 14·5 volumes of nitrogen, 62·3 of carbonic acid, and 23·2 of oxygen; whilst in venous blood (charged with the products of combustion of the body) the same gases are found in the proportion of 13·1, 71·6, and 15·3 volumes respectively.

Amongst the other most important animal fluids may be mentioned *Gastric Juice*, a clear liquid secreted by the lining membrane of the stomach and the active agent in effecting the digestion and solution of the albuminous parts of the food. It has an acid reaction due to the presence of free lactic and hydrochloric acids. The *Bile*, a liquid secreted in the liver and poured out into the duodenum; this substance contains several peculiar nitrogenized acids, viz., taurocholic acid $C_{26} H_{45} N SO_7$, and glycocholic acid, $C_{26} H_{43} N O_6$. A peculiar substance termed *taurin*, $C_2 H_7 N SO_3$, is obtained by the action of acids on bile; this body, which is isomeric with the compound of aldehyde ammonia with sulphurous acid, can be prepared artificially by heating ammonium-isethionate, $H_4 N C_2 H_5 S O_4$, which parts with $H_2 O$, and forms taurin.

Milk. The composition of this important secretion varies considerably in different animals, but each kind contains all the materials needed for the formation of the body of the young animal; thus it contains casein (a body having nearly the same composition as flesh), fats (butter), and milk sugar, together with those inorganic salts, especially the alkaline chlorides, and calcium phosphates, needed for the formation of bone. The following gives the average composition of milk of different animals:—

	Woman.	Cow.	Goat.	Ass.	Bitch.
Water	88·6	87·4	82·0	90·5	66·3
Butter	2·6	4·0	4·5	1·4	14·8
Milk Sugar and Soluble Salts	4·0	5·0	4·5	6·4	2·9
Casein and Insoluble Salts	3·9	3·6	9·0	1·7	16·0

The specific gravity of milk varies from 1·03 to 1·04.

The Urine. It is in the urine that a large portion of the waste nitrogenous portions of the body pass off as urea and uric acid; the urine is secreted by the kidneys from the arterial blood. Healthy urine contains, in 1,000 parts, 957 parts of water, 14 of urea, 1 of uric acid, 15 of other organic matter, and 13 of inorganic salts.

Functions of Animals and Plants.

The general characteristics of *animal* and *vegetable* life may be stated as follows: the animal lives upon *organized* materials, taking up oxygen and evolving carbonic acid and other oxidized products; the plant lives upon *unorganized* materials, especially carbonic acid, water, ammonia, and salts, organizing them and evolving oxygen. The chemical function of the animal is oxidation, that of the plant reduction. The food of the plant serves merely to increase its bulk; that of the animal is employed (after it has attained its full growth) to replace the material worn out by all the active operations of life. The animal obtains the energy necessary for its existence from the oxidation of its own body; the plant obtains the energy necessary for the organization of its food directly from the sun.

Respiration and Animal Heat. The process of respiration, essential to the life of all animals, consists in the aerating of the blood circulating through the lungs or similar apparatus, by means of the oxygen of the air. The blood does not come into actual contact with the air, but is separated by a large surface of very thin membrane, through which the exchange of gases takes place by solution and diffusion. Not only does the blood gain in oxygen (*see* Blood, p. 340), but it loses the products of combustion with which it is charged, and is thus rendered fit again to circulate and carry away used-up material. The volume of air thrown out of the human lungs at each ordinary expiration amounts to from 350 to 700 cubic centimeters: this, however, by no means empties the lungs, whose capacity is much greater; the number of respirations amounts to about fifteen in each minute. The expired air differs remarkably from the inspired air, as it contains from 3 to 6 per cent. of carbonic acid, and will not support the combustion of a candle.

Under different circumstances of health or disease, activity or repose, sleeping or waking, after a meal or fasting; according to the temperature, pressure of the air, and from other varying conditions, the quantity of exhaled carbonic acid varies considerably. The determination of the quantity of carbonic acid exhaled by an animal under the above circumstances is a subject of the highest importance, but one which is surrounded by numerous experimental difficulties. We may assume, as the result of the best experiments, that a man gives off 19·8 litres of carbonic acid (at 0° and 760 mm.) each hour; this amounts to about 40 grms. of carbonic acid, or 11 grm. of carbon per hour: the heat which is always evolved by the combustion of this carbon goes to keep up the temperature of the body. It is difficult to determine with accuracy how far the whole of the animal heat can be accounted for by the combustion of this carbon, as the chemical changes which go on in the body are of a very complicated nature, and as yet little understood. Considering, however, the subject in a general point of view, there cannot be much doubt that the whole of the animal heat is derived from the combustion of the materials of the body ; thus we find that in birds, whose temperature is higher than that of mammalia, the quantity of carbonic acid evolved is more than half as much again as in larger animals ; whilst in cold climates, where the loss of animal heat is great, men find it necessary to eat enormous quantities of fat, this doubtless serving to maintain the temperature of the body.

The effect of starvation on the quantities of carbonic acid and urea, taken as representing the rate of change going on in the body, is very remarkable; in a dog, the quantity of carbonic acid was reduced by fasting for ten days to one-third, and the urea to one twenty-second part of the amount given off on full diet; whereas in a man the carbonic acid was nearly reduced to one-third by starvation. An interesting fact has been observed, viz., that hydrogen and marsh-gas are evolved in small quantities from the skin and lungs under certain conditions. This subject is quite in its infancy, and demands careful experimental investigation, as it is by such patient research alone that we can hope to form any real estimate of the income and expenditure of the body. The special study of the chemistry of the body has been made a separate branch science, termed *Physiological Chemistry.*

Food of Plants. Animals, as we have seen, are unable to produce the complicated chemical compounds which they need for their structure ; plants are, however, able to do this, and from the elementary constituents to build up their various parts. This function of plants is entirely dependent upon the sunlight; without sunlight the green coloring matter of the leaves of plants cannot decompose the atmospheric carbonic acid, and, therefore, without sunlight, the plant cannot grow.

In order to separate the atoms of carbon and oxygen, an expenditure of force is necessary; this force is derived from the rapidly vibrating solar rays; it is they which tear asunder the carbon and oxygen atoms, and thus enable the leaves to take up and assimilate the carbon, throwing out the oxygen into the air for the subsequent use of animals. When vegetable matter is ignited, it burns to carbonic acid, and generates exactly the same amount of force, as the vibrations of heat, which were needed, in the form of vibrations of light, originally to decompose the atmospheric carbonic acid. Hence when coal burns, the light and heat evolved may truly be said to be that of the sun; and, as animals depend for their existence upon vegetables, and these in their turn cannot live without the solar radiations, animals may with truth be called children of the sun.

The bodies of plants may be considered to be composed of two kinds of substances, *organic*, such as starch, vegetable fibre, &c.; and *inorganic* salts, constituting the ash of the plant. The carbon, needed for the first of these materials, the plant obtains mainly from the atmosphere; the nitrogen, hydrogen, and oxygen, which the organic substances contain, the plant takes up both by its leaves and by its roots; whilst the whole of the inorganic salts are absorbed from the earth by the roots, which act as the mouth of the plant, whilst the leaves may be compared to the lungs of animals. Every plant has in the atmosphere an unlimited supply of carbon and water; but for the supply of inorganic materials, the plant is dependent upon the nature of the particular soil upon which it grows. Plants possess the peculiar power of selection, by the roots, of the mineral constituents of food, as well as the subsequent chemical elaboration of the materials. Of the causes of the changes which thus go on we know nothing: thus we cannot explain why an acorn turns out always to be an oak, or why of two seeds sown in the same soil and exposed to the same sunlight and air, one evolves a poisonous and the other a wholesome plant.

Concerning the growth of plants a large amount of information has been amassed, but we are far from possessing even an approach to a knowledge of the laws which regulate this important subject. For an account of the interesting facts which have been ascertained respecting the questions of manuring, fertility of the soil, &c., we must refer the reader to books on the branch science of *Agricultural Chemistry*.

By Act of Parliament (27 and 28 Vict. cap. 117, 29th July, 1864) the use of the Metrical System of Weights and Measures is rendered legal. The weight of the Kilogramme is settled by this Act to be equal to 15432.3487 *English grains.*

COMPARISON OF THE METRICAL WITH THE COMMON MEASURES. BY DR. WARREN DE LA RUE.

MEASURES OF LENGTH.

	In English Inches.	In English Feet = 12 Inches.	In English Yards = 3 Feet.	In English Fathoms = 6 Feet.	In English Miles = 1,760 Yards.
Millimetre	0.03937	0.0032809	0.0010936	0.0005468	0.0000006
Centimetre	0.39371	0.0328090	0.0109363	0.0054682	0.0000062
Decimetre	3.93708	0.3280899	0.1093633	0.0546816	0.0000621
Metre	39.37079	3.2808992	1.0936331	0.5468165	0.0006214
Decametre	393.70790	32.8089920	10.9363310	5.4681655	0.0062138
Hectometre	3937.07900	328.0899200	109.3633100	54.6816550	0.0621382
Kilometre	39370.79000	3280.8992000	1093.6331000	546.8165500	0.6213824
Myriometre	393707.90000	32808.9920000	10936.3310000	5468.1655000	6.2138244

1 Inch = 2.539954 Centimetres.
1 Foot = 3.0479449 Decimetres.
1 Yard = 0.91438348 Metre.
1 Mile = 1.6093149 Kilometre.

MEASURES OF SURFACE.

	In English Square Feet.	In English Sq. Yards = 9 Square Feet.	In English Poles = 272.25 Sq. Feet.	In English Roods = 10,890 Sq. Feet.	In English Acres = 43,560 Sq. Feet.
Centiare or sq. metre	10.7642993	1.1960333	0.0395383	0.000988457	0.0002471143
Are or 100 sq. metres	1076.4299342	119.6033260	3.9538290	0.098845724	0.0247114310
Hectare or 10,000 sq. metres	107642.9934183	11960.3326020	395.3828959	9.884572398	2.4711430996

1 Square Inch = 6.4513669 Square Centimetres.
1 Square Foot = 9.2899683 Square Decimetres.
1 Square Yard = 0.83609715 Square Metre or Centaire.
1 Acre = 0.404671021 Hectare.

MEASURES OF CAPACITY.

	In Cubic Inches.	In Cubic Feet = 1,728 Cubic Inches.	In Pints = 34·65923 Cubic Inches.	In Gallons = 8 Pints = 277·27384 Cubic Inches.	In Bushels = 8 Gallons = 2218·19075 Cubic Inches.
Millilitre, or cubic centimetre	0·061027	0·0000353	0·001761	0·00022010	0·000027512
Centilitre, or 10 cubic centimetres	0·610271	0·0003532	0·017608	0·00220097	0·000275121
Decilitre, or 100 cubic centimetres	6·102705	0·0035317	0·176077	0·02200967	0·002751208
Litre, or cubic decimetre	61·027052	0·0353166	1·760773	0·22009668	0·027512085
Decalitre, or centistere	610·270515	0·3531658	17·607734	2·20096677	0·275120846
Hectolitre, or decistere	6102·705152	3·5316581	176·077341	22·00966767	2·751208459
Kilolitre, or stere, or cubic metre	61027·051519	35·3165807	1760·773414	220·09667675	27·512084594
Myriolitre, or decastere	610270·515194	353·1658074	17607·734140	2200 96676750	275·120845937

1 Cubic Inch = 16·3861759 Cubic Centimetres. 1 Cubic Foot = 28·3153119 Cubic Decimetres.
1 Gallon = 4·543457969 Litres.

MEASURES OF WEIGHT.

	In English Grains.	In Troy Ounces = 480 Grains.	In Avoirdupois Lbs. = 7,000 Grains.	In Cwts. = 112 Lbs. = 734,000 Grains.	Tons = 20 Cwt. = 15,620,000 Grains.
Milligramme	0·015432	0·000032	0·0000022	0·00000002	0·000000001
Centigramme	0·154323	0·000322	0·0000220	0·00000020	0·000000010
Decigramme	1·543235	0·003215	0·0002205	0·00000197	0·000000098
Gramme	15·432349	0·032151	0·0022046	0 00001968	0·000000984
Decagramme	154·323488	0·321507	0·0220462	0·00019684	0·000009842
Hectogramme	1543·234880	3·215073	0·2204621	0·00196841	0·000098421
Kilogramme	15432·348800	32·150727	2·2046213	0·01968412	0·000984206
Myriogramme	154323·488000	321·507267	22·0462126	0·19684118	0·009842059

1 Grain = 0·064798950 Gramme. 1 lb. Avd. = 0·45359265 Kilogr.
1 Troy oz. = 31·103496 Gramme. 1 Cwt. = 50·80257689 Kilogr.

QUESTIONS AND EXERCISES UPON THE FOREGOING LESSONS.

IN order to enable the pupil to master the principles of the science, he must conscientiously write out answers to the Questions, and work out the Exercises given in illustration of each Lesson.

LESSON I. *Introduction.*

1. Describe an experiment to prove that when a candle burns, the materials are not annihilated.
2. Distinguish between a chemical element and a compound.
3. What is the construction and use of a chemical balance?
4. Name a few important elements.
5. Is it likely that we are now acquainted with all the elements existing on the earth, and why?
6. Describe some cases in which chemical actions occur.

LESSON II. *Oxygen and Hydrogen.*

1. How did Priestley first prepare oxygen gas?
2. Describe the process now adopted for obtaining this gas.
3. Whence is the name oxygen derived?
4. State the action produced (1) by animals, (2) by plants, on the air.
5. Learn by heart the composition by weight of potassium chlorate.
6. I want 100 pounds of oxygen; how many pounds of potassium chlorate must I take?
7. What is meant by the combining weights of the elements? Give an example.
8. How can hydrogen be obtained from water?

9. Mention the chief properties of hydrogen.

10. What is formed when hydrogen burns in the air? How can this be exhibited?

11. 65·2 parts by weight of zinc in decomposing water yield 2 parts by weight of hydrogen. How much zinc must be employed to obtain 100 pounds of hydrogen?

12. What is the derivation of the word hydrogen?

LESSON III. *Chemical Calculations, &c.*

[It will generally be found necessary to divide this into several lessons, and to familiarize the pupil to the subject by a much larger number of exercises than those here given.]

1. Describe shortly the metrical system of weights and measures.

2. How many cubic centimetres are contained in 1 cubic metre?

3. How is a thermometer made and graduated?

4. Describe the three thermometric scales now in use. Work out the following:

How many degrees C. and R. correspond to + 42° and — 32° Fah.?
" C. and F. " + 327° and — 2° R.?
" F. and R. " + 78° and — 172° C.?

5. If 273 volumes of gas be at the temperature of 0° C., to what temperature must they be heated in order to expand to 295 volumes?

6. What volume will 1,063 litres of hydrogen at -2° occupy when heated to 100°?

7. State Boyle's Law of Pressures.

8. What volume will 1,000 cbc. of oxygen at 0° and 760 mm. become at a temperature of 16°·5, and under a pressure of 735 mm.?

9. Learn by heart the weight in grammes of one litre of hydrogen at 0° and under 760 mm. pressure.

10. What simple method is used for calculating the weight

of one litre of the elementary gases at the standard temperature and pressure ?

11. How many cubic centimetres of oxygen gas measured at 10° and under 745 mm. pressure can be got by heating 20 grammes of potassium chlorate ?

12. What is the weight in grammes of 516 litres of hydrogen gas measured at —20° and under 770 mm. pressure ?

LESSON IV. *Water.*

[It may be necessary to divide this into two or more lessons.]

1. How did Cavendish determine the composition of water ?

2. Describe the most exact methods of determining the composition of water (1) by volume, (2) by weight.

3. What is meant by the latent heat of water ? How is this determined ?

4. How can you show that when a liquid solidifies, heat is given out ?

5. Describe the changes in bulk which water undergoes when heated from 0° to 100°.

6. When does water boil ?

7. How is the latent heat of steam determined ?

8. Define the term "thermal unit."

9. How is the tension of aqueous vapor measured ?

10. Why must the barometric pressure be noticed when graduating a thermometer ?

11. How is pure water obtained ?

12. What is the composition of hydric peroxide ?

LESSON V. *The Atmosphere.*

1. How may pure nitrogen gas be prepared ?

2. What is the mean height of the barometer at the sea's level ?

3. Why does water boil at a lower temperature than 100° on a mountain top?

4. What reasons have we for believing that the air is a mechanical mixture, and not a chemical combination of nitrogen and oxygen?

5. Describe the mode of making a eudiometric analysis of the air.

6. How may we determine the composition of the air by weight as regards nitrogen and oxygen?

7. Draw and describe an apparatus for estimating the quantity of carbonic gas contained in the air.

8. What important part does this carbonic acid play as regards vegetation?

9. How is rain formed?

10. Explain the formation of dew and hoar frost.

11. What is the use of hygrometers?

12. Name other constituents of the atmosphere.

Lesson VI. *Nitric Acid and Oxides of Nitrogen.*

1. Give the composition by weight of the five oxides of nitrogen.

2. Explain what is meant by chemical combination in multiple proportions.

3. State the principles of Dalton's atomic theory.

4. What happens when electric sparks are passed through the air?

5. Learn by heart the combining weights of oxygen, hydrogen, nitrogen, chlorine, potassium, sulphur; and the formulæ of nitric acid, sulphuric acid, nitre, and potassium sulphate.

6. Write out in symbols the decomposition occurring in the preparation of nitric acid, and explain the meaning of these symbols.

7. I want 500 grms. pure nitric acid; how many grms. of nitre and sulphuric acid shall I need, and how many grms. of hydric potassium sulphate will remain?

8. Mention the tests for nitric acid.

9. How is nitric pentoxide prepared ?

10. 100 parts by weight of this substance contain 25·93 parts of nitrogen, and 74·07 parts of oxygen. Show that the formula of the substance is $N_2 O_5$.

Lesson VII. *Oxides of Nitrogen and Ammonia.*

1. Name the chief properties of laughing gas.

2. How many grms. of nitrous oxide and water can be obtained from 213 grms. ammonium nitrate ?

3. How is the composition by volume of nitrous oxide determined ?

4. I want 100 litres of nitric oxide gas when the temperature is 0° and the pressure 760 mm. What weight in grms. of copper and nitric acid must I take ?

5. Point out the relation between nitric pentoxide and the nitrates, and nitric trioxide and the nitrites.

6. Give the formulæ representing three different modes by which ammonia can be produced.

7. How many litres of ammonia measured at 10° and under a pressure of 755 mm. can be obtained from 100 grms. of sal ammoniac.

8. Describe the principles of Carré's freezing machine.

9. How is the composition by volume of ammonia ascertained ?

10. How can ammonia be liquefied ?

Lesson VIII. *Carbon and Carbon Dioxide.*

1. Name the three allotropic modifications of carbon. State their chief peculiarities.

2. Give a short description of the nature of coal. What changes have occurred in the passage of wood into coal ?

3. Required 562 grms. carbonic dioxide ; how will you obtain it, and what weight of materials will you need to use ?

4. What law regulates the absorption of this gas in water?

5. How can carbonic dioxide be obtained in the liquid and in the solid state? What peculiar property does this liquid exhibit?

6. Explain the mode adopted for obtaining very low temperatures by means of solid carbonic acid.

7. Describe, with a drawing, the apparatus used to determine the composition of carbonic acid.

8. State the results of this determination.

9. How many litres of carbonic dioxide measured at 300°, and under a pressure of 740 mm. can be obtained by burning one kilogramme of Wigan cannel (No. 5 on page 69)?

10. Show, by describing an experiment, that carbonic dioxide contains its own volume of oxygen.

Lesson IX. *Carbon Monoxide and Hydrocarbons.*

1. How many grms. of carbon will be needed to convert 100 litres of carbonic dioxide at 0° and 760 mm. into carbonic oxide, and how many litres of this latter gas will be formed?

2. Find the volume in litres at 10° and 740 mm. of carbonic oxide which can be obtained from 100 grms. of oxalic acid and formic acid respectively.

3. What is formed when caustic potash and carbonic oxide are heated together?

4. How is the composition of carbonic oxide ascertained by eudiometric analysis?

5. What is the composition of marsh-gas and fire-damp?

6. How is olefiant gas prepared?

7. State shortly the properties and composition of coal gas.

8. How is the illuminating power of coal gas ascertained?

9. Describe the construction of a Bunsen's burner.

10. Explain the principles of the Davy lamp.

11. How many litres of carbonic acid are formed by the combustion of one litre of olefiant gas?

12. How is cyanogen gas prepared?

13. I want 50 grms. pure hydrocyanic acid; how many grms. of potassium cyanide and sulphuric acid shall I need to use?

LESSON X. *Chlorine.*

1. Write down as an equation the decompositions which occur in the preparation of chlorine from rock salt.
2. I want 100 litres of chlorine gas at 10°, and under the pressure of 735 mm.; how many grms. of the materials, viz., $Na\,Cl$, $H_2\,SO_4$, and $Mn\,O_2$ shall I require?
3. Describe experiments proving the power of chlorine to combine with hydrogen.
4. Explain the bleaching action of chlorine.
5. How many kilos. of salt and sulphuric acid must be taken to yield 100 kilos. of aqueous hydrochloric acid containing 20·22 per cent. of the gas?
6. How is the composition of hydrochloric acid determined?
7. Write out the formulæ of the oxides of chlorine and the corresponding acids.
8. Describe the action of water upon hypochlorous oxide, nitric pentoxide, and carbonic dioxide.
9. What is the composition of bleaching-powder?
10. How is potassium chlorate prepared?
11. Show from the composition of the salt that the formula of potassium chlorate is $K\,Cl\,O_3$.
12. Show that the aqueous perchloric acid containing 72·3 per cent. of $H\,Cl\,O_4$ does not correspond to any definite compound of this acid with water.

LESSON XI. *Bromine, Iodine, and Fluorine.*

1. Describe the mode of obtaining pure bromine.
2. What is the composition of bromic and perbromic acids?
3. Write out in an equation the decompositions occurring in the manufacture of iodine from potassium iodide.

4. How is hydriodic acid gas prepared?

5. Show that the aqueous hydriodic acid, boiling at a constant temperature, and containing 57 per cent. of H I does not correspond to a definite hydrate.

6. How would you detect iodine, bromine, and chlorine when present in solution together?

7. How can fluorine be prepared?

8. Mention the most remarkable property of H F.

9. State the general relations which Cl, Br, I, and F exhibit amongst themselves.

LESSON XII. *Sulphur and Sulphurous Acid.*

1. State the different compounds in which sulphur is met with in nature.

2. Name some of the chief properties of sulphur.

3. Write down the names and symbols of the compounds of sulphur, oxygen, and hydrogen.

4. How is sulphuric dioxide prepared? How can it be liquefied?

5. How many cubic centimetres of sulphuric dioxide at 0° and 760 mm. can be got by the use of 12 grms. of copper, and how many grms. of sulphuric acid will be needed?

6. How is real sulphurous acid formed from sulphuric dioxide? Explain the constitution of the salts termed sulphites.

7. How does sulphurous acid act as a bleaching agent?

LESSON XIII. *Sulphuric Acid and Sulphuretted Hydrogen.*

1. How is sulphuric trioxide prepared, and what are its properties?

2. Describe the decompositions by which sulphuric acid is prepared in the leaden chamber.

3. How many tons of chamber-vitriol, containing 70 per

cent. of real acid (H_2SO_4), can be prepared from 250 tons of pyrites, containing 42 per cent. of sulphur?

4. What is meant by the term "molecule" as distinguished from "atom?"

5. How many grms. of oxygen can be obtained by the decomposition of 450 grammes H_2SO_4 at a red heat?

6. Explain what is meant by the terms "monatomic" and "di-atomic."

7. How would you detect the presence of sulphuric acid?

8. What is the composition of sodium hyposulphite?

9. How is sulphuretted hydrogen prepared?

10. Explain how this gas may be used for the separation of the metals into groups.

11. Point out the relations existing between the oxygen and sulphur compounds.

LESSON XIV. *Silicon, Boron, &c.*

1. Mention the chief properties of selenium and tellurium.
2. How is silicon prepared?
3. What names does the substance SiO_2 go by?
4. How can we obtain (1) soluble and (2) insoluble silica?
5. Explain the terms "dialysis," "colloid," and "crystalloid."
6. How is silicic tetrafluoride prepared?
7. Where does boracic acid occur?
8. What is the composition of borax?

LESSON XV. *Phosphorus Compounds.*

1. Whence do animals ultimately get the phosphorus which they need?
2. How is phosphorus prepared from bone-ash?
3. Describe the different modifications of phosphorus.
4. What weight of phosphorus pentoxide can be obtained by burning one kilo. of phosphorus?
5. How is trihydric phosphate prepared?

6. Write down the formulæ of the tribasic sodium phosphates.

7. How many grms. of sodium metaphosphate can be got by heating 100 grms. of microcosmic salt?

8. Write down the decomposition which occurs when we mix solutions of hydric di-sodium phosphate and silver nitrate ($Ag\ NO_3$).

9. $4\ (H_3\ PO_3) = 3\ (H_3\ PO_4) + PH_3$. Describe this decomposition, and give the properties of the substances formed.

10. How are the chlorides of phosphorus prepared?

Lesson XVI. *Arsenic Compounds.*

1. How is arsenic separated from its ores?

2. Name the oxides of arsenic.

3. What are the peculiar characteristics of the arsenites and arsenates?

4. How does ferric oxide act as an antidote to the poisonous properties of the arsenites and arsenates?

5. What is the composition and mode of preparation of arseniuretted hydrogen?

6. Name the tests by which arsenic can be detected with certainty.

7. Point out the general chemical relations of the arsenic, phosphorus, and nitrogen compounds.

8. Explain fully what is meant when we say that chlorine is *monad*, oxygen is *dyad*, nitrogen is *triad*, and carbon is *tetrad*.

9. Give examples of compound radicals belonging to the monad, dyad, and triad groups.

10. How is the *atomicity* of an element or radical denoted?

Lesson XVII. *The General Properties of the Metals.*

1. Name the metals which are lighter than water.

2. At what temperature does mercury boil and freeze?

3. Describe the modes in which the metallic ores generally occur.

4. State some of the peculiar properties of the alloys.

5. Under what typical form may all the oxides be classed?

6. What is meant by a metallic salt?

7. To what types do the substances $K_2 SO_4$, $Ba SO_4$, $Al_2 3SO_4$, $Ba N_2O_6$, Ag_3PO_4, belong?

LESSON XVIII. *Crystallography.*

1. Give the chief characteristics of crystalline structure.

2. Distinguish between amorphous and cellular structure.

3. How is the cube derived from the regular octohedron?

4. Explain what is meant by the axes of a crystal.

5. What are the distinguishing characteristics of the six systems of crystallography?

6. How is the rhombohedron derived from the double six-sided pyramid?

7. What is the meaning of isomorphous and of dimorphous bodies?

LESSON XIX. *Metals of the Alkalies.*

1. How was potassium first prepared, and how is it now manufactured?

2. State the sources of the potassium compounds.

3. How is caustic potash obtained?

4. Describe what happens when gunpowder is burnt.

5. Supposing that the decomposition is a simple one, how many cbc. of (1) carbonic oxide and (2) of nitrogen gas at 0° and 760 mm. will be given off by burning one gramme of English musketry powder?

6. Name the characteristic tests for potassium salts.

7. What are the sources of the sodium compounds?

8. Describe the salt-cake process.

9. How many tons of vitriol containing 72 per cent. of H_2

SO_4 will be needed to convert 100 tons of salt into salt-cake, and how many tons of this latter will be formed?

10. How many tons of aqueous hydrochloric acid containing 30 per cent. of HCl will be formed in the preceding reaction?

11. Describe the decompositions by which salt-cake is converted into soda ash.

12. Required 500 tons of soda-crystals, what will be the weight of salt and pure sulphuric acid needed?

13. How were the two new alkaline metals discovered?

14. Explain the analogy in constitution existing between the potassium and ammonium salts.

LESSON XX. *Metals of the Alkaline Earths and Aluminium.*

1. What is the composition of slaked lime?
2. Describe the uses of lime in agriculture.
3. How can temporarily hard water be softened?
4. Name the commonest minerals containing barium and strontium.
5. How can oxygen gas be prepared from barium dioxide, and how can this process be rendered continuous?
6. Mention the distinguishing reactions of the compounds of calcium, strontium, and barium.
7. How is metallic aluminium prepared?
8. What is the meaning of a mordant?
9. Calculate the percentage composition of common alum?
10. Give a short account of the composition and properties of the different kinds of glass.
11. How are colored glasses obtained?
12. How is common earthenware glazed?

LESSON XXI. *Magnesium, Zinc, Manganese.*

1. Find the formula of a salt having the following percentage composition:

Magnesium	9·76
Sulphur	13·01
Oxygen	26·01
Water	51·22
	100·00

2. How can these magnesium salts be distinguished and separated from those of calcium ?

3. State the method employed to extract zinc from its ores.

4. How many grms. of crystallized zinc sulphate can be got from 1,000 grms. of blende ?

5. State the composition of the several manganese oxides.

6. How many litres of oxygen at 12° and under the pressure of 750 mm. can be got (1) by heating 500 grms. of manganese dioxide, and (2) by treating the same weight of the same oxide with sulphuric acid ?

7. What tests would you employ to detect the presence of the compounds of zinc, cadmium, and manganese ?

Lesson XXII. *Iron.*

1. Mention some of the most important physical properties of iron.

2. How is ferrous sulphate obtained ? How many tons of crystals can be obtained by the slow oxidation of 230 tons of pyrites containing 37·5 per cent. of sulphur ?

3. What is the composition of red hæmatite and specular iron ore ?

4. How can the ferrous and ferric salts be distinguished ?

5. Describe the manufacture of cast-iron from clay ironstone.

6. What chemical changes go on in the processes of "refining" and "puddling ?"

7. How do cast-iron, steel, and wrought-iron differ in their composition ?

8. Describe (1) the common method for making steel, and (2) that known as Bessemer's method.

9. 3·285 grms. of pure iron wire are burnt in excess of oxygen and in chlorine gases; required the weight (1) of oxide, and (2) of chloride formed.

10. What is the cause of difference in the appearance and properties of "mottled" and "white" cast-iron?

LESSON XXIII. *Cobalt, Nickel, Chromium, Tin, &c.*

1. Mention some of the chemical characteristics of cobalt.
2. How can cobalt and nickel be distinguished by the blow-pipe?
3. Give the formulæ and names of the chromium oxides.
4. How can we pass from chromium sesquioxide to the tri-oxide, and *vice versâ?*
5. Write down the formulæ of the potassium chromates.
6. What is the constitution and mode of preparation of chlorochromic oxide?
7. In what form does tin occur?
8. How can tin compounds be distinguished?
9. What weight of crystallized "tin salts" can be prepared from one ton of metallic tin?

LESSON XXIV. *Antimony, Bismuth, Lead, Thallium.*

1. Write down the formulæ of the corresponding oxides of arsenic and antimony.
2. How are the two chlorides of antimony prepared?
3. How much manganese dioxide, salt, and sulphuric acid will furnish chlorine enough to convert 100 grms. of antimony into the trichloride?
4. Point out the chief distinguishing properties of the bismuth compounds.
5. Mention the decompositions which occur in the process of lead smelting.
6. Describe the action of lead upon water.
7. How is white lead manufactured?

8. 100 grms. of lead oxide when reduced to the metallic state in a current of hydrogen lost 7·1724 grms. Calculate the combining weight of lead.

9. 4·9975 grms. of lead chloride needed 3·881 grms. of metallic silver for complete precipitation. Required the combining weight of lead, those of silver and chlorine being given.

LESSON XXV. *Copper and the Noble Metals.*

1. How is copper obtained from copper pyrites?

2. Calculate the percentage of water contained in crystallized copper sulphate.

3. What is the density of mercury vapor? Does it obey the usual law of densities?

4. What weight of mercury and corrosive sublimate must be taken to yield three kilos. of calomel?

5. How is the silver extracted from argentiferous lead?

6. 100 parts by weight of silver yield 132·84 parts of silver chloride. Given the combining weight of chlorine, required that of silver.

7. What decomposition does silver chloride undergo in the light?

8. Describe the method used for the extraction of gold.

9. How can platinum-ore be worked into coherent metal?

10. Give the distinguishing tests for copper, mercury, silver, and gold.

LESSON XXVI. *Spectrum Analysis.*

1. Describe the phenomenon observed when a source of white light is examined by means of a prism.

2. What peculiarity is observed in the spectra of colored flames?

3. How does the spectrum of a glowing solid differ from that of a glowing gas?

4. Mention some facts to show the extreme delicacy of the spectrum analytical methods.

5. How can the spectra of the metals be obtained?

6. Describe the construction and mode of use of the spectroscope.

7. Draw a rough sketch of the spectra of the following:—sodium, potassium, rubidium, lithium, and strontium (see Frontispiece).

8. Explain what is meant by Fraunhofer's lines.

9. Describe shortly an experiment to show the reversion of the bright line of sodium.

10. Why does Kirchhoff conclude that iron exists in the solar atmosphere?

11. How do we know that the fixed dark solar lines are not caused by absorption in the earth's atmosphere?

12. How can we learn the composition of the atmospheres of the fixed stars, and why are we in ignorance about the composition of the planets?

13. State the results of Mr. Huggins' observations upon the spectra of the nebulæ.

LESSON XXVII. *Introduction to Organic Chemistry.*

1. Name the two chief peculiarities of the carbon compounds.

2. Give examples of monad, dyad, triad, and tetrad elements.

3. Explain what is meant by saturated and non-saturated carbon compounds.

4. Name the chlorine substitution products of marsh-gas.

5. Explain, with a drawing, the constitution of the mono-, di-, and tri-carbon series of saturated compounds.

6. What is the constitution of the hydrides, chlorides, and alcohols of the first three series of carbon compounds?

7. What is meant by an organic radical, and by the term polyatomic radicals?

8. Show that the constitution of the saturated compound benzol, $C_6 H_6$, is different from that of the alcohol group of bodies.

9. Give examples to show the distinction between an atom and a molecule.

10. How can all chemical changes be represented as double decompositions?

11. Give examples of bodies (organic and inorganic) formed according to the types—

$$\left.\begin{matrix}H\\H\end{matrix}\right\},\quad \left.\begin{matrix}H\\H\end{matrix}\right\}O,\text{ and }\left.\begin{matrix}H\\H\\H\end{matrix}\right\}N.$$

12. To what types do the following substances belong: magnesium chloride, sulphuric acid, tribasic phosphoric acid, nitric acid, sulphuretted hydrogen, arseniuretted hydrogen?

LESSON XXVIII. *Organic Analysis, &c.*

[It will be necessary for the pupil to work out many more exercises on the lesson than are here given.]

1. Describe shortly the process adopted for the estimation of the carbon and hydrogen contained in organic compounds.

2. 0·3059 grm. of a body containing carbon, hydrogen, and oxygen, yielded on combustion 0·6000 grm. carbonic acid, and 0·3040 grm. water. Required the relation between the number of atoms of the component elements.

3. What is the molecular weight of an acid (mono-chloracetic) whose silver salt contains 53·6 per cent. of this metal?

4. 0·305 grm. of an acid yielded on combustion 0·761 grm. of carbonic acid and 0·136 grm. water; 0·391 grm. of the silver salt contained 0·184 grm. silver. Required the formula of the acid containing carbon, hydrogen, and oxygen.

5. How does the determination of the vapor density of an organic body serve as a means of ascertaining its molecular weight?

6. What is the density of ammonia, marsh gas, olefiant gas, methyl alcohol, ethyl alcohol?

7. Describe the two methods employed for determination of vapor density.

8. Required from the following numbers the vapor density of a hydrocarbon of the marsh gas series :—

Globe filled with air	at	16°·5	7·566 grm.
———————— vapor	"	140°	7·783 "

Capacity of globe 115·5 cbc.

LESSON XXIX. *Monatomic Alcohol Group.*

1. Explain the analogy in constitution existing between the ethyl and potassium compounds.

2. Write down the formula for ethyl alcohol, ether, acetyl-acetate, aldehyde, acetamide.

3. Give the names of the following :

$$\left.\begin{matrix} N\,(C_2\,H_5)_4 \\ H \end{matrix}\right\} O, \qquad \left.\begin{matrix} C_2H_3O \\ C_2H_3O \\ C_2H_5 \end{matrix}\right\} N, \qquad \left.\begin{matrix} C_2H_5 \\ C_2H_5 \\ C_2H_5 \end{matrix}\right\} P.$$

4. What is the chemical change which occurs in the passage from an alcohol to the corresponding acid ?

5. Write down a list of the first eight alcohols with their derived acids.

6. Name the properties and mode of preparation of methyl alcohol.

7. What is the action of sulphuric acid upon methyl alcohol ?

8. By what reactions are we enabled to pass from the methyl to the ethyl series ?

LESSON XXX. *Di-carbon or Ethyl Series.*

1. How can alcohol be prepared from its inorganic materials ?

2. How many grms. of alcohol can be completely burnt by 1,000 litres of oxygen at 0° and 760 mm ?

3. Give the formulæ for potassium ethylate, potassium-ethyl-sulphate, and ether.

4. Describe the continuous etherification process.

5. Write down in formulæ the decomposition by which ethyl cyanide yields propionic acid.

6. How many grms. of ethylamine can be prepared from 100 of ethyl-cyanate, and how many grms. of potassium carbonate will be produced?

7. Write down the formulæ of propyl alcohol, propionic acid, propyl chloride, butyl alcohol.

8. How can amyl alcohol be prepared from $C_5 H_{11} H$?

9. What is the action of chlorine upon $\left.\begin{matrix} C_5 H_{11} \\ C_5 H_{11} \end{matrix}\right\}$?

10. How can the higher alcohols be prepared from American petroleum?

Lesson XXXI. *Compound Ammonias.*

1. Mention the reactions by which the compound alcoholic ammonias can be prepared.

2. Required the percentage of platinum contained in $2 (N (C_2 H_5)_3 H Cl) + Pt Cl_4$.

3. What is the molecular weight and possible formula of a double platinum salt yielding on heating 29·4 per cent. of metallic platinum?

4. How is tetra-ethyl ammonium hydrate prepared?

5. Give examples of primary, secondary, and tertiary monamines.

6. How would you determine the *constitution* of a compound ammonia of the composition $C_3 H_9 N$?

7. What is the formula of tri-ethyl phosphine, and how is this substance prepared?

8. What is the composition of cacodyl and cacodylic acid?

9. How is zinc-ethyl prepared, and what are its chief properties?

10. Na ($C_2 H_5$) + CO_2 = $C_3 H_5 Na O_2$; explain this reaction.

Lesson XXXII. *Oxidized Derivatives of the Alcohols.*

1. Mention the chief reactions by which the fatty acids can be formed.
2. How many grms. of potassium formate can be got from 500 litres of carbonic oxide at 15° and 745 mm.?
3. Required 100 kilos. of $C H_2 O_2$; how many kilos. of oxalic acid are needed?
4. What is the formula of formamide?
5. How can aldehyde be produced from acetic acid, and how can aldehyde be reduced to alcohol?
6. Explain what is meant by the acetous fermentation.
7. What is the composition of red- and iron-liquors?
8. How many grms. of glacial acetic acid can be obtained from 25 kilos. of potassium acetate?
9. How is acetyl acetate (acetic anhydride) prepared?
10. Name some of the chlorine substitution products of acetic acid.
11. Give the formulæ and mode of preparation of thiacetic acid, acetyl peroxide, acetamide, acetone, acetylene.
12. Show that by substituting hydrogen in the radical of acetic acid by methyl and ethyl we obtain (1) propionic and (2) butyric acids.
13. Point out several methods by which we can pass from the di- to the tri-carbon series.

Lesson XXXIII. *Diatomic Alcohols.*

1. What is meant by a diatomic alcohol?
2. Mention the chlorine substitution products of ethylene.
3. Why is ethylene regarded as a non-saturated compound?
4. How is glycol prepared?
5. What are the products of oxidation of glycol?

6. How is ethylene oxide distinguished from aldehyde ?

7. How many grms. of oxygen are required to burn completely 100 grms. of tri-ethylene glycol ?

8. Write down a list of the olifines with their formulæ.

9. What is the name of $\left.\begin{matrix} C_6 H_{12} \\ H_2 \end{matrix}\right\} O_2$?

10. Write out the formulæ of some ethylene di-amines.

Lesson XXXIV. *Diatomic Acids.*

1. How are the acids of (1) the lactic series and (2) those of the oxalic series derived from the corresponding glycols ?

2. Show that hydrated carbonic acid is the first term of the first series.

3. Write the formula of di-methyl sulpho-carbonate.

4. How many grms. of oxygen are required to oxidize 100 grms. of glycolic to oxalic acid ?

5. Describe the manufacture of oxalic acid from saw-dust.

6. Show that lactic acid can be formed from chlor-propionic acid.

7. In what important respect, as regards the formation of salts, do lactic acid and its homologues differ from oxalic acid and the higher terms of its series ?

8. How can malic and tartaric acid be obtained from succinic acid ?

9. Describe the several varieties of tartaric acid.

10. What is the action of hydriodic acid upon tartaric acid ?

Lesson XXXV. *Cyanogen Compounds.*

1. Write down the typical formulæ for the most important cyanogen compounds.

2. Describe the tests for hydrocyanic acid.

3. What are the chief points of relationship between the cyanogen and the oxalic acid groups of compounds ?

4. How much yellow prussiate of potash, manganese dioxide, and ammonium sulphate can yield 500 grms. urea ?

5. 50 grms. of urine yielded on analysis 475 cbc. of nitrogen at 11° and 754 mm. Required the percentage of urea contained.

6. Write the formulæ for cyanuric acid, di-ethyl urea, sulphocyanic acid, and cyanamide.

Lesson XXXVI. *Triatomic and Hexatomic Alcohols.*

1. Show the relation in composition existing between propyl alcohol, propylene glycol, and glycerine.

2. Explain the process of saponification.

3. Explain by typical formula the composition of the chlorhydrines.

4. $\left.\begin{matrix} C_3H_5 \\ (C_{18}H_{35}O)\,H_2 \end{matrix}\right\} O_3$; $\left.\begin{matrix} C_3H_5 \\ (C_{18}H_{35}O)_2\,H \end{matrix}\right\} O_3$;

$$\left.\begin{matrix} C_3H_5 \\ (C_{18}H_{35}O)_3 \end{matrix}\right\} O_3.$$

Name the above bodies.

5. How is allyl alcohol prepared from glycerine ?

6. What is the composition of acroleine ?

7. What is the constitution of mannite ; and what reasons have we for supposing it to be a hexatomic alcohol ?

8. How are the so-called iso-alcohols prepared ?

Lesson XXXVII. *Sugars and Glucoses.*

1. Give a short description of the preparation and refining of cane sugar.

2. Write down the formula of sucroses and glucoses.

3. What is meant by right- and left-handed rotatory power ?

4. What is the action of yeast and dilute sulphuric acid upon cane sugar ?

5. How is dextrose prepared?

6. Give a short account of the principal phenomena of fermentation.

LESSON XXXVIII. *Starch, Gums, and Glucosides.*

1. How does starch differ in constitution from glucose?

2. What weight of dextrine and dextrose can be obtained from 1 kilo. of starch by the action of diastase?

3. What is the composition of gun cotton, and what advantages does it offer over gunpowder?

4. $C_{20} H_{27} N O_{11} + 2 H_2O = C_7 H_6 O + HCN + 2C_6 H_{12} O_6$. Explain the above equation.

5. What is the composition of ink?

6. State some of the general characteristics of a glucoside.

LESSON XXXIX. *Group of Aromatic Compounds.*

1. How do we suppose the carbon atoms in benzol are arranged?

2. Write down the formulæ for benzol, toluol, xylol, and cumol.

3. What substances are formed by the replacement of one atom of hydrogen in benzol by $N O_2$, $N H_2$ and $H O$?

4. Describe the methods employed for preparing aniline.

5. Required the volumes at 0° and 760 mm. of nitrogen and carbonic dioxide obtained by the combustion of 216 grms. of aniline.

6. How is rosaniline prepared?

7. How can oil of bitter almonds be converted into benzoic acid and *vice versâ*?

8. Explain the following :—

$$\left.\begin{matrix} C_7 H_5 O \\ K \end{matrix}\right\} O + C_7 H_5 O Cl = \left.\begin{matrix} C_7 H_5 O \\ C_7 H_5 O \end{matrix}\right\} O + K Cl.$$

9. Explain the relation of leucaniline to rosaniline, and of white to blue indigo.

LESSON XL. *Turpentines and Vegeto-alkaloids.*

1. What is meant by physical isomerism?
2. What is the general composition of essential oils?
3. State the composition and chief properties of naphthalin.
4. What is the chief coloring matter of madder?
5. Name the chief peculiarities of the group of vegeto-alkaloids.
6. What tests may be used to ascertain the presence of morphine, brucine, and strychnine?
7. Point out the chief properties of quinine, cinchonine, and its isomers.
8. How is theobromine connected with theine?
9. Required the molecular weight of an alkaloid whose hydrochlorate contains 11·0 per cent. of chlorine.

LESSON XLI. *Albuminous Substances.*

1. In what chemical characters do the albuminous bodies differ from chemical compounds?
2. How may fibrin, albumin, and casein be separated?
3. Describe shortly the composition and properties of blood, milk, and bile.
4. Distinguish between animal and vegetable life.
5. From what source do animals obtain the energy necessary for existence, and whence do plants draw the energy needed for the organization of their food?

APPENDIX.

NOTE.—*On the Absolute Weights of Hydrogen, Oxygen, and the other Gases.*

IN the foregoing pages we have taken the density of Oxygen to be 16 (Hydrogen = 1), and we have chosen as the standard of absolute weight Regnault's number 1·429802 as the weight in grammes of 1 litre of oxygen gas, weighed in the latitude of Paris at 0° C., and under a pressure of 760 mm. of mercury, this being generally received as the most reliable of this great experimenter's numbers. Hence we adopt $\frac{1}{16}$ of this, or 0·0893626, as the absolute weight of 1 litre of Hydrogen measured under the same circumstances. Regnault's experimental number for the weight of 1 litre of Hydrogen is, however, 0·089578, and if we accept both of these experimental numbers as correct, the density of Oxygen becomes 15·96 instead of 16.

In his recent classical researches on the combining weights of the elements (the most reliable and accurate determinations which we now possess), Stas concludes that the true combining weight, and therefore the true density, of Oxygen is 15·960 when H = 1, and thus a remarkable confirmation of the accuracy of Regnault's experimental results is obtained.

If we wish to calculate the weights of the gases with the greatest possible degree of precision, from the best experimental data, we must adopt $\frac{1.429802}{15.96}$ or 0·089586 as the unit (or *crith*), and in order to find the weight of 1 litre of the following gases, or their gaseous compounds, we must multiply this number by the densities of the respective gases, as given in (or obtained from) the accompanying table, and derived from the combining weights, as determined by Stas.

In the third column the experimental densities of the gases are found as they have been determined by exact observation; it will be seen that in some cases these numbers closely agree with those obtained by Stas, whilst in others (especially those of the older experimenters) the difference becomes more apparent.

	I. Density calculated. (Stas).	II. Weight of 1 litre at 0° & 760 mm.	III. Density Experimental.	
Hydrogen	1·000	0·089586	1·000	
Oxygen	15·960	1·429802	15·960	Regnault.
Nitrogen	14·009	1·255010	14·025	"
Chlorine	35·368	3·168478	35·343	Bunsen.
Bromine	79·750	7·144483	79·978	Mitscherlich.
Iodine	126·533	11·335586	125·83	Dumas.
Steam	8·980	0·804482	9·001	Gay Lussac and Thenard.
Ammonia	8·504	0·761839	8·614	Biot and Arago.
Hydrochloric Acid	18·184	1·629071	18·008	" "

INDEX.

—o—

D.

E.

F.

G.

H

I.

K.

L.

M.

Q.

R.

S.

www.ingramcontent.com/pod-product-compliance
Lightning Source LLC
LaVergne TN
LVHW020105110826
845151LV00001B/63